Advanced Transmission Techniques in WiMAX

Advanced Transmission Techniques in WiMAX

Editor

Dan Hallmen

scitus academics

Advanced Transmission Techniques in WiMAX
Edited by **Dan Hallmen**

ISBN: 978-1-68117-277-4
Library of Congress Control Number: 2016937449

www.scitusacademics.com
Box No. 4766, 616 Corporate Way,
Suite 2, Valley Cottage,
NY 10989

Printed in United States of America on Acid Free Paper

Preface

WiMAX (Worldwide Interoperability for Microwave Access) is a long-range standard for wireless networking. WiMAX equipment exists in two basic forms–base stations installed in the coverage area by customer installed service providers and receivers. WiMAX supports multiple networking models. WiMAX refers to the interoperability of the IEEE 802.16 Wireless Network Standards Family through the WiMAX forum. Wi-Fi also refers to IEEE 802.11 Wireless LAN standards certified by the Wi-Fi Alliance. WiMAX Forum certification enables vendors to sell WiMAX certified fixed or mobile products, thus ensuring interoperability with other certified products as long as they fit the same profile. WiMAX does not replace Wi-Fi home networking or wireless hotspot technology mainly because of its much higher costs. The WiMAX connectivity devices are known as subscriber stations (SS). Portable units include handsets (similar to cell phones); PC peripherals (PC cards or USB dongles); and Wi-Fi embedded laptop devices. In addition, operators put a great deal of emphasis on consumer electronics equipment such as gaming consoles, MP3 players and the like. WiMAX is Wi-Fi more like any 3G cell technology.

This book has been prepared to introduce the latest WiMAX technology. The book covers theoretical and technical information to WiMAX engineers and researchers with an in-depth review of the latest developments in the field.

Table of Contents

1

CHAPTER

MULTI-CRITERIA VERTICAL HANDOVER COMPARISON BETWEEN WIMAX AND WIFI

Maroua Drissi [1,*] and Mohammed Oumsis [1,2]

[1] *GSCM-LRIT Associated Unit to CNRST - URAC n'29, Faculty of Science, Mohammed V University, Rabat 10 000, Morocco*

[2] *High School of Technology, Mohammed V University, Sale 11 000, Morocco*

ABSTRACT

In next generation wireless networks, the most tempting feature is the ability of the user to move smoothly over different access networks regardless of the network access technology. In this paper we study the benefit of Multiple Attribute Decision Making (MADM) strategies for network selection. We compare three of these methods naming Simple Additive Weighting (SAW), Multiplicative Exponential Weighting (MEW) and Technique for Order Preference by Similarity to an Ideal Solution (TOPSIS) in a realtime ns-3 simulation. Analytic Hierarchy Process (AHP) provides the weights of attributes which allow the comparison in different types of applications. Therefore, we propose a performance evaluation model with a reconfiguration of AHP parameters used in the literature. Simulation results show that the proposed parameters provide an improvement of Delay and offer allowable Packet loss in different types of applications.

Keywords: 4G; Multiple Attribute Decision Making (MADM); Analytic Hierarchy Process (AHP); heterogeneous network; network selection; vertical handover; WIMAX; WLAN; NS-3

1. INTRODUCTION AND MOTIVATION

The most auspicious Next Generation Networks (NGN) are the heterogeneous networks. They are based on the coexistence and interoperability of different types of Radio Access Technologies (RAT), including the current and the forthcoming networks, intended to fit the requirements for various applications in the future. 4G came with the features “Always Best Connected (ABC)”, “Anytime Anywhere” and “seamless communications”. In [1], the authors assert that a terminal supports the ABC feature means that it is not only always connected, but also connected through the best available network and access technology at all times. The ABC concept includes virtually all types of access technologies; ABC achieve a win-win partnership because it considers user's and operator's benefits. In the mean time, 4G provides a larger coverage area, faster data transfer, low latency and low data transfer cost. The main issue for the 4G heterogeneous networks is Network Selection, i.e., a flawless and proficient handover scheme that allows the roaming of mobile devices from one wireless system to another. Therefore, with ABC functionalities, terminals are able to select the best access network to fit various Quality of Service (QoS) requirements; terminals escape a network with high traffic load to avoid congestion and reduce costs by using network selection, this is known as Vertical handover, and vertical handover decision strategies are dressed specifically for this purpose. In generic terms for all wireless standards [2] proposed a handover decision scheme based on multi-criteria decision-making modeling providing a continuous connection in heterogeneous wireless networks, unlike traditional approaches which are mainly concerned with specific scenarios.

Being the key for resource management in a wireless heterogeneous network, dynamic network selection algorithms intend to provide users with the adequate QoS in terms of attributes and user’s preferences. Both are aforethought during the process of network selection (see Figure 1). Hence, this technology is a hot research topic in the field of wireless communication. Network selection can be programmed by terminal or can depend on measurements of link quality of the network when multiple alternatives are available at the same time. The terminals are equipped more and more by new technologies; every terminal has no less than two network interfaces, such as WLAN, GPRS, HSPA, WIMAX or LIFI [3]. In the optimal cases, the user can move from one wireless network to a different wireless network seamlessly with no deterioration of the connection.

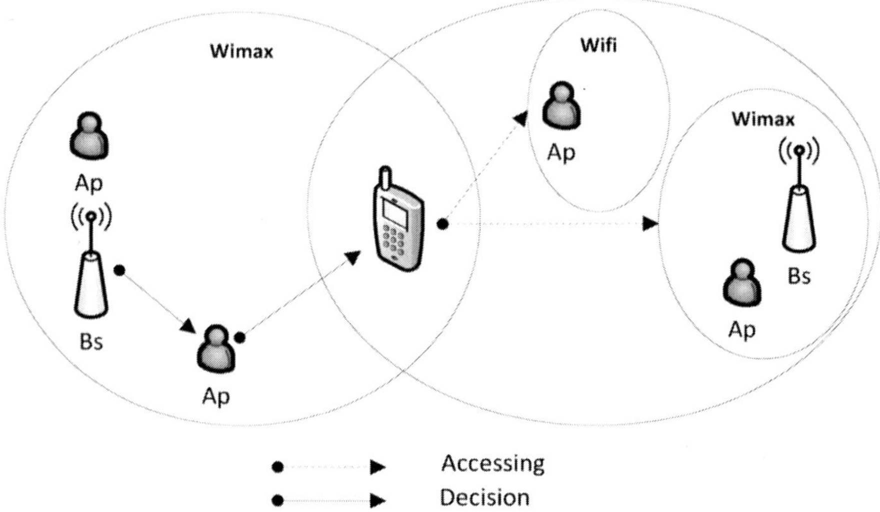

Figure 1. Network Selection Process.

The remainder of this paper is organized as follows. Related works about vertical handover decision (VHD) strategies are summarized in Section 2. In Section 3, the authors introduce the performance evaluation model and the new configuration of AHP weights and the handover algorithms are compared. Section 4 presents the simulations and the results discussion. Finally, Section 5 concludes the work and gives some perspectives.

2. RELATED WORKS

Vertical Handover is an important step to acomplish the best QoS in an heterogeneous environment that needs a dynamic selection of the best convenient network, and the decision phase is the most crucial in this case. It is about assembling the performances of each candidate network, and ranking them in purpose to select the best network. It is leading a revolution in the internet by carrying an improved Quality of Experience (QoE) for users of wireless services.

In the literature, different approaches have been proposed to handle the vertical handover decision problem: good network performance, user satisfaction, flexibility, efficiency, and multi-criteria solution. In the same context of heterogeneous environment, authors of [4] compared classical handover decision strategies (RSSbased), and came to the conclusion that they are not sufficient to make a vertical handover decision. They do not consider the current context or user preferences. Indeed, vertical handover decision strategy involves complex considerations and compromises. It needs to be flexible and efficient considering the useful criteria and reasonable policies or rules applicable to both the user's professional and personal communications. In this context, [5] conclude that the mobility management of heterogeneous networks not only depends on network parameters, but also on terminal's velocity, battery power, location information, user's preferences, service's capabilities and QoS requirements. [4,6] presented a panorama of vertical Handover Decisions. Using cost or utility functions; utility/cost refers to the satisfaction that a good or service provides to the decision maker. In the same context, [6] specified that other studies on the network selection issue could also evaluate networks based on utility/cost functions which combine multiple attributes. Furthermore, authors of [7], as well, proposed an algorithm for network selection based on averaged received signal strength, outage probability and distance.

In [8], an intelligent handover decision mechanism has been proposed based on the context-aware strategy and Media Independent Handover in order to reduce the latency, to enable seamless vertical handover and always guarantee the best connected concept. However, some studies focus also on other mathematical models such as fuzzy logic methods which can be used in VHD. The authors of [9] proposed an utility-based Radio Access Technology (RAT) selection optimization in heterogeneous wireless networks, they introduced three utility-based optimization functions based on the type of application that users request. They formulated the terminal assignment problem as an optimization problem, which is recognized as NP-hard.

The goal in vertical handover process is that a terminal should join the best Access Point to address the needs of the users. Interface selection turns into a "decision making" problem with multiple options -Alternative networks - and attributes -QoS Parameters- user preferences and application requirements. Among several approaches that have been proposed for decision making and network selection, MADM is one of the most promising methods.

In [10], MADM refers to making preference decision over the available alternatives that are characterized by multiple attributes. MADM problems have several common characteristics such as Alternatives, Multiple attributes, Decision matrix, Attribute weights and Normalization. Most MADM algorithms that have been studied in [11,12,13,14,15] for the network selection problem are compensatory algorithms which means they combine multiple attributes to find the best alternatives, versus non compensatory algorithms used to find acceptable alternatives that satisfy the minimum cutoff.

The most popular classical MADM methods according to [4] are SAW, MEW and TOPSIS which are compensatory algorithms. The mathematical modeling of these methods is given in the next section. The authors of [15] proposed giving weights to know the relative importance of each metric or attribute in the conseidered QoS Class. They used AHP to calculate the vectors of weight. Their results showed the importance of those weights in the decision.

Therefore, in this paper we focus on investigating in how to generate the importance weight according to new access technologies.

3. EVALUATION APPROACH

3.1. System Model

As in [16], our contribution consists of a handover decision phase; decision makers must choose the best network from available networks. The goal of this contribution is to analyze the weights given to the metrics, so that, as reported in the Algorithm block diagram shown in Figure 2, simulation provides the system with the metrics in realtime and a pairwise comparison is applied according to each QoS class: Conversational, Streaming, Interactive and Background. Then a weighting method (AHP in this work) is used to calculate the weight vectors in order to obtain the relative importance of each metric or attribute. Consequently, SAW, TOPSIS and MEW are applied to the weighted matrices to obtain the final decision for each method. The handover decisions are made in real-time and repeatedly, in our case the process is repeated every 5 s.

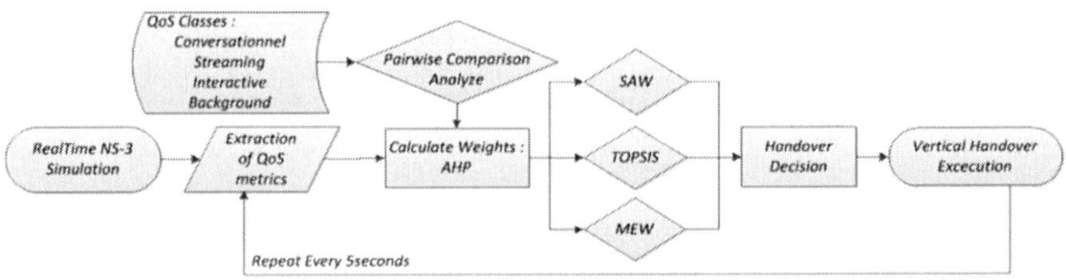

Figure 2. Algorithm Block.

In [17], authors explain that a determined set of weights produces certain quality or merit degree for each network; these merit values change if we consider another set of weights. The goal is to obtain the best merit value, which will correspond to the selected network for the VH decision phase. Accordingly, the more combinations of weights, the more possibilities to get better merit values we will have. In this work, and to compare SAW, TOPSIS and MEW algorithms, we have been concentrated on the weighting phase with AHP, we have reconfigured the weights done previously in the literature, considering Throughput, Delay, Jitter and Bit Error Rate (Ber) of the participating access networks to make the handover decisions. The vertical handoff decision problem can be formulated as a matrix form Q, where each row i corresponds to the candidate network and each column j corresponds to an attribute. The matrix of alternative networks is settled conforming to the attributes.

$$Q_{N,M} = \begin{array}{c} \\ Network_1 \\ Network_2 \\ \vdots \\ Network_N \end{array} \begin{array}{c} \begin{array}{cccc} Attribute_1 & Attribute_2 & \dots & Attribute_M \end{array} \\ \begin{pmatrix} q_{11} & q_{12} & \cdots & q_{1M} \\ q_{21} & q_{22} & \cdots & q_{2M} \\ \vdots & \vdots & \ddots & \vdots \\ q_{N1} & q_{N2} & \cdots & q_{NM} \end{pmatrix} \end{array}$$

As all three methods grant the evaluated criteria to be expressed in different measurement units, it is necessary to convert them into normalized values.

$$n_{ij} = \frac{q_{ij}}{\sqrt{\sum_{i=1}^{i=N} q_{ij}^2}}, j = 1, ..., M \tag{1}$$

In addition, weight factor is assigned conveniently to each criterion to report its importance which is determined by (AHP).

3.2. Analytic Hierarchy Process : AHP

The concept of weight involved in each method, is solved by the use of the AHP method. The authors of [18] proposed this process for decision-making in multi-criteria problems, they introduced AHP as a method of measurement with ratio scales, AHP allows comparison and a choice of preset options. It is based on the comparison of pairs of options and criteria.

In our approach, AHP is used to calculate the weight vector w, which represents the importance of each metric with respect to different QoS classes for use in SAW, MEW and TOPSIS methods. AHP provide as results *wj*>0, which represents the weight or importance of the *jth* attribute, given that

$$\sum_{j=1}^{j=M} w_j = 1$$

Weight computing needs answering to a sequence of comparisons between a pair metrics. The common way to ask a question is to consider two elements, and find out which one satisfies the criterion more. The answers are given by using the fundamental 1-9 AHP scale [18,19]. The values in the answers correspond to the scale in Table 1 below.

Table 1. The scale of importance [18].

Importance	Definition	Explanation
1	*Equal importance*	Two parameters contribute equally
3	*Moderate importance of one over another*	Experience favored 3 times one than another
5	*Strong importance*	Experience favored 5 times one than another
7	*Very Strong importance*	A parameters is favored and dominant in practice
9	*Extreme importance*	The evidence favoring one activity over another is of the highest possible order of affirmation
2,4,6,8	*Intermediate values*	When Compromise is needed

Table 2 presents the answers of the questions asked about the relative importance between each pair, for example, in Conversational Class, the first comparison is (Ber, Jitter), the question is : How much more is Ber preferred over Jitter in conversational Class? Indeed, Jitter is 7 times more important than Ber, so the value in matrix is 1/7, and accordingly 7 is put in the opposite side (symmetrical to the diagonal). The other values can be obtained from Table 1.

Table 2. Analytic Hierarchy Process (AHP) Matrices for Each Traffic Class.

Conversational	Ber	Jitter	Delay	Throughput
Ber	1	1/7	1/7	3
Jitter	7	1	3	7
Delay	7	1/3	1	7
Throughput	1/3	1/7	1/7	1

Interactive	Ber	Jitter	Delay	Throughput
Ber	1	3	7	3
Jitter	1/3	1	5	3
Delay	1/7	1/5	1	1/7
Throughput	1/3	1/3	7	1

Streaming	Ber	Jitter	Delay	Throughput
Ber	1	1/3	1/7	1/7
Jitter	3	1	1/3	1/3
Delay	7	3	1	1/3
Throughput	7	3	3	1

Background	Ber	Jitter	Delay	Throughput
Ber	1	7	7	9
Jitter	1/7	1	3	5
Delay	1/7	1/3	1	5
Throughput	1/9	1/5	1/5	1

In [15] the authors explain that if C is defined as an AHP comparison matrix as in Table 1, then by solving the system : *U.w=nmax* (where *nmax* is the largest eigenvalue of U), the priority or importance vector w can be obtained. Thus, the weights rely on the QoS prerequisite of the traffic classes. We use the eigenvector method used by the AHP to figure out the weights prensented in Table 3.

Table 3. Importance Weights Per Class.

Traffic Class	Ber	Jitter	Delay	Throughput
Conversational	0.07968	0.55464	0.31956	0.04610
Streaming	0.05104	0.13444	0.29493	0.51957
Interactive	0.50385	0.27509	0.04608	0.17496
Background	0.68037	0.17644	0.10390	0.03926

3.3. Vertical Handover Decisions

In this section we present the mathematical models of the methods SAW, TOPSIS and MEW. To calculate the score of the available networks given by the methods in the four QoS classes, the problem is also expressed as:

$$Q_{N,M} = \begin{array}{c} \\ \\ Network_1 \\ Network_2 \\ \vdots \\ Network_N \end{array} \begin{pmatrix} \overset{Attribute_1}{w_1} & \overset{Attribute_2}{w_2} & \overset{\dots}{\dots} & \overset{Attribute_M}{w_M} \\ \\ n_{11} & n_{12} & \dots & n_{1M} \\ n_{21} & n_{22} & \dots & n_{2M} \\ \vdots & \vdots & \ddots & \vdots \\ n_{N1} & n_{N2} & \dots & n_{NM} \end{pmatrix}$$

3.3.1. SAW

The overall score of each candidate network is determined by the weighted sum of all attribute values. The score *SSAW* of each candidate network *Ci* is obtained by summing the contributions of each *nij* normalized metric multiplied by the assigned importance weight *wj*. The selected network is:

$$S_{SAW} = \max_i \sum_{j=1}^{M} w_j \cdot n_{ij} \tag{2}$$

Such that, *wj* is the weight vector taken from Table 3. *nij* is the value of normalized attribute j of network i. N and M are respectively the number of candidates and the number of network attributes considered.

3.3.2. TOPSIS

The selected network is the one that is closest to the ideal solution and farthest from the worst case solution. These networks are defined as the network with the best and worst values for each metric. For a metric of performance, the best value is the largest. For a cost metric, the best value is the lowest. The steps of TOPSIS are cited in [20]:

Step 1: Calculate the weighted normalized decision matrix as follows:

$$v_{ij} = w_{ij} \cdot n_{ij} \tag{3}$$

Step 2: Determine the "positive ideal" and "negative ideal" alternatives. They are given by:

$$A^{+} = \{v_1^{+}, v_2^{+}, ...\}, v_j^{+} = \max_i(v_{ij})$$
$$A^{-} = \{v_1^{-}, v_2^{-}, ...\}, v_j^{-} = \min_i(v_{ij}) \quad (4)$$

Step 3: Calculate the distances. The separation of each alternative *Ai* from the ideal solution *A*+ and the separation of each alternative *Ai* from the ant-ideal solution *A*− are calculated as:

$$d_i^{+} = \sqrt{\sum_{j=1}^{M} |v_i^{+} - v_{ij}|} \; d_i^{-} = \sqrt{\sum_{j=1}^{M} |v_i^{-} - v_{ij}|} \quad (5)$$

Step 4: Calculate the relative proximity to the ideal solution. The relative proximity of *Ai* with regards to *A*+ and *A*− is given by *Ri* and can be expressed as:

$$R_i = \frac{d_i^{+}}{d_i^{+} + d_i^{-}} \quad (6)$$

where *Ri* is the best while it is most proximate to 0; *Ri*∈[0,1].

3.3.3. MEW

The score *SMEW* of network i is determined by the weighted product of the attributes (or metrics):

$$S_{MEW} = \prod_{j=1}^{M} n_{ij}^{w_j} \quad (7)$$

Note that in this eq, *wj* is a positive power for gain metrics $n_{ij}^{w_j}$, and a negative power for cost metrics $n_{ij}^{-w_j}$. Since the score of a network obtained by MEW does not have an upper limit, it is proper to compare each network with the score of the positive ideal network *A+* which is defined by the first part of Equation (4). The value ratio *Ri* between network i and the positive ideal is calculated by:

$$R_i = \frac{\prod_{j=1}^{M} n_{ij}^{w_j}}{\prod_{j=1}^{M} (n_{ij}^{*})^{w_j}} \tag{8}$$

where $Ri \in [0,1]$. Hence, the selected network is : $A_{MEW} = \max_i(R_i)$

4. PERFORMANCE EVALUATION

4.1. Simulations

In this simulation, to explain Vertical Handover schemes we use the same scenario for different mathematical theories.

On the network side, we consider two available networks (WLAN and Wimax). Both are ideal partners for operators to deliver convenient and affordable services, they are open IEEE wireless standards built from scratch for IP-based applications. We used the same scenario as in [16]. On the user side, we consider four types of applications with different QoS requirements including Conversational, Streaming, Interactive and background. Each traffic class is combined with four different QoS parameters or attributes: Throughput, e2e Delay, Jitter, and Ber. The four traffic classes have different QoS requirements. Therefore, different weights have been assigned to the AHP method for the same attribute with different traffic classes. We use our own weights which are shown in Table 3. Although the simulation is maintained for 10 min, 100,000 packets are supposed to be sent and the decision is made every 5 s in realtime.

In case of N different available networks, the terminal needs to apply the methods for the four QoS classes. For each method, the terminal perform 4 x N computations to select the optimal network, the total of $3 \times 4 \times N$ computations in the case of all decision schemes.

To calculate the time T the terminal will require selecting the optimal network, we translated the complexity of the algorithms

$\mathcal{O}(n^2) = max(\mathcal{O}(SAW), \mathcal{O}(TOPSIS), \mathcal{O}(MEW))$.Typically, for high quality and seamless handover, the packet delay must not exceed 150 ms. Assuming the terminal performs one billion operations per second, the approximate running time of each algorithm where $n<5$ is 250 ns. Accordingly, the time to find the rank of all the available network is ~1 ms for the four types of applications

The following section details the development of the network through the simulation.

4.2. Results

In some point of the simulation, we give Wifi more importance with the weighting vectors, but in all of the following figures Wimax is selected, indeed the choise is based on multiple factors, bandwidth occupation for example. Therefore, to compare the performance of each decision we choose the number of lost packets and the end-to-end delay during the simulation. Those parameters vary within the change of performance of the methods and considering the time taken by each method to be excecuted and the packets dropped all along the vertical handover excecution.

The Figure 3 and Figure 4 describe the end to end delay and the Packet Loss respectively of each traffic class for each handover decision algorithm, also, it presents the difference of the performance between the weights given in [15] and presented with our scenario in [16] and our weights with the same scenario.

In terms of delay, the weights offer a better performance for the four type of application against the packet loss, the new weight helped in increasing the number of lost packets, even though the number of lost packets is very low considering the number of packets that should be transmitted (100,000).

In Conversational Class, the results show that TOPSIS and SAW maintained the same performance in terms of delay and packet loss. In Streaming Class, SAW presented a slightly low delay as did TOPSIS for packet loss. On the Interactive Traffic, MEW showed the best performance in terms of delay and SAW provides the best performance in terms of packet loss. Finally, in the Background Class, MEW has better loss rate whereas both MEW and TOPSIS maintained the same performance.

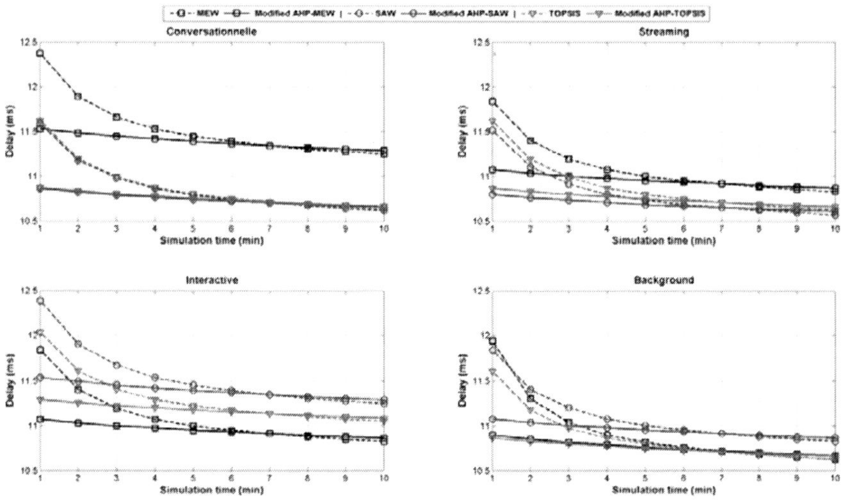

Figure 3. Behavior of Delay over Time.

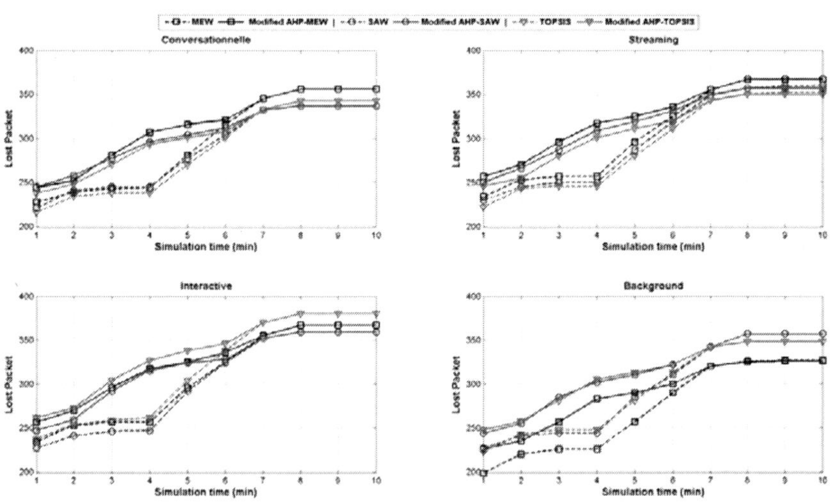

Figure 4. Behavior of Packet Loss over Time.

5. CONCLUSIONS

In terms of resource management in the heterogeneous wireless network, dynamic network selection algorithms provide users with the appropriate QoS in terms of operator's and user's preferences.

Therefore, in this paper, we presented the mathematical models of three different vertical handoff decision algorithms, SAW, TOPSIS and MEW and the comparison between their performances in four types of applications: conversational, streaming, interactive and background. In addition, we investigated the weights provided to those methods in order to figure out how to establish, objectively, the vertical handover without decision makers mediation required by AHP. The realtime simulation includes Wifi and Wimax. The attributes considered are throughput, delay, jitter, and Ber. Results showed that the proposed parameters provide an improvement of Delay and offer allowable Packet loss in different types of application.

For the future, we intend to enhance AHP with fuzzy logic to achieve the objectivity needed in vertical handover. Also, we consider improving the simulation with sufficient performance comparison using realtime data acoording to each type of applications.

AUTHOR CONTRIBUTIONS

Maroua Drissi: Main writing and also designing the total system, analyzing and improving the proposed system; Mohammed Oumsis: Total supervision of the paperwork, review, and comments, etc. Both authors have read and approved the final manuscript.

CONFLICTS OF INTEREST

Authors declare no conflict of interest.

REFERENCES

1. Gustafsson, E.; Jonsson, A. Always best connected. Wirel. Commun. IEEE **2003**, 10, 49–55.
2. Khan, M.; Han, K. A Vertical Handover Management Scheme based on Decision Modelling in Heterogeneous Wireless Networks. IETE Tech. Rev. **2015**, 31, 1–12.

3. Xu, B.; Yu, G.; Dai, J.; Zhu, X. Li-Fi: Light fidelity-a survey. Wirel. Netw. **2015**, 19, 1–11.

4. Kassar, M.; Kervella, B.; Pujolle, G. An overview of vertical handover decision strategies in heterogeneous wireless networks. Comput. Commun. **2008**, 31, 2607–2620.

5. Chandavarkar, B.; Reddy, G. Survey paper: Mobility management in heterogeneous wireless networks. Proc. Eng. **2012**, 30, 113–123.

6. Wang, L.; Kuo, G.S. Mathematical Modeling for Network Selection in Heterogeneous Wireless Networks 2014; A Tutorial. Commun. Surv. Tutor. IEEE **2013**, 15, 271–292.

7. Ahuja, K.; Singh, B.; Khanna, R. Network selection algorithm based on link quality parameters for heterogeneous wireless networks. Opt. Int. J. Light Electron. Opt. **2014**, 125, 3657–3662.

8. Maaloul, S.; Afif, M.; Tabbane, S. A new vertical Handover Decision Based Context Awareness for Ubiquitous Access. In Proceedings of the 2012 International Conference on Communications and Information Technology (ICCIT), Hammamet, Tunisia, 26–28 June 2012; pp. 231–236.

9. Kosmides, P.; Rouskas, A.; Anagnostou, M. Utility-based RAT selection optimization in heterogeneous wireless networks. Pervasive Mob. Comput. **2014**, 12, 92–111.

10. Rao, R.V. Decision Making in the Manufacturing Environment: Using Graph Theory and Fuzzy Multiple Attribute Decision Making Methods; Springer-Verlag: Berlin, Germany, 2007.

11. Cechich, A.; Piattini, M.; Vallecillo, A. Component-Based Software Quality: Methods and Techniques; Springer-Verlag: Berlin, Germany, 2003.

12. Karam, F.W.; Jensen, T. Performance Analysis of Ranking for QoS Handover Algorithm for Selection of Access Network in Heterogeneous Wireless Networks. In Proceedings of the 2012 21st International Conference on Computer Communications and Networks (ICCCN), München, Germany, 30 July–2 August 2012; pp. 1–6.

13. Savitha, K.; Chandrasekar, C. Trusted network selection using SAW and TOPSIS algorithms for heterogeneous wireless networks. Int. J. Comput. Appl. **2011**, 26, 22–29.

14. Liu, S.; Su, P.; Xu, M. An Improved TOPSIS Vertical Handoff Algorithm for Heterogeneous Wireless Networks. In Proceedings of the 2010 12th IEEE International Conference on Communication Technology (ICCT), Nanjing, China, 11–14 November 2010; pp. 750–754.

15. Stevens-Navarro, E.; Wong, V.W. Comparison between Vertical Handoff Decision Algorithms for Heterogeneous Wireless Networks. In Proceedings of the VTC 2006-Spring 63rd Vehicular Technology Conference, Melbourne, Australia, 7–10 May 2006; Volume 2, pp. 947–951.

16. Drissi, M.; Oumsis, M. Performance Evaluation of Multi-Criteria Vertical Handover for Heterogeneous Wireless Networks. In Proceedings of the Intelligent Systems and Computer Vision (ISCV), Fez, Morocco, 25–26 March 2015; pp. 1–5.

17. Jaraiz-Simon, M.D.; Gomez-Pulido, J.A.; Vega-Rodriguez, M.A. Embedded intelligence for fast QoS-based vertical handoff in heterogeneous wireless access networks. Pervasive Mob. Comput. **2015**, 19, 141–155.

18. Saaty, R. The analytic hierarchy process—What it is and how it is used. Math. Model. **1987**, 9, 161–176.

19. Saaty, T.L. The Analytic Hierarchy Process: Planning, Priority Setting, Resource Allocation; McGraw: New York, NY, USA, 1980.

20. Vega, A.; Aguarón, J.; García-Alcaraz, J.; Moreno-Jiménez, J.M. Notes on Dependent Attributes in TOPSIS. Proc. Comput. Sci. **2014**, 31, 308–317.

2

CHAPTER

A DOUBLE-NEGATIVE METAMATERIAL-INSPIRED MOBILE WIRELESS ANTENNA FOR ELECTROMAGNETIC ABSORPTION REDUCTION

Touhidul Alam [1], Mohammad Rashed Iqbal Faruque [1,*] and Mohammad Tariqul Islam [2]

[1] *Space Science Centre (ANGKASA), Universiti Kebangsaan Malaysia, UKM, Bangi 43600, Selangor, Malaysia*

[2] *Department of Electrical, Electronic and Systems Engineering, Universiti Kebangsaan Malaysia, UKM, Bangi 43600, Selangor, Malaysia*

ABSTRACT

A double-negative metamaterial-inspired antenna is presented for mobile wireless applications. The antenna consists of a semi-circular radiating patch and a 3 × 4 hexagonal shaped metamaterial unit cell array in the ground plane. The antenna is fed with a 50 Ω microstrip feed line. The electric dimensions of the proposed antenna are 0.20λ × 0.26λ × 0.004λ, at the low-end frequency. The proposed antenna achieves a −10 dB impedance with a bandwidth of 2.29 GHz at the lower band and 1.28 GHz at the upper band and can operate for most of the mobile applications such as upper GSM bands, WiMAX, Bluetooth, and wireless local area network (WLAN) frequency bands. The focused novelties of the proposed antenna are its small size, multi-standard operating bands, and electromagnetic absorption reduction at all the operating frequencies using the double-negative metamaterial ground plane.

Keywords: antenna; double negative metamaterial; electromagnetic absorption; wireless communication

1. INTRODUCTION

Following the recent advent of artificial metamaterials, there has been much interest in microwave applications. Metamaterials are engineered materials that are usually formed by embedding periodic unit cells to produce exotic electromagnetic properties that are naturally unavailable, such as inverted Snell's law or a negative refractive index. Some of these materials can have either negative permittivity or negative permeability at some frequencies, referred to as single negative (SNG) materials. If both negative permittivity (ε) and negative permeability (μ) are found at a certain frequency, the composite material exhibits a negative refractive index (η) property, and it is usually referred to as a double-negative (DNG), negative refractive index (NRI), or left-handed material (LHM). These unconventional properties of metamaterials are used in many current applications, such as microwave component design, antenna design, electromagnetic absorption reduction, contactless measurement, and invisibility cloaking [1,2,3,4].

Much study on the human health risk due to electromagnetic (EM) field radiation from wireless devices is in progress. Many short- and long-term effects of EM radiation on human health, such as disorders in sleep, cognitive function, heart rate, blood pressure, headaches, and brain tumors, are being studied by various health organizations like the World Health Organization (WHO). Now, several international organizations [5,6] have established guidelines for radio frequency exposure from wireless devices. The electromagnetic absorption limit recommended by the International Commission on Non-Ionizing Radiation Protection (ICNIRP) and IEEE C95.1:2005 guideline is 1.6 W/kg averaged over 1 gram of tissue volume in the shape of a cube and 2.0 W/kg average over any 10 grams of continuous tissue.

In recent years, extensive research efforts have been devoted to electromagnetic absorption reduction from mobile handset antennas. Different methods have been used to reduce EM absorption, such as embedding ferrite sheets [7,8], parasitic elements [9], artificial magnetic conductors, electromagnetic band gaps [10], and metamaterials [11,12,13]. In [14], the author presents the SAR reduction using metamaterial, but did not provide detailed information. Tay et al. proposed a reflector with a dipole to reduce the electromagnetic absorption in [15]. The drawback of this technique is the use of an additional reflector together with the main antenna, resulting in increased manufacturing cost and device dimensions. Kitra et al. investigated the EM absorption reduction upon the inclusion of ferrite in a material-loaded antenna and succeeded in reducing the EM absorption by 88% compared to conventional phones [8]. Though the ferrite material has special properties of permittivity and permeability to reduce EM absorption, it increases the manufacturing cost. In [9], Zhan et al. combined PIFA and a side-mounted inverted "F" antenna (IFA) for multifunctional applications as commercially needed and compared the SAR value with that of a conventional PIFA antenna. Although a reduction of 30% was achieved by combining a PIFA with a long IFA as the parasitic element, a large space is required to mount with its wireless devices. Sultan et al. proposed an EBG structure embedded antenna to reduce the maximum SAR [10]. In [11], Rashed et al. proposed a DNG metamaterial structure, which can be attached to the PCB to reduce the EM absorption. The major drawback of this technique is that the metamaterial structure needs additional space to mount with the PCB.

Antenna researchers are also extensively researching the minimization of the antenna size and cost, together with increasing the bandwidth to cover multiband. Chang et al. developed a Penta-band printed PIFA antenna for WLAN operation in a mobile phone [16] that can operate in two wide bands at approximately 900 MHz and 1900 MHz. In [17], Jie et al. presented a printed octaband monopole antenna for mobile phones sized at 15 × 26 mm^2, which can operate in GSM850 (824–894 MHz), GSM900 (880–960 MHz), DCS (1710–1880 MHz), PCS (1850–1990 MHz), UMTS (1920–2170 MHz), and WiMAX (3400–3600 MHz). Chen et al. proposed a modified T-shaped planar antenna for wireless mobile applications that can operate in the DCS, UMTS, and lower and higher WLAN frequency bands [18]. The proposed antenna size was quite larger for mobile applications, which was 65 × 40 mm^2. In [19], a crescent-shaped mobile wireless antenna was presented. The presented antenna can cover the frequency bands of 1.7 to 3.1 GHz, with antenna dimensions of 57 × 37.5 × 0.8 mm^3. Sung et al. presented a modified L-shaped feed antenna that achieved an impedance bandwidth of 3.51 GHz (1.21–4.72 GHz) [20]. The antenna dimension was also larger than convenient for mounting on mobile devices.

In this paper, a metamaterial-loaded microstrip patch antenna is proposed for mobile wireless communication systems. The hexagonal metamaterial structure is embedded on the ground plane to reduce the maximum electromagnetic radiation of the proposed antenna. Moreover, the antenna performance has been investigated. This paper is structured as follows. Section 2 describes the structural design of the proposed antenna and unit cell array. Metamaterial characterization is included in Section 3. The proposed antenna performance is discussed in Section 4. The specific absorption rate analysis is discussed in Section 5, and Section 6 concludes the paper.

2. DESIGN OF THE PROPOSED ANTENNA AND UNIT CELL

The proposed metamaterial antenna and unit cell structure is presented in Figure 1. A hexagonal shaped metamaterial unit cell array is designed and fabricated on a 0.8 mm thick FR-4 substrate. The proposed antenna is also printed on a 0.8 mm thick FR-4 substrate of dimensions 45 × 35 mm^2. The antenna is incorporated with a semi-circular patch and a hexagonal shaped metamaterial array in the ground plane. The semi-circular patch is printed on the top layer, and the metamaterial array is printed on the bottom layer of the substrate material. The antenna and unit cell design specifications are listed in Table 1.

Table 1. Antenna design and unit cell specifications.

Parameter Name	Value (mm)	Parameter Name	Value (mm)
L	45	L3	12
W	35	L4	13
Lf	17	L5	32
Wf	1.25	L6	28
L1	30	a	8
L2	15	g	0.8

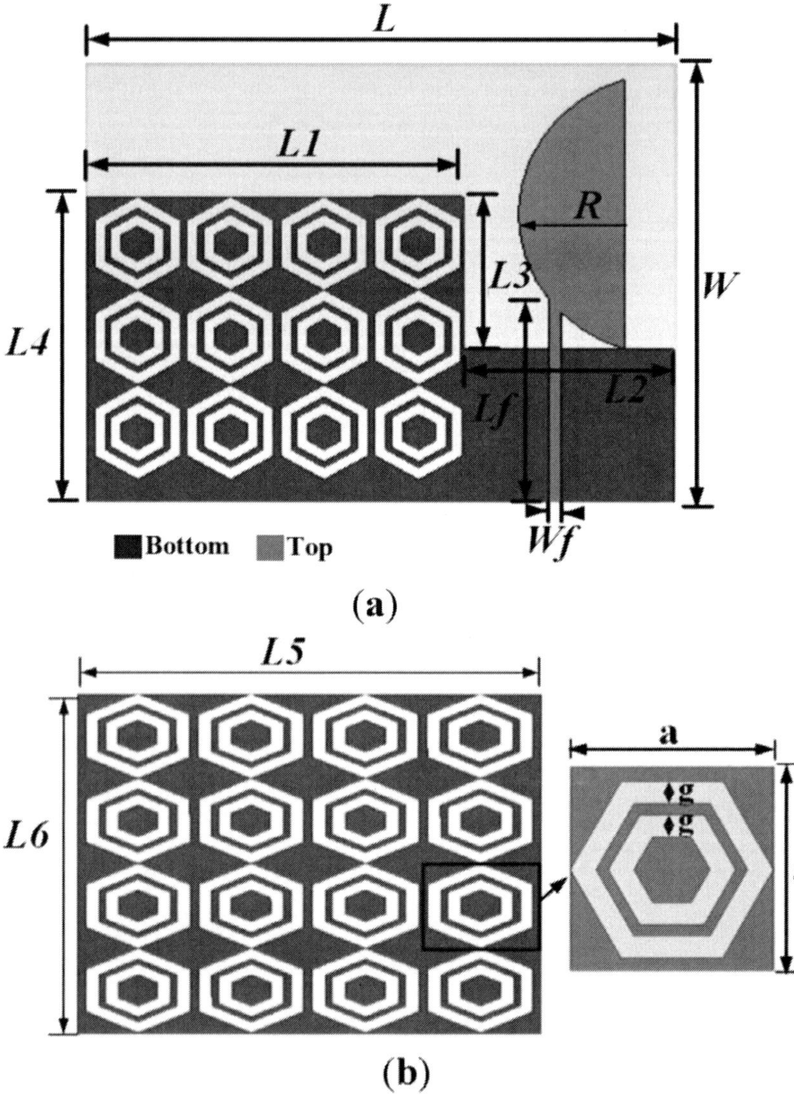

Figure 1. (**a**) Schematic diagram of the antenna; (**b**) Unit cell array and unit cell configuration.

3. DNG METAMATERIAL CHARACTERIZATION

The metamaterial structure interacts with electromagnetic waves and shows some special properties. For characterizing the metamaterial, the array structure was positioned between two waveguide ports on the negative and positive x-axis and excited by a transverse electromagnetic (TEM) wave. The perfect electric conductor (PEC) boundary and the perfect magnetic conductor (PMC) boundary were defined along the y and z axes, respectively, as shown in Figure 2a. A frequency solver with a tetrahedral mesh was used for simulation. The normalized impedance was set to 50 Ω. The simulation was run in the frequency range of 1–6 GHz for both the metamaterial and antenna performance investigations. The constitutive parameters of the metamaterial were retrieved using scattering parameters, the method used in [21] and presented in Figure 3. The measured and simulated spectral analyses of the proposed metamaterial structure are illustrated in Figure 3a. It may be observed from Figure 3 that there are two resonance points at 1.963 GHz and 5.03 GHz where the DNG characteristics of the metamaterial have been found. It is shown from Figure 3b that the the retrieved negative permittivity regions of the the structure are found 1.97–3.03GHz and 5.0–6.0 GHz. Moreover, the retrieved negative permeability regions are 1.96–3.5 GHz and 5.05–6 GHz. Similarly, refractive index regions are obtained at 1.68–3.43 GHz and 5.04–6.0 GHz. Therefore, the metamaterial structure achieves double-negative medium of about 1.50 GHz at the lower band and about 0.95 GHz at the upper band. The magnetic resonance behavior can be assumed by observing simulated current distributions, as shown in Figure 2b, and comparing them with the existing behavior of the metamaterials [22,23]. It is shown from Figure 2b that the resonant electric current oscillates along the finite conductor. The parallel finite conductor can be considered as an LC resonant circuit. The inductance of the structure is formed by self- and mutual inductance of the conductors and capacitance is introduced between the gaps. Moreover, the periodic arrangement of the unit cells has an additional coupling between adjacent unit cells.

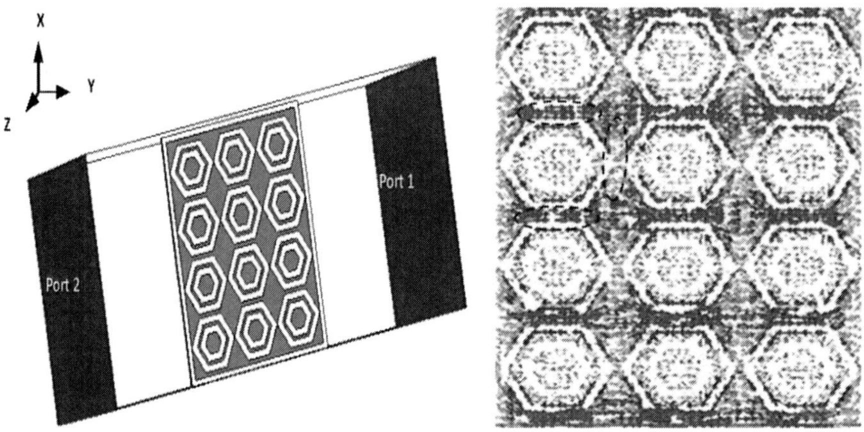

(a) (b)

Figure 2. (**a**) Simulation arrangement of a unit cell array of metamaterial characteristics; (**b**) Surface current distribution at 1.97 GHz.

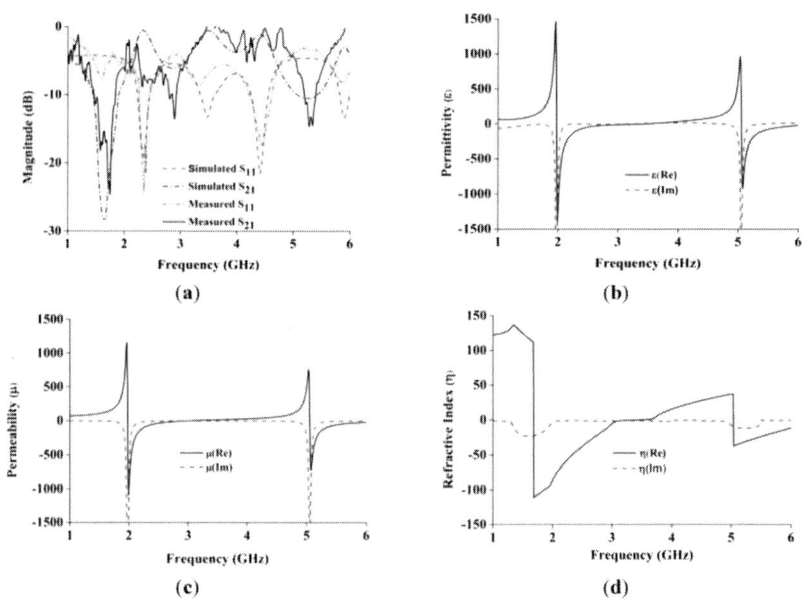

Figure 3. (**a**) Spectral response of the metamaterial structure; (**b**) Real and imaginary values of effective permittivity (ε) vs. frequency; (**c**) Real and imaginary values of effective permeability (μ) vs. frequency; (**d**) Real and imaginary values of refractive index (η) vs. frequency.

4. ANTENNA PERFORMANCE ANALYSIS

A prototype of the antenna has been fabricated using an LPKF Laser and Electronics machine and is shown in Figure 4. The reflection coefficient of the proposed antenna has been measured using a PNA network analyzer, presented in Figure 5. The proposed antenna achieved measured impedance bandwidths of 2.29 GHz (1.66–3.95 GHz) and 1.28 GHz (4.45–5.73 GHz), enabling it to operate in the frequency bands of GSM (1800, 1900, 2100), WiMAX (3.2–3.6 GHZ), Bluetooth (2.4 GHz), and WLAN (5.47–5.9 GHz). Although slight disagreement is found between the measured and simulated reflection coefficients, the two results are most likely identical. The main reasons for the disagreement between the two results are fabrication tolerance and deficient soldering effects of the SMA connector.

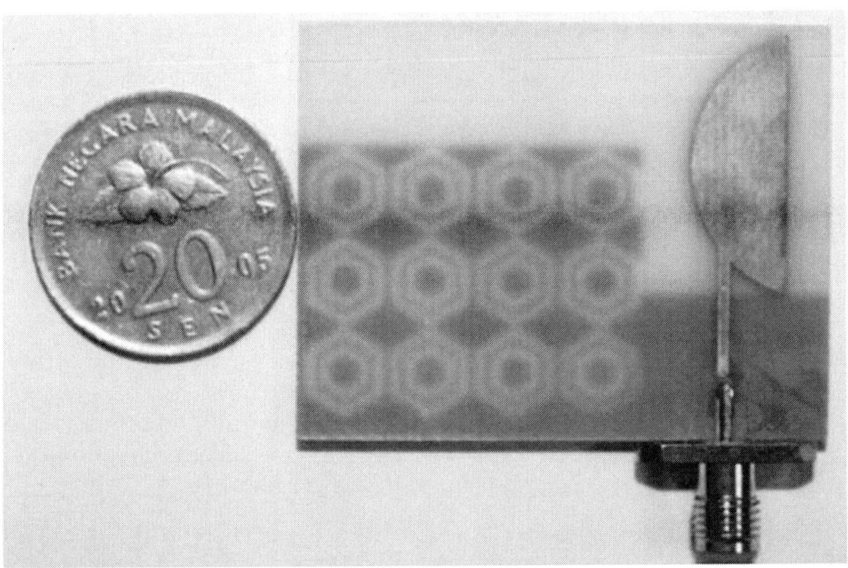

Figure 4. Proposed antenna fabricated prototype.

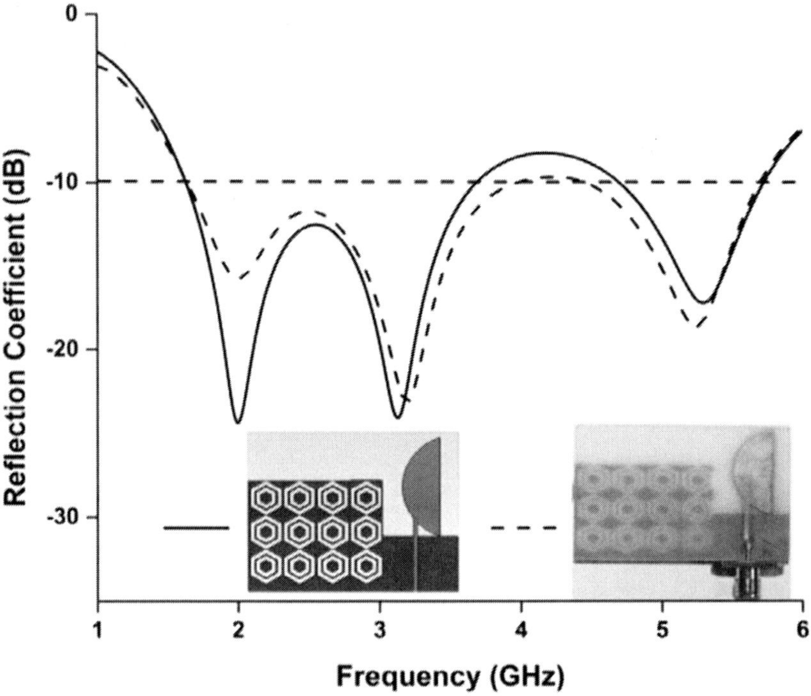

Figure 5. Simulated and measured reflection coefficients of the proposed antenna.

To observe the physical phenomenon of the proposed antenna, the current distribution at different frequencies is analyzed. The surface current distribution is obtained from simulation software for different frequencies, as shown in Figure 6. A stronger surface current distribution is observed along the metamaterial ground plane and near the feed line.

The radiation pattern of the proposed antenna has been measured using the Satimo nearfield measurement system (Satimo Starlab). The measured radiation patterns at 1.8 and 2.4 GHz are demonstrated in Figure 7 where both Phi = 0° and Phi = 90° are included. It is seen from Figure 7a,b that the radiation patterns at Phi = 90° are nearly omnidirectional for $E\phi$. According to the experimental result, it is seen that for the overall antenna volume, the proposed antenna with the compact size of $37 \times 47 \times 0.8$ mm^3 has an antenna size at least 33% less than [18], 18.5% less than [19], and 80% less than [20], and shows better antenna performances.

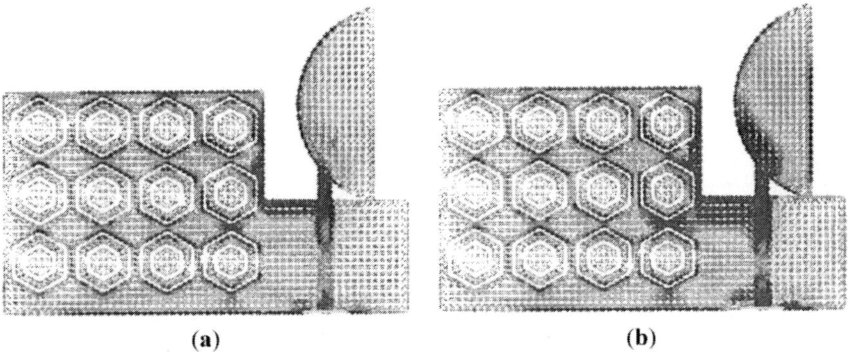

Figure 6. Surface current distribution of the proposed antenna at (**a**) 1.8 GHz; (**b**) 2.4 GHz.

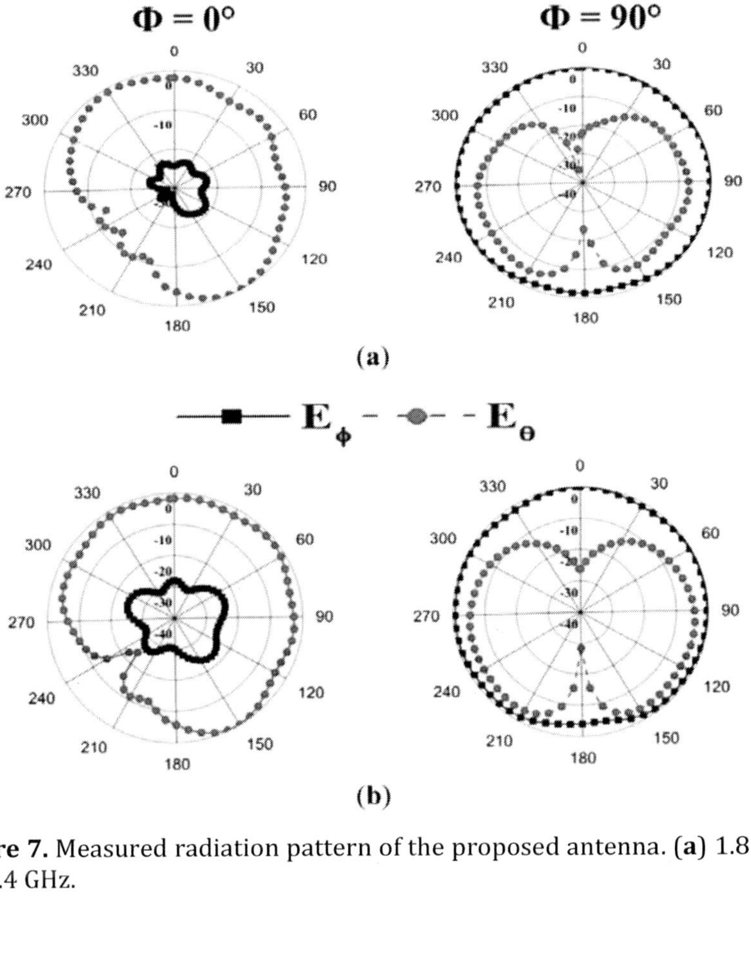

Figure 7. Measured radiation pattern of the proposed antenna. (**a**) 1.8 GHz; (**b**) 2.4 GHz.

5. ELECTROMAGNETIC ABSORPTION ANALYSIS

The specific absorption rate of the proposed antenna has been studied using a commercially available finite-difference time-domain (FDTD) method-based CST microwave studio. The simulation arrangement was set up according to IEEE and Federal Communications Commission (FCC) guidelines. The input power was set to 500 mW, and the distance between the head phantom and the mobile phone was approximately 2 mm. The SAM head phantom consists of head equivalent liquid ($\varepsilon_r = 40$, $\sigma = 1.4$) and shell ($\varepsilon_r = 5$, tangent delta = 0.05). The simulated 1 g SAR at 1.8 GHz and 2.4 GHz has been analyzed and is presented in Figure 8. It is shown in Figure 8 that the metamaterial antenna shows 1 g SAR values at 1.8 GHz and 2.4 GHz of 0.708 W/Kg and 0.484 W/Kg, respectively. The simulated SAR values of the proposed metamaterial-loaded antenna are much lower than the standard safety guidelines. Here, the metamaterial structure plays the most important role in reducing the SAR values. The metamaterial structure has high electromagnetic surface currents and acts as a perfect magnetic conductor (PMC) in a specified frequency range. Moreover, the stop band characteristics of the metamaterial structure can control the radiation characteristics of the antenna. These characteristics of the metamaterial can reduce the undesirable EM waves that travel to the human head without degrading the antenna performance.

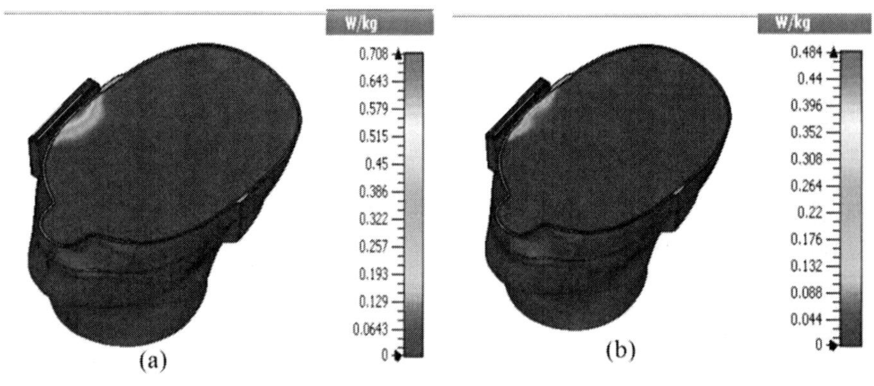

Figure 8. Simulated 1 g SAR values of the proposed antenna (**a**) at 1.8 GHz and (**b**) at 2.4 GHz.

The SAR values of the proposed antenna have been measured using the Satimo COMOSAR measurement system. The system consists of a robot to move the field probe, head phantom, and test zig, as shown in Figure 9. The field probe is connected to the system computer. The head phantom is filled with liquid, which maintains the equivalent dielectric properties of the human head. The metamaterial antenna-loaded mobile phone was placed in a test zig and connected with an input power supply set at 27 dBm (500 mW). The distance between the head phantom and the mobile phone was approximately 6 mm. The measurement was performed at 1.8 and 2.4 GHz. The measured 1 g SAR value of the proposed antenna is shown in Figure 10, and the simulated and measured results are listed in Table 2. It is seen from Table 2 that the proposed antenna has succeeded in a large-scale reduction of SAR values as compared to reported antennas.

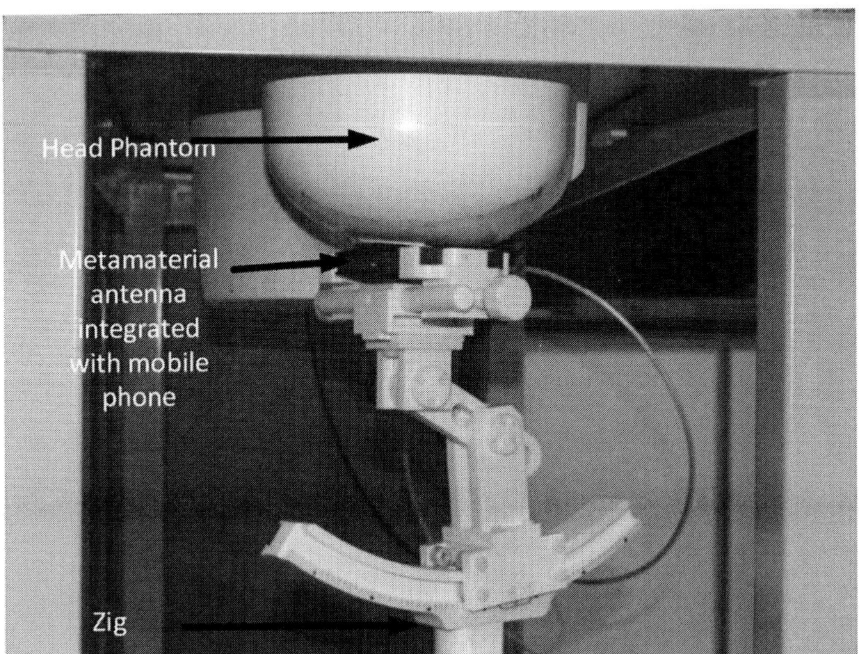

Figure 9. SAR measurement in the Satimo SAR measurement lab.

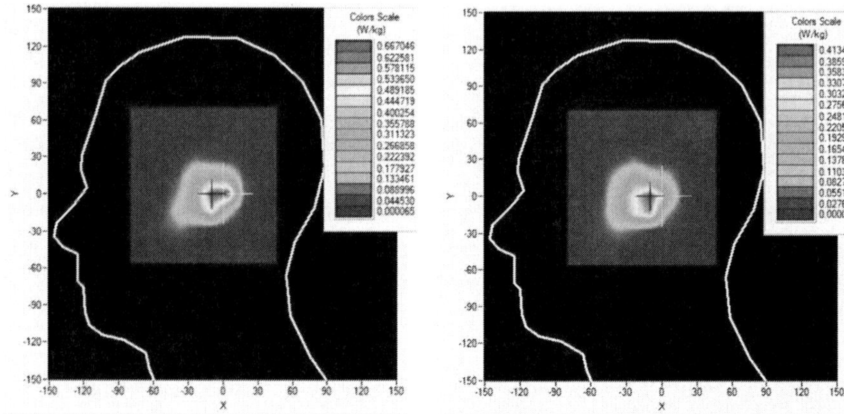

Figure 10. 1 g SAR measurements of the proposed antenna at (**a**) 1.8 GHz and (**b**) 2.4 GHz.

Table 2. SAR values of the proposed antenna.

Type	1g SAR (W/Kg)			
	Condition	**Frequency (GHz)**	**SAR values (W/Kg)**	**S_{11} (dB)**
Metamaterial antenna	simulated	1.8	0.708	−16
	measured	1.8	0.667	−14.8
	simulated	2.4	0.484	−13.5
	measured	2.4	0.413	−14.2

The equivalent isotropic radiated power (EIRP) is a very important criterion that all wireless equipment and devices must satisfy to minimize the exposure of human beings to electromagnetic fields. The EIRP is related to the power transmitted (Pt), cable losses (Lc), and the antenna gain (Ga), and its expression is presented in Equation 1. The EIRP of the proposed antenna has been calculated and is presented in Figure 11. It is seen from Figure 11 that the antenna satisfies the EIRP limit for wireless applications.

$$ETRP\ [dBm] = Pt\ [dBm] - Lc\ [dB] + Ga\ [dBi] \quad (1)$$

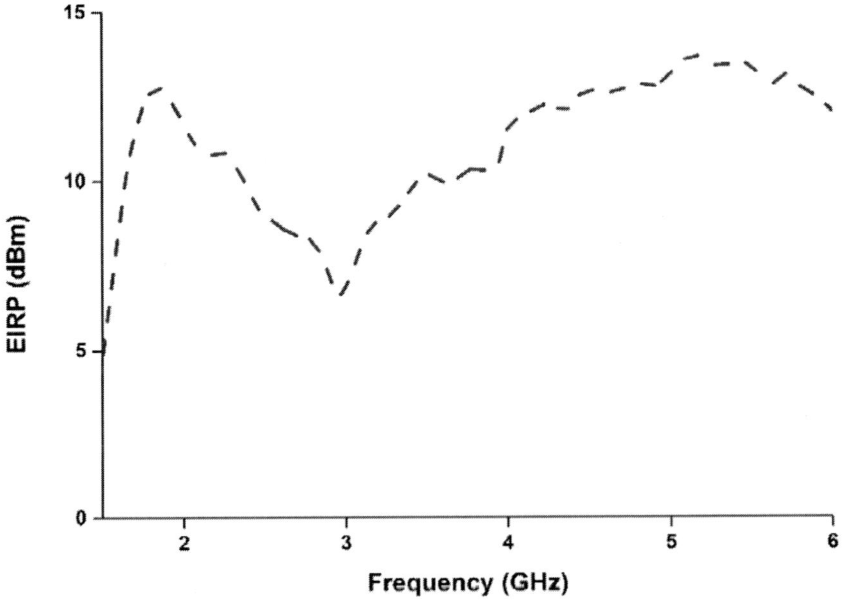

Figure 11. Equivalent isotropic radiated power (EIRP) of the proposed antenna.

6. CONCLUSIONS

A low-profile metamaterial antenna has been presented for low electromagnetic absorption mobile applications. The proposed antenna with a metamaterial structure was found to reduce the peak SAR values without degrading the antenna performance. The measured 1 g SAR values of the proposed antenna were 0.667 W/kg and 0.413 W/kg at 1.8 GHz and 2.4 GHz, respectively, which are 58.31% and 74.19% lower than the standard safety guidelines. Therefore, the human body can be sheltered from the hazardous effects of the electromagnetic radiation using the proposed antenna.

ACKNOWLEDGMENTS

This work is supported by the Ministry of Education Malaysia (MOE) under grant no. FRGSTOPDOWN/2014/TK03/UKM/01/1 and research university grant Dana Lonjakan Penerbitan-DLP-2014-003.

AUTHOR CONTRIBUTIONS

Touhidul Alam has designed the antenna and investigated the antenna performance analysis and optimization. Mohammad Rashed Iqbal Faruque and Mohammad Tariqul Islam have contributed in revising the article.

CONFLICTS OF INTEREST

The authors declare no conflict of interest.

REFERENCES

1. Islam, M.M.; Islam, M.T.; Samsuzzaman, M.; Faruque, M.R.I.; Misran, N.; Mansor, M.F. A miniaturized antenna with negative index metamaterial based on modified SRR and CLS unit cell for uwb microwave imaging applications. Materials **2015**, 8, 392–407.

2. Wu, B.-I.; Wang, W.; Pacheco, J.; Chen, X.; Grzegorczyk, T.M.; Kong, J.A. A study of using metamaterials as antenna substrate to enhance gain. Prog. Electromagn. Res. **2005**, 51, 295–328.

3. Lee, H.-M.; Lee, H. A metamaterial based microwave absorber composed of coplanar electric-field-coupled resonator and wire array. Prog. Electromagn. Res. C **2013**, 34, 111–121.

4. Yahiaoui, R.; Chantalat, R.; Chevalier, N.; Jouvet, M.; Lalande, M. Metamaterial-based highly directive antenna: Application in a monochromatic wave radar for a contactless measurement of the breathing activity. Prog. Electromagn. Res. C **2013**, 44, 185–195.

5. IEEE Standard for Safety Levels with Respect to Human Exposure to Radio Frequency Electromagnetic Fields,3 KHz to 300 GHz; IEEE Std C95.1–2005 (Revision of IEEE Std C95.1–1991); Institute of Electrical and Electonics Engineers: New York, NY, USA, 2006.

6. International Non-Ionizing Radiation Committee of the International Radiation Protection Association. Guidelines on limits on exposure to radio frequency electromagnetic fields in the frequency range from 100 KHz to 300 GHz. Health Phy. **1988**, 54, 115–123.

7. Islam, M.T.; Faruque, M.R.I.; Misran, N. Design analysis of ferrite sheet attachment for SAR reduction in human head. Prog. Electromagn. Res. **2009**, 98, 191–205.

8. Kitra, M.I.; Panagamuwa, C.J.; McEvoy, P.; Vardaxoglou, J.; James, J.R. Low SAR ferrite handset antenna design. IEEE Trans. Antennas Propag. **2007**, 55, 1155–1164.

9. Zhan, L.; Rahmat-Samii, Y. Optimization of PIFA-IFA combination in handset antenna designs. IEEE Trans. Antennas Propag. **2005**, 53, 1770–1778.

10. Sultan, K.; Abdullah, H.; Abdallah, E.; Hashish, E. Low-SAR, miniaturized printed antenna for mobile, ISM, and WLAN services. IEEE Antennas Wirel. Propag. Lett. **2013**, 12, 1106–1109.

11. Faruque, M.R.I.; Islam, M.T. Design of miniaturized double-negative material for specific absorption rate reduction in human head. PLoS ONE **2014**, 9.

12. Manapati, M.; Kshetrimayum, R. Sar reduction in human head from mobile phone radiation using single negative metamaterials. J. Electromagn. Waves Appl. **2009**, 23, 1385–1395.

13. Faruque, M.R.I.; Islam, M.T. Novel triangular metamaterial design for electromagnetic absorption reduction in human head. Progr. Electromagn. Res. **2013**, 141, 463–478.

14. Alam, T.; Faruque, M.; Islam, M. Specific absorption rate reduction of multi-standard mobile antenna with double-negative metamaterial. Electron. Lett. **2015**, 51, 970–971.

15. Tay, R.Y.-S.; Balzano, Q.; Kuster, N. Dipole configurations with strongly improved radiation efficiency for hand-held transceivers. IEEE Trans. Antennas Propag. **1998**, 46, 798–806.

16. Chih-Hua, C.; Kin-Lu, W. Printed λ/8-PIFA for penta-band wwan operation in the mobile phone. IEEE Trans. Antennas Propag. **2009**, 57, 1373–1381.

17. Ma, J.; Yin, Y.Z.; Guo, J.L.; Huang, Y.H. Miniature printed octaband monopole antenna for mobile phones. IEEE Antennas Wirel. Propag. Lett. **2010**, 9, 1033–1036.

18. Chen, S.-B.; Jiao, Y.-C.; Wang, W.; Zhang, F.-S. Modified T-shaped planar monopole antennas for multiband operation. IEEE Trans. Microwave Theory Tech. **2006**, 54, 3267–3270.

19. See, C.H.; Abd-Alhameed, R.A.; Zhou, D.; Lee, T.H.; Excell, P.S. A crescent-shaped multiband planar monopole antenna for mobile wireless applications. IEEE Antennas Wirel. Propag. Lett. **2010**, 9, 152–155.

20. Sung, Y. A printed wide-slot antenna with a modified L-shaped microstrip line for wideband applications. IEEE Trans. Antennas Propag. **2011**, 59, 3918–3922.

21. Islam, S.S.; Faruque, M.R.I.; Islam, M.T. The design and analysis of a novel split-H-shaped metamaterial for multi-band microwave applications. Materials **2014**, 7, 4994–5011.

22. Ramakrishna, S.A. Physics of negative refractive index materials. Rep. Prog. Phys. **2005**, 68.

23. Koschny, T.; Zhou, J.; Soukoulis, C.M. Magnetic response and negative refractive index of metamaterials. Proc. SPIR **2007**, 6581.

3

CHAPTER

LTE AND WIMAX: COMPARISON AND FUTURE PERSPECTIVE

Ismat Aldmour

Department of Computer Engineering and Science, Al Baha University, Al Baha, Saudi-Arabia

ABSTRACT

Two wireless technologies, WiMAX based on IEEE standards and LTE standardized by 3GPP, are two competing technologies, nevertheless, are very technically similar. This competition started with the advent of their pre-4G versions (802.16e for Mobile WiMAX and 3GPP release 8 for LTE) and continued with the advent of their 4G versions (WiMAX 2.0 based on IEEE 802.16 m and LTE-Advanced standardized by Release 10). It looks that the competition ended with the advantage of LTE. Plans are set for WiMAX to migrate/integrate with LTE in a multiple heterogeneous access technology mode. This article addresses the technical similarities and differences that advantage one technology over the other technology in order to determine which of these factors might have contributed to LTE winning. Nontechnical factors of commercial and historical nature which might also advantage one technology over the other one are also explored. Finally, current activities in the standardization of both WiMAX and LTE are presented with a perspective on the prospects of both technologies.

Keywords: LTE-Advanced; WiMAX 2; 3GPP Release 10; IEEE 802.16m; WiMAX-Advanced; LTE vs WiMAX

1. INTRODUCTION

WiMAX (Worldwide Interoperability for Microwave Access); is a technology standardized by IEEE. IEEE issued a series of standards, IEEE 802.16 series of standards, starting 2000, which aimed to provide a metropolitan area data access called Wireless MAN standards. The first in the series that found real applicability was IEEE 802.16d in 2004 [1]. This standard aimed to provide high throughput wireless data, last-mile broadband, to fixed users, which formed a real competitor to DSL and cable data providers. IEEE 802.16e in 2005 [2] formed the basis to what is known as Mobile WiMAX, or WiMAX R1.0. Lately, IEEE 802.16m, standardized in March, 2011, was considered for WiMAX Release 2.0. Release 2.0 offers many folds higher data rates than Release 1.0 and was lately officially recognized as 4G in 2012 [3]. 4G technologies shall satisfy the IMT-Advanced of the ITU [4] aiming at peak data rates in the order of 1 Gbps for low mobility users and 100 Mbps for high mobility users on the downlink to support advanced services and applications [4].

On the other hand, LTE (Long Term Evolution), a mobile telecommunication technology standardized by 3GPP, is the biggest jump on the evolution path from 3G UMTS and CDMA2000 towards 4G, with ambitious requirements for data rates, capacity and latencies [5]. An advanced version of LTE, LTE-Advanced based on 3GPP UMTS Rel 10 in 2011, is also a 4G recognized mobile technology [3].

The two technologies, WiMAX and LTE, competed with each other starting their pre-4G versions and continued with their 4G versions while having much in common. It looks like that finally WiMAX gave up the competition and selected to harmonize and integrate with LTE in its future harmonized WiMAX advanced standard supporting multiple access technologies. This work addresses the technical similarities and differences between the two technologies trying to pinpoint those differences that advantage one technology over the other one. Other factors, commercial, historical, political, etc. which might advantage one technology over the other technology are also exploited. Finally, the article reviews plans of WiMAX and LTE future with alternatives to WiMAX discussed.

The rest of the paper is organized as follows: Section 2 presents the evolution of both LTE and WiMAX standards. Section 3 presents salient features of both. Section 4 discusses some of the main technical differences between the two technologies, while Section 5 discusses other non-technical factors that favor one or the other technology. Section 6 discusses the future of both technologies followed by the conclusions in Section 7.

2. STANDARDS EVOLUTION

Figure 1 below shows the evolution paths of LTE and WiMAX technologies. LTE originated from the growth path of 1G, 2G, 3G mobile generations of the telecom companies and their 3GPP and 3GPP2 associations. 1G is characterized as being analogue with systems standardized by companies like TACS in Europe and AMPS in North America. 2G systems of GSM and cdma-one came as digital solution with circuit switched voice capacity as objective. Moderate data rates were enabled using 2.5G packet switched enhancements like GPRS and EDGE while voice remained circuit switched. The IMT-2000 initiative of the ITU put the requirements for 3G systems with video and data as targets. WCDMA was the adopted system of 3GPP in Europe and many other parts of the world. CDMA2000 was the 3G system of North America (NA). Both of these two used a pair of separate frequency bands to transmit and receive in what is called Frequency Division Duplex (FDD) mode. A third 3G system adopted in China called the synchronous CDMA (SCDMA) uses a single band on Time Division Duplex (TDD) basis. 3G systems continued as both circuit and packet switched based. They received a number of enhancements enabling higher data rates such as High Speed Packet Access (HSPA) enhancement of WCDMA. The NA's counterpart, CDMA2000, received similar enhancements as well.

The advent of OFDM based all IP LTE in 3GPP release 8 with TDD and FDD modes seemed to incite migration of NA's systems to it targeting a unified telecom standard. Plans for Ultra Mobile Broadband (UMB); a suggested OFDM system of NA's 3GPP2, were cancelled. Moreover, TD-SCDMA migration is planned toward TDD mode of LTE (TD-LTE). LTE is considered a 3G system but it provides many folds higher rates than the basic 3G systems. It is therefore called pre-4G or sometimes 3.9G. Commercially it is considered 4G.

IMT-Advanced initiative by the ITU in 2003 [4] set the framework and overall objectives of 4G systems and accelerated the work toward these systems by setting October 2009 as a deadline date for submitting proposals. It sets an ambitious high data rate of 1.0 Gbps peak rate for fixed services and 100 Mbps for mobile services with mobility up to 500 km/hr. It also sets more requirements on packet and handover latency and VOIP efficiency. Additionally, it sets peak and average cell spectral efficiencies and spectral efficiency requirements at cell edges to ensure high throughputs all over. The advanced version of LTE, LTE-Advanced based on 3GPP UMTS Rel 10 in 2011, is an officially recognized 4G technology satisfying IMT-Advanced.

On the other hand WiMAX[2] grew out from IEEE standards adopted by data companies with deep roots in wireline technologies (e.g. the Ethernet standards of IEEE 802.3, etc.) and WiFi wireless technologies (IEEE 802.11). Early attempts to provide metropolitan wireless broad band were based on fixed wireless access techniques that require line of sight and rooftop antennas and operated at high frequencies that can be affected by rain and atmospheric conditions. These attempts had little success. The first in the series that found real applicability was IEEE 802.16d in 2004 [1]. This standard aimed at providing high throughput wireless data, last-mile broadband, to fixed users, which formed a real competitor to DSL and cable data providers. IEEE 802.16e in 2005 [2] formed the basis to what is known as Mobile WiMAX, or WiMAX R1.0. WiMAX was officially recognized as 3G technology in 2007 [6] many years after the recognition of the initial 3G technologies in 2000. Lately, IEEE 802.16m, standardized in March, 2011, was considered for WiMAX Release 2.0. The aim of IEEE 802.16m is to develop an advanced air interface to meet the requirements for IMT-Advanced while being compatible with previous 802.16 standards. Release 2.0, which is the other official 4G technology, offers many folds higher data rates than Release 1.0 satisfying the IMT-Advanced of the ITU [4].

Both LTE-Advanced and WiMAX 2.0 are not yet around. LTE-Advanced is expected to be the main one as most WiMAX operators have put plans to migrate to LTE. Nevertheless, both camps have plans for the future of their technology as shown by the rightmost part of Figure 1 (Beyond 4G/Evolved 4G/Evolved 3G). These will be discussed later in the paper.

3. FEATURES OF LTE/WIMAX

Both WiMAX and LTE are all-IP, all packet technology with a packet network core. This makes them best for bursty data traffic with good VoIP support. Both use OFDMA, Orthogonal Frequency Division Multiple Access, a multiple access technology which is a form of FDM in which the subcarriers are made orthogonal to each other. This enables compacting more subcarriers in the spectrum provided giving rise to higher spectral efficiency. The small subcarrier separation results in large symbol size. This helps mitigating ISI (Inter-Symbol Interference) and reduces the need for

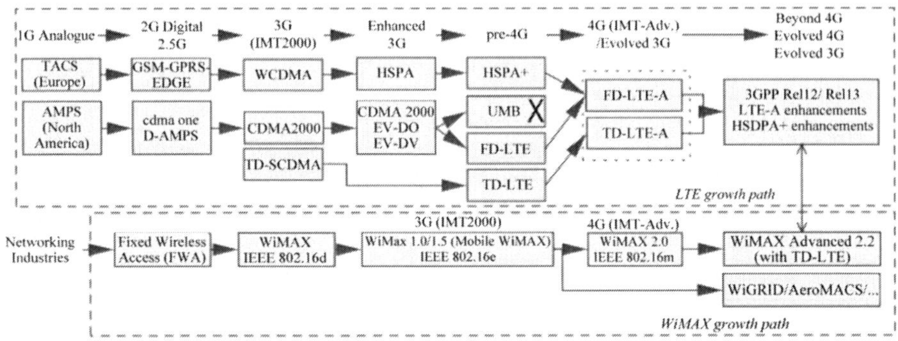

Figure 1. Growth paths for LTE and WiMAX.

complex adaptive equalization needed in single carriers wideband systems. OFDM is robust against frequency selective burst errors and narrow band interference. In OFDMA, the connection is scheduled in both time and frequency by having many connections sharing multiple carriers whereas this sharing can be re-worked periodically to maximize performance [7].

Some of the other features associated with LTE and WiMAX are the following features [8]:

- **Sub-channelization and permutation:** In the allocated spectrum, some subcarriers are used for data while some others are used as guard bands and pilots. Data carriers and pilots are randomly allotted to different sub-channels periodically. In other words, the channels are hopping. This is similar to hopping in Wifi. However, only one channel is hopping in WiFi, while in these, all the sub-channels in the spectrum, say 30, are hopping. This results in interference averaging giving rise to less error corrections and restoring system capacity [9]. Subcarriers are subdivided into groups and only some of these groups can be used in any cell. This is called partial usage of subcarriers (PUSC). PUSC reduces neighboring cells interference therefore improving performance. An alternative is the use of fractional frequency re-use (FFR). In FFR users in the cell close to cell centre use all the frequencies while those toward the cell boundary use frequencies different to those used toward boundaries of neighboring cells in order to reduce inter-cell interference.
- Both LTE and WiMAX (The mobile version of WiMAX, i.e. WiMAX 1.0, and the later WiMAX 2.0) use a version of OFDMA called scalable OFDMA (SOFDMA). In this scheme, whenever the bandwidth allotted to service provider changes, the number of subcarriers also changes so as to keep the inter-carrier spacing fixed. Therefore, the Doppler effect on

performance is kept the same for mobile users. WiMAX 16e may use any available spectrum width from 1.25 MHz to 28 MHz while LTE R8 can use 1.25, 2.5, 5, 10, 15, 20 MHz.

- **Both LTE and WiMAX use AMC** (adaptive modulation and coding) for link adaptation. In this scheme, the connection that is currently using modulation like 16 QAM can be re-scheduled to say, QPSK, a more robust modulation whenever the user's signal is fading, e.g. the user is moving away from base station therefore struggling with smaller signal to noise ratios. This ensures that the connection is kept at an acceptable quality thereafter increasing the range. On the other hand, an approaching user connected at say 16 QAM who is enjoying improved signal to noise ratio can be switched to a higher order modulation like 64 QAM which can provide higher bps/Hz, thereafter increasing capacity. AMC when combined with multicarrier OFDM will result in more advantageous results. This is because adapting a narrowband channel to noise conditions is more efficient than adapting to averaged noise in a wide band channel [10]. Other features of LTE/WiMAX are the use of Hybrid ARQ (HARQ) technique for error detection and multiple antennas to further augment performance and data rates.

The 4G version of WiMAX; WiMAX 2.0, supports a number of modifications on the physical layer that target higher throughputs and improved performance. Amongst these are the following:

- Framing: WiMAX R1.0, suffered from high latencies due to its relatively long frames of 5 ms compared to the later LTE which has a subframe of 1 ms long. Therefore, WiMAX, in its new release subdivided the 5 ms frame into 8 subframes (now the subframe is 5/8 ms). They also kept the 5 ms frames for compatibility purposes with legacy WiMAX R1.0. Furthermore, a super frame of 20 ms is introduced. It aims to reduce framing overheads by incorporating the common headers and control bits in it. Thus, the new release framing structure is a 3 level framing structure. The new framing structure offers persistent allocations to insure QoS for recurring (VOIP) transmissions. Similarly, LTE has a similar 3 level framing structure with basic slot of 0.5 ms duration. The frames structure for both LTE-Advanced and WiMAX 2.0 are shown in Figure 2.
- Data rates: To attain high peak data rates requirements of IMT-advanced, LTE-Advanced and WiMAX R2.0 work on increasing the transmission bandwidth utilized up to 40/100 MHz bandwidth respectively. As it is not likely that such large bandwidths be available in one band, subcarriers are spread over multiple frequency bands. This is called multicarrier or carrier aggregation. Subcarriers in any one

channel can be contiguous in the same band or selected from separate bands.

- Cell edge throughputs: Besides peak data rates, IMT-Advanced concentrated on average as well as cell edge throughputs. This issue has been lightly tackled in previous systems Many of the WiMAX R2.0/LTE-Advanced specifications exceed those of IMT advanced requirements [11,12]; for example, WiMAX 2.0 shall offer a downlink spectral efficiency of 2.6 and 0.09 bit/s/Hz/sector for cell and cell edge users respectively while IMT advanced requires 2.2 and 0.06 respectively.
- Other features: LTE-Advanced and WiMAX R2.0 support other measures to further improve data rates (e.g. [11,13,14]) such as:

    - Femto cells support for high throughputs in small cells;
    - Higher order MIMOs. with up to 8 × 8 MIMO in LTE-Advanced;
    - The use of coordinated multipoint base stations (CoMP) in which two base stations or more can cooperate on reaching the user equipment especially of those at cell edges;
    - The use of relay nodes to improve coverage and cell edge throughput;
    - Self-Organizing Networks (SON) to reduce costs of operation and maintenance and optimize performance.

4. TECHNICAL DIFFERENCES

As can be deduced from the previous section, there are lots of technical similarities between the two technologies in architecture and targets. Both uses OFDMA with flat-IP architecture and both are meant to meet or even surpass IMT-Advanced requirements with similar enabling technologies. Nevertheless, a number of technical differences exist. Some of these differences with mobile WiMAX (WiMAX 1.0) compared to LTE and WiMAX 2.0 compared to LTE-Advanced are the following:

Duplex mode: Both LTE and WiMAX provide for both TDD and FDD. However, FDD was the focus of all telecom companies and continued throughout the different generations. TD-LTE is gaining popularity as migration path of the synchronous CDMA of China 3G. WiMAX, on the other hand, had TDD focus throughout. Future WiMAX embracing LTE will most likely head to TD-LTE.

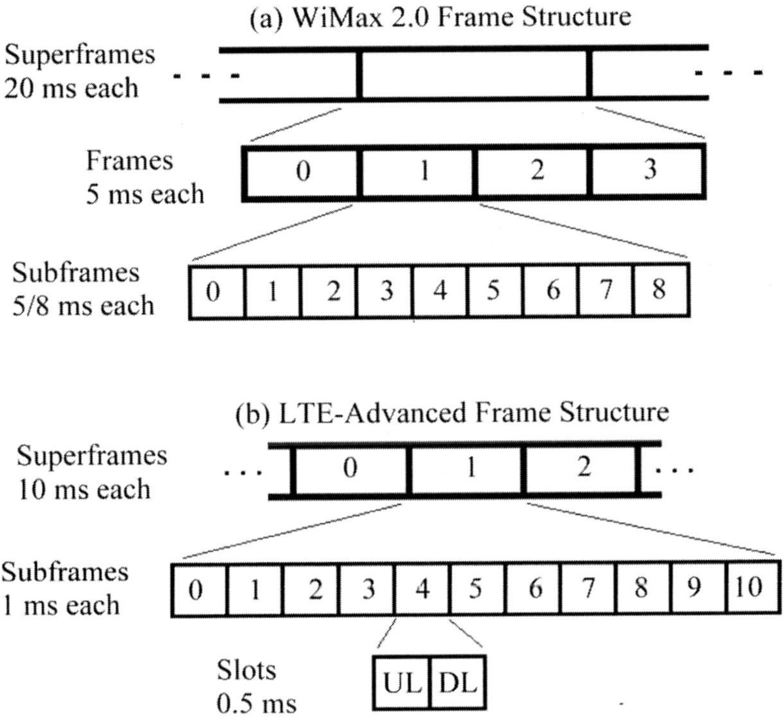

Figure 2. Frame (a) WiMAX 2.0; (b) LTE-Advanced.

Spectrum: Legacy LTE and LTE advanced use Licensed IMT-2000 Bands at bands like 700, 900, 1800, 2100, and 2600 MHz while legacy WiMAX is Licensed & unlicensed, at 2.3, 2.5, 3.5 and 5.8 GHz. Thus LTE is generally available at preferred low frequency band which gives it coverage advantage. This enhanced its opportunities to serve as public wide area network. With WiMAX heading to LTE some operators started trying LTE in some of the WiMAX bands they already posses.

Intercarrier spacing: LTE uses a standard 15 KHz intercarrier spacing while WiMAX 2.0 uses 10.94 KHz. The larger intercarrier spacing the higher the immunity against Doppler spread. LTE can handle mobility speeds up to 350 Km/hr while WiMAX can support speeds of the order of 120 Km/hr with WiMAX 2.0 up to 350 Km/hr.

Access technology: LTE-Advanced access technology for its downlink (OFDMA) is different to its uplink. In the uplink Single Carrier FDMA (SC-FDMA) is used. SC-FDMA reduces Peak-to-Average-Power-Ratio (PAPR) by

3 - 5 dB giving rise to uplink improvements that can be utilized to improve coverage or throughputs of cell edge users. 802.16m uses SOFDMA for both uplink and downlink. In fact, the major problem in extending GSM TDMA and wideband CDMA to broadband systems is increased receiver complexity with multipath signal reception. The main advantage of OFDM, as is for SCFDMA, is its robustness against multipath signal propagation, which makes it suitable for broadband systems [15].

Comparing 4G WiMAX and LTE from other aspects are covered in the literature as well. For example [16] compares LTE and WiMAX when used in enterprise environment from two aspects security and integration into enterprise IT network. It was found that both WiMAX and LTE can be hosted and deployed by the enterprise as the next generation of mobile enterprise network and WiMAX can meet enterprise security naturally because of the WiMAX's authentication protocols.

Generally, it can be concluded that LTE design seems to be superior especially concerning mobility, data throughputs and capacity. However, these factors, though important may not be the only factors that impact the popularity of one technology compared to the other. This is discussed in the coming section.

5. OTHER DIFFERENCES AND FACTORS

Beside the technological constrains other regional, operator and regulatory factors and constraints, lead to the preference of LTE over WiMAX or WiMax over LTE.

WiMAX proceeded LTE in appearance and deployment. WiMAX was backed in the USA by Clearwire and later Clearwire/Sprinit partnership. It was also adopted by major companies in Korea, Russia, Japan and other countries. Many other developing countries like India found WiMAX, even the non-mobile version, suitable for its broadband requirements [17] due to the lack of proper fixed line infrastructure.

WiMAX is a TDD technology which does not require paired spectrum with flexibility in sharing the time frame between the uplink and the downlink. This made it appear more appropriate for data as the viable wireless alternative to wired DSL. On the other hand, phone companies, adopting 3GPP/3GPP2 standards of 2G, 3G and LTE use one band of frequencies for the uplink and another band for the downlink. This makes these systems more costly in terms of spectrum and equipment. However, the introduction of TD-LTE, a TDD version of LTE with single band for its operation wipes out one main advantage of WiMAX over LTE.

The set of IEEE standards, on which WiMAX releases are based, are modular stand-alone standards offering high performance. The 4G version of WiMAX (WiMAX 2.0) have no support for legacy 3GPP devices, which means no handovers possible to and from 2G (GSM) and 3G (UMTS) etc. On the other hand, 3GPP provided a clear evolution path towards LTE for 2G and 3G standards of Europe, North America and China. LTE-Advanced is backward compatible with all previous standards. Hence, operators worldwide who have already deployed their networks based on 3GPP standards find it a good business case with easy upgrade and with possibility that they re-use their already possessed paired spectrum of abandoned technologies of 2G or so for the more efficient LTE.

The 3G TD-SCDMA of China will take a home-made evolution path aiming at a smooth migration to LTEAdvanced [18]. To save investment and make full use of the network infrastructure available, the design of TDLTE takes into account the features of TD-SCDMA, and keeps TD-LTE backward compatible with TD-SCDMA systems to ensure smooth migration. At the same time, WiMAX has limited spectrum options for deployment making it difficult for existing operators (networks based on 3GPP legacy standards) to migrate to it.

LTE success is the result of being backward compatible to the abundant customer base built over the different generations. This is said though WiMAX showed superiority in early stages. Table 1 below provides a comparison of 3GPP track that led to LTE and IEEE 802.16 track that led to current WiMAX. The table shows superiority of IEEE standards as it used OFDMA, provided high data rates, all IP and flat architecture from the beginning. It managed later on to provide mobility and cater for voice servicing. On the other hand, 3GPP targeted wide coverage and ubiquitous service while at the same time gradually staffed some salient IEEE features like OFDMA, TDD, all IP, flat architecture and higher data rates which together contributed to LTE success later on.

6. FUTURE OF LTE AND WIMAX

WIMAX had the precedence advantage over LTE in bringing to light much of the themes currently adopted such as the flat architecture, all IP network and TDD structure. 3GPP on the other hand, moved from all circuit switched of 2G, through half packetized of 2.5G and 3G and finally to all IP in LTE and LTE-Advanced. The objective of telecom companies from the start was big customer base, public networks and wide coverage while WiMAX headed to serve vertical segments requesting broadband. This, together with the

Table 1. 3GPP track and IEEE 802.16 track compared.

Characteristic	3GGP Track	IEEE 802.16 Track
All IP vs Circuit Switched	Started Circuit switched, moved to half IP (2.5/3 G) and finally "All IP" (LTE)	All IP from the beginning
Architecture	Centric architecture, gradually moving to flat architecture	Flat architecture from the beginning
Mobility	High mobility from the beginning	Started as fixed Wireless, moved quickly to mobile WiMAX
Voice versus Data	Started voice centric gradually moved to data centric	Started as data centric gradually serving voice
Mode of operation	FDD is the main mode with increased interest in TDD recently	TDD mode mainly
Access [redacted]	Different access technologies like TDM/FDM and Spread [redacted]	OFDMA was considered at early stages by IEEE [redacted]

[redacted]

previously discussed factors, made WiMAX people realize that the ecosystem of WiMAX as a stand-alone technology cannot continue to compete with 3GPP stream and its attractive LTE. WiMAX forum had to set plans to proceed. One of which is to harmonize and integrate with LTE rather than continue rivaling it as is discussed below.

6.1. Future of WIMAX

The future of WiMAX can be seen in three different directions:

1. WiMAX integration with LTE and other multiple access technologies. This will keep it a mobile operator solution in a robust ecosystem. In late 2012, WiMAX forum approved requirements for WiMAX coexistence and harmonization with LTE networks [19]. In WiMAX R2.2, expected to be finalized by the end of 2013, focus will be on coexistence of multiple radio access technologies and on features such as link aggregation and load balancing. Networks that support both standards and more multiple access technologies may emerge in 2014.
2. WiMAX technology as private network in vertical segments like utilities, aviation and similar segments which require reliable networking to manage their operations. This is called the wireless Ethernet.
3. WiMAX operators who hold TDD spectrum and built WiMAX networks will continue to grow within some of its traditional markets in Japan, Korea, Malaysia and in the USA. It is expected that it will be there for some time. There are even ongoing plans for launching WiMAX 2.0 (IEEE 802.16m) in Japan by its large WiMAX operator (UQ) and in Malaysia by YTEL.

Wireless technology can be an optimal solution in environments such as the energy utilities and aviation. This provided an alternative for WiMAX; i.e. to grow in private networks serving dedicated segments. The WiGRID, is a new wide-area networking technology based on IEEE 802.16e with requirements announced early this year [20]. It is a technology for energy utilities and smart grid industrial applications of telemetry, measurements and managements of critical systems that require realtime and high security [21]. For such services, WiMAX shall be optimized for uplink (uplink biased) with reduced latencies and increased range [20]. Frequency bands that various utilities hold, namely 1.4 GHz, 1.8 GHz, 2.3 GHz, 3.65 GHz and 5.8 GHz, can be utilized [20]. Public WiMAX, LTE and other 3GPP solutions will not provide the same service with reliability due to congestions. Public cellular networks on the other hand can be used for smart metering applications because of good coverage and less requirements of realtime and security [22]. WiMAX based private wireless system is suggested for monitoring power distribution stations. For example, in [22] cognitive spectrum WiMAX sharing with SCADA system in 223 - 225 MHz is suggested. The solution requires few modifications on MAC and PHY layer to support interference management [22].

AeroMACS is another WiMAX technology to support aviation industry in airports communications infrastructure. Other similar segments of interest to WiMAX's are transportation and oil and gas industries. Other uses of the WiMAX wireless Ethernet can be video distribution techniques for m-Health and similar applications.

The migration of WiMAX to LTE aims to open broader the ecosystem of radio access technologies and devices beyond WiMAX Release 1.0 and 2.0 [23]. The harmonized WiMAX Advanced 2.2 standard for network evolution will grant operators the flexibility to leverage their all-IP broadband data network capabilities and support multiple broadband wireless access technologies including TD-LTE devices (dual-mode WiMAX/LTE devices). The impact of TD-LTE on WiMAX operators is best described by the WiMAX forum's president saying [24]: "I don't see the advent of TD-LTE networks, ...putting those WiMAX networks out of business. I think there will be a complementary parallel network paradigm for some period of time. How long that is, I can't tell you."

The migration to LTE path will necessitate lots of software and hardware upgrades to network components, core network, backhaul and devices to serve the multimode heterogeneous harmonized set of access technologies. The cost incurred can be understood if this will give the WiMAX operators, who have small customer base compared to 3GPP, access to the much larger base of 3GPP technologies, customers, and ecosystem. On the contrary it can be an opportunity for 3GPP big operators to acquire WiMAX customers. Regardless of some recent growth of WiMAX technology in certain Asian countries, its continuity will largely depend on its success in vertical markets like aviation, utility and transportation rather than on reliance on public networking. Success in harmonization with LTE means that specialized segments can be served through dedicated WiMAX networks while at the same time offering them the public service. This looks like a unique combination in which embracing the competition is elected as the approach instead of the competition approach itself.

6.2. Future of LTE

3GPP provided a smooth path of transition of its current technologies to LTE and LTE advanced. Lots of efforts are in progress to issue new 3GPP releases in 2013 and 2014. The upcoming Release 12 and beyond will be as significant as ever in the industry's quest to extend mobile broadband availability, provide much more consistent service quality, and economically satisfy demand for spiraling data growth in face of spectrum scarcity [25].

Current broadband is characterized as being patchy in coverage and erratic in data rates. 3GPP future releases (Rel-12 and Rel-13) aims at providing a stable broadband; i.e. broadband anytime and anywhere while satisfying the continuously exponential rise of traffic demand. Keith Mallinson, in his article "2020 Vision for LTE" [26] following 3GPP TSG meeting in Slovenia in 2012 heralded an improvement figures of $3 \times 6 \times 56 = 1008$ as a result of the new releases. These are: 3× increase in spectrum employed, 6× improvement in spectral efficiency and 56× higher average cell density. Future tools [10,11,26,27] enabling such massive improvements include:

- Multiplication of current peak, average, and cell edge data rates due to the factors of more bandwidth, higher order 3D-MIMO, higher order QAMs, small cells etc.
- Macro cell will continue to exist as the upper layer in a multi-tier network shadowing underneath it large number of randomly distributed micro, pico, and femto cells that may run at higher frequency bands than those used in macro-cells. The resulting heterogeneous structure will be self organized with enhanced mobility between the different cell types and efficient Inter-Cell Interference Coordination (ICIC) procedures. This architecture will help in offloading high data traffic to small cells thereby increasing total throughput. It will also help lessening coverage holes and provide high traffic indoors. WLAN (WiFi) will also play an increasing role in 3GPP operator's networks to support broadband in hotspots and indoors alongside other technologies like femto cells.
- More reliance on TDD mode for LTE as a result of further enhancements in DL-UL interference management and traffic adaptation.
- UL performance improvements, e.g. deployment of 8 Rx antennas at eNodeB, use of cross-polarized antennas, and higher carrier frequency, etc.
- 3D MIMO and massive antenna beamforming. arrays of as many as 64 antenna elements enable additional frequency reuse within cell sectors.

7. CONCLUSIONS

This paper aimed at comparing LTE with WiMAX wireless technologies. It also discusses the factors that led to LTE winning as the technology for near future public networks. The article also foresees future directions of both technologies and the alternatives that WiMAX technology has.

For theses purposes, the article first reviewed the evolution of the two technologies; LTE coming from the generations path of wireless mobile phone standards and WiMAX coming from data networking industry. Mobile technology path started voice centric and progressed gradually toward data while WiMAX started with broadband data in focus on progress toward serving the voice users. Salient common features of the two technologies were also revised. The two technologies, which were competing with each other, have much in common. Both are all IP with flat architecture and similar enabling technologies. Technical differences reviewed include spectrum allocation, intercarrier spacing, frames/subframes and access technology on the uplink. Generally, the impact of each of these on the prospects of the two technologies made that LTE provides more throughputs and capacity and a better mobility.

Factors that influenced the competition are not only technological. Other factors influenced the fate of the competition. The final outcome is that WiMAX people realized that it is necessary to harmonize and integrate with LTE rather than continue rivaling it. This formed the future migration strategy for WiMAX. Alternatively, WiMAX is progressing toward providing networking service to serve privately specialized segments like energy utilities, aviation and transportation. The resort of such segments on private networking of WiMAX enables avoiding the congestion that may occur in public cellular networks serving large customer base. Integration of WiMAX with LTE enables its companies to serve their specialized segments with private networks while at the same time offering them the public service. WiMAX forum is working to complete the standards that will integrate multiple WiMAX and LTE technologies. Lots of software and hardware changes/additions need to be applied to devices, base stations and core networks to enable them to operate in multimode with a possible need for backhaul capacity increase. This gives less opportunities for WiMAX to continue as public wide area network. WiMAX's continuity will largely depend on its success in vertical markets like aviation, utility and transportation rather than on reliance on public networking.

3GPP, on the other hand, is actively progressing to enhance the capabilities of LTE and other current technologies. Two new releases, R12 and R13, are under preparation. The objectives of these updates are to improve coverage, to provide much higher stable data rates, and to meet the expected high traffic demand. A number of enhancement and new enabling technologies are suggested in order to achieve these objectives such as 3D sectorization, small cells, efficient ICIC, more TDD etc. It can be said that LTE technology will prevail as one standard in the near future for serving public networks while WiMAX has good opportunities in dedicated segments.

REFERENCES

1. IEEE, "IEEE Standard for Local and Metropolitan Area Networks Part 16: Air Interface for Fixed Broadband Wireless Access Systems," IEEE Standard 802.16-2004, 2004.
2. IEEE, "Amendment for Physical and Medium Access Control Layers for Combined Fixed and Mobile Operation in Licensed Bands," IEEE Standard 802.16e-2005, 2006.
3. ITU Press Release, "IMT-Advanced Standards Announced for Next-Generation Mobile Technology," 2012. http://www.itu.int/net/pressoffice/press_releases/2012/02.aspx
4. ITU, "Framework and Overall Objectives of the Future Development of IMT-2000 and Systems beyond IMT- 2000," ITU-R M.1645, 2003.
5. 3GPP, "Requirements for Evolved UTRA (E-UTRA) and Evolved UTRAN (E-UTRAN) Release 7," 3GPP TS 25.913, 2007.
6. ITU Press Release, "ITU Radiocommunication Assembly Approves New Developments for Its 3G Standards," 2007. http://www.itu.int/newsroom/press_releases/2007/30.html
7. C. Gessner, A., Roessler and M. Kottkamp, "UMTS Long Term Evolution (LTE) Technology Introduction," Rohde and Schwarz, Application Note, IMA111_3E, 2012.
8. A. Zreikat, I. Aldmour and K. Al-Begain, "Performance Model of a WIMAX 2.0 All-IP 4G System," Wireless Personal Communications, Vol. 72, No. 1, 2013, pp. 191- 210.
9. L. Korowajczuk, "LTE, WiMAX and WLAN Network Design, Optimization and Performance Analysis," Wiley, 2011.

10. C. Y. Wong, R. S. Cheng, K. B. Lataief and R. D. Murch, "Multiuser OFDM with Adaptive Subcarrier, Bit, and Power Allocation," IEEE Journal on Selected Areas in Communications, Vol. 17, No. 10, 1999, pp. 1747-1758.

11. S. Ahmadi, "An Overview of Next-Generation Mobile WiMAX Technology," IEEE Communications Magazine, Vol. 47, No. 6, 2009, pp. 84-98.

12. T.-T. Tran, Y. Shin and O.-S. Shin, "Overview of Enabling Technologies for 3GPP LTE-Advanced," EURASIP Journal on Wireless Communications and Networking, Vol. 2012, No. 1, 2012, pp. 1-12.

13. R. Irmer, H. Droste, P. Marsch, M. Grieger, G. Fettweis, S. Brueck, H.-P. Mayer, L. Thiele and V. Jungnickel, "Coordinated Multipoint: Concepts, Performance, and Field Trial Results," IEEE Communications Magazine, Vol. 49, No. 2, 2011, pp. 102-111.

14. K. Etemad, "Overview of Mobile WiMAX Technology and Evolution," IEEE Communications Magazine, Vol. 46, No. 10, 2008, pp. 31-40.

15. IXIA, "SC-FDMA: Single Carrier FDMA in LTE," IXIA White Paper, November 2009.

16. L. Yi, K. Miao and A. Liu, "A Comparative Study of WiMAX and LTE as the Next Generation Mobile Enterprise Network," Proceedings of the 13th International Conference on Advanced Communication Technology (ICACT), Gangwon-do, South Korea, 13-16 February 2011, pp. 654-658.

17. "WiMAX: The Quintessential Answer to Broadband in India," 2009. http://resources.wimaxforum.org/sites/wimaxforum.org/files/wimax_in_india_protiviti_paper_0.pdf

18. M. Peng and W. Wang, "Technologies and Standards for TD-SCDMA Evolutions to IMT-Advanced," IEEE Communications Magazine, Vol. 47, No. 12, 2009, pp. 50-58.

19. WiMAX Forum, "Requirements for WiMAX Coexistence with LTE Networks," WMF-T31-132-v02, December 2012.

20. WiMAX Forum, "WiMAX Forum System Profile Requirements for Smart Grid Applications—Requirements for WiGRID," WMF T31-002-R010-v01, February 2013.

21. Z. Brabec, J. Holecek and T. Hruza, "Overview of Applicability of ICT Technologies for Smart Grids," Proceedings of 15th International Symposium on MECHATRONIKA, Prague, 5-7 December 2012, pp. 1-5.

22. Q. Wang, J. Wang, Y. Lin, J. Tang and Z. Zhu, "Interference Management for Smart Grid Communication under Cognitive Wireless Network," Proceedings of IEEE Third International Conference on Smart Grid Communications (SmartGridComm), Tainan, 5-8 November 2012, pp. 246- 251.

23. B. Ayvazian and R. Schwartz, "WiMAX Technology Roadmap: The Path to Harmonization of 4G Technology Standards," 2012. http://www.wimaxforum.org/LiteratureRetrieve.aspx?ID=177981

24. T. Parker, "WiMAX Forum President describes Roadmap Wherein WiMAX, LTE Are Complementary," 2012. http://www.fiercebroadbandwireless.com/node/14529/print

25. S. Parkvall, E. Dahlman, A. Furuskar, Y. Jading, M. Olsson, S. Wanstedt and K. Zangi, "LTE-Advanced-Evolving LTE towards IMT-Advanced," Proceedings of IEEE 68th Vehicular Technology Conference (VTC 2008-Fall), Calgary, 21-24 September 2008, pp. 1-5.

26. K. Mallinson, "2020 Vision for LTE," 2012. http://www.3gpp.org/IMG/pdf/wiseharbor.pdf

27. Y. Yang, H. Hu, J. Xu and G. Mao, "Relay Technologies for WiMAX and LTE-Advanced Mobile Systems," IEEE Communications Magazine, Vol. 47, No. 10, 2009, pp. 100-105.

4

CHAPTER

PERFORMANCE STUDY OF MOBILE TV OVER MOBILE WIMAX CONSIDERING DIFFERENT MODULATION AND CODING TECHNIQUES

Jamil Hamodi[1,2], Ravindra Thool[1], Khaled Salah[3], Anwar Alsagaf[2], Yousef Ilolba[2]

[1]*Information Technology Department, SGGS Institute of Engineering and Technology, Nanded, India*

[2]*Computer Engineering Department, Hodeidah University, Hodeidah, Yemen*

[3]*Computer Engineering Department, Khalifa University of Science Technology and Research, Sharjah, UAE*

ABSTRACT

With the advent of the wide-spread use of smart phones, video streaming over mobile wireless networks has suddenly taken a huge surge in recent years. Considering its enormous potential, mobile WiMAX is emerging as a viable technology for mobile TV which is expected to become of key importance in the future of mobile industry. In this paper, a simulation performance study of Mobile TV over mobile WiMAX is conducted with different types of adaptive modulation and coding taking into account key system and environment parameters which include the variation in the speed of the mobile, path-loss, scheduling service classes with the fixed type of modulations. Our simulation has been conducted using OPNET simulation. Simulation results show that dynamic adaptation of modulation and coding schemes based on channel conditions can offer considerably more enhanced QoS and at the same time reduce the overall bandwidth of the system.

Keywords: Mobile TV; QoS; Adaptive Modulation and Coding; WiMAX; OPNET

1. INTRODUCTION

The primary challenge for present and future communication systems is the ability to transport multimedia content over a variety of networks in an energy efficient manner at different channel conditions and bandwidth capacities with various requirements of QoS [1]. Worldwide Interoperability for Microwave Access (WiMAX) technology is one of the future communications capable of offering high QoS at high data rates for IP networks. The high data rate and Quality of Service (QoS) assurance provided by this standard has made it commercially viable to support multimedia applications such as video telephony, video gaming, and mobile TV broadcasting. System architecture to support high definition video broadcasting (like MPEG-X, H.264/AVC and SVC) that provides a mobility of 30 kmph in an urban and sub-urban environment has been developed [2,3].

Mobile TV is a technology that enables users to transmit and receive TV program data through IP-based wired and wireless networks. Users can enjoy IPTV services anywhere and anytime with mobile devices. Four kinds of mobile TV technology approach, namely: mobile TV over IP that was discussed in this work, IPTV over mobile device, cellular IPTV, internet IPTV. However, with rapid adaptation to the user's requirements, Mobile TV may be preeminent in the future. It is convenient for users who can access an IPTV service through various wireless networks with mobile devices.

In cellular communication, the quality of a signal received by user equipment depends on a number of factors, namely: the distance between the desired user and interfering base stations, path loss exponent, log-normal shadowing, short term Rayleigh fading and noise. In order to improve system capacity, peak data rate and coverage reliability, the signal transmitted by a particular user is modified to account for the signal quality variation through a process commonly referred to as link adaptation.

Traditionally, CDMA systems have used fast power control as the preferred method for link adaptation. Recently, Adaptation Modulation and Coding (AMC) has offered an alternative link adaptation method that promises to raise the overall system capacity. Based on channel conditions, the scheme can be changed on per frame and per user basis. In order to maximize throughput in a channel varying in time, Adaptive and Modulation Coding can be used [4]. AMC provides the flexibility to match the modulation-coding scheme to the average channel conditions for each user. With AMC, the power of the transmitted signal hold constant over a frame interval, and the modulation and coding format are changed to match the current received signal quality or channel conditions. In a system with AMC, users close to the BS are typically assigned higher order modulation with higher code rates (e.g. 64 QAM with R = 3/4 turbo codes), but the modulation-order and/or code rate will decrease as the distance from BS increases [5].

Our work in this paper is geared towards investigating the performance study of Mobile TV (VoD) over Mobile WiMAX networks when considering different adaptive modulation and coding schemes using simulation software OPNET Modeler. OPNET provides comprehensive development of network models including all the necessary parameters that need to be reflected in the design procedure of PHY and/or MAC layers. A series of simulation scenarios under OPNET for broadband wireless communication are developed. In sharp contrast to prior work [6], the research work and results presented in this paper focus mainly on the use of real-time audio/video movies coded by MPEG-4 for modeling and simulation Mobile TV deployment over Mobile WiMAX. This paper, in the main, aims to establish a comparative study of performance for Mobile TV (VoD) over Mobile WiMAX under varying speeds of the mobile nodes and using different path-loss models under both AMC and fixed types of modulation techniques, and to identify the factors affecting the video performance.

The rest of this paper is organized as follows: Section 2 presents a brief overview of WiMAX. Section 3 presents performance metrics of video over WiMAX. Video streaming over WiMAX is given in Section 4. Section 5 describes the practical steps to be taken prior to simulation. Simulation results and analysis obtained are provided in Section 6. Section 7 is naturally the conclusion of my findings as a whole and a summation of my modest research endeavor.

2. BACKGROUND AND PRELIMINARIES

This section provides relevant background and preliminaries pertaining to QoS support and the physical layer on WiMAX modulation and coding schemes. The WiMAX physical layer is based on orthogonal frequency division multiplexing (OFDM). OFDM is the transmission scheme of choice to enable high-speed data, video, and multimedia communications and is used by a variety of commercial broadband systems, including DSL, Wi-Fi, Digital Video Broadcast-Handheld (DVB-H), and MediaFLO, besides WiMAX. The standard defined within OFDM is an elegant and efficient scheme for high data rate transmission in a non-line-of-sight or multipath radio environment. Five PHY layer interfaces are in the IEEE 802.16 standard. These physicals interfaces are Wireless MAN-SC is known as SC operating in LOS condition at a frequency between 10 - 66 GHz, Wireless MAN-SC is known as SCa operating in NLOS condition at a frequency below 11 GHz, Wireless MAN-OFDM known as OFDM operating in NLOS condition at a frequency below 11 GHz, Wireless MAN-OFDMA known as OFDMA operating in NLOS condition at a frequency below 11 GHz, and last PHY interfaces is a Wireless HUMAN with frequency below 11 GHz. Among these PHY interface standards Wireless-MAN OFDM, a 2.5 GHz base frequency using 5 MHz has been used in our simulation that provides a 512-point TDD-based OFDM PHY for point-to-multipoint operations in NLOS conditions on frequencies between 2 GHz and 11 GHz [7].

WiMAX supports a variety of modulation and coding schemes and allows the scheme to change on a burst-byburst basis per link, depending on channel conditions. Using the channel quality feedback indicator helps the mobile provide the base station with feedback on the downlink channel quality. For the uplink, the base station can estimate the channel quality based on the received signal quality. The base station scheduler can take into account the channel quality of each user's uplink and downlink and assign a modulation and coding scheme that maximizes the throughput for the available signalto-noise ratio. Adaptive modulation and coding significantly increases the overall system capacity, as it allows real-time trade-off between throughput and robustness on each link. The modulation used in WiMAX is the orthogonal frequency division multiplexing (OFDM). WiMAX OFDM features multiple subcarriers ranging from a minimum of 256 up to 2048, each modulated with QPSK, 16 QAM, or 64 QAM modulation, where 64 QAM is optional on the uplink channel. The advantage of orthogonality is that it minimizes self-interference, a major source of error in receiving signals in wireless communications.

Channel coding schemes are used to help reduce the SINR requirements by recovering corrupted packets that may have been lost due to burst errors or frequency selecting fading. These schemes generate redundant bits to accompany information bits when transmitting over a channel. Coding schemes include convolution coding (CC) at various coding rates (1/2, 2/3 and 3/4) as well as conventional turbo codes (CTC) as various coding rates (1/2, 2/3, 3/4, and 5/6). The coding rate is the ratio of the encoded block size to the coded block size. The available coding rates for a given modulation scheme with the minimum signal to noise rate and the peak UL and DL data rates for 5 MHz channel mobile WiMAX with different information bits/symbol are listed in Table 1 [8,9].

The key feature of adaptive modulation is that it increases the range over which a higher modulation scheme can be used, since the system can flex to the actual fading conditions, as opposed to having a fixed scheme that is budgeted for the worst case conditions [10]. Two sets of AMC schemes, AMC-1 and AMC-2, are considered to be used in DL in our simulation. These AMC profiles are same as in [11,12]. Each AMC is basically characterized by two threshold parameters, one is mandatory and the other is the minimum entry threshold for different modulation schemes. The mandatory exit threshold is the SINR at or below where this burst profile can no longer be used and where a change to a more robust (but also less frequency-use efficient) burst profile is required and the minimum entry threshold is the minimum SINR required to start using this burst profile when changing from a more robust burst profile [13]. Table 2 shows the AMC profile used in our simulation. Here, AMC-2 is conservative AMC as it is using lower order MCS most of the time.

WiMAX supports different signal bandwidths ranging from 1.25 to 20 MHz to facilitate transmission over longer ranges in different multipath environments. In wireless communication systems, information is transmitted between the transmitter and the receiver antenna by electromagnetic waves. During propagation, electromagnetic waves interact with the environment, thereby causing a reduction of signal strength. Another limiting factor for higher sustained throughput in wireless communications, especially when the terminal nodes have the mobility, is caused by reflections between a transmitter and receiver, viz, a propagation path between the transmitter and the receiver is regarded. The propagation path between the transmitter and the receiver may vary from simple line-of-sight (LOS) to very complex one due to diffraction, reflecting and scattering [14]. To estimate the performance of mobile TV over WiMAX channels, propagation models in [6] are often used. Path loss models describe the signal attenuation between a transmitting and a receiving antenna as a function of the propagation distance and other parameters

Table 1 . Mobile WiMAX PHY data rate and SINR for 5 MHz channel.

Modulation scheme	Coding	Information bits/symbol	Minimum SINR (dB)	Dwon-link rate (Mbps)	Up-link rate (Mbps)
QPSK	1/2	1	5	3.17	2.28
	3/4	1.5	8	4.75	3.43
16 QAM	1/2	2	10.5	6.34	4.57
	3/4	3	14	9.5	6.85
64 QAM	1/2	3	16	9.5	6.85
	2/3	4	18	12.6	9.14
	3/4	4	20	14.26	10.28

Table 2. AMC profiles.

Modulation and coding schemes	AMC-1		AMC-2	
	Mandatory exit threshold (dB)	Minimum entry threshold (dB)	Mandatory exit threshold (dB)	Minimum entry threshold (dB)
QPSK 1/2	−20	2.0	−20	2.0
QPSK 3/4	5.0	5.9	11	11.9
16 QAM 1/2	8.0	8.9	14	14.9
16 QAM 3/4	11	11.9	17	17.9
64 QAM 1/2	14	14.9	20	20.9
64 QAM 2/3	17	17.9	23	23.9
64 QAM 3/4	19	19.9	25	25.9

which provide details of the terrain profile required to estimate the attenuating signals. Path loss models represent a set of mathematical equations and algorithms which apply to radio signal propagation prediction in certain environments [15]. Pathloss is highly dependent on the propagation model, the common propagation models namely Free Space, Suburban Fixed (Erceg), Outdoor to Indoor and Pedestrian Environment and Vehicular Environment are given in Table3 These models are used in mobile WiMAX performance evaluation through OPNET simulation.

IEEE 802.16 MAC defines up to five separate service classes to provide QoS for various types of applications. The service classes include: Unsolicited Grant Scheme (UGS), Extended Real Time Polling Service (ertPS), Real Time Polling Service (rtPS), Non Real Time Polling Service (nrtPS) and Best Effort Service (BE). Each service class has its own QoS parameters such as the way to request bandwidth, minimum throughput requirement and delay/jitter constraints. These service classes are [16] [17] described as follows: UGS is designed for constant bit rate (CBR) real-time traffic flows generating fixedsize data packets such as E1/T1 circuit emulation. It provides a fixed periodic bandwidth allocation. Once the connection is setup, there is no need to send any other requests. The main QoS parameters are maximum sustained rate, maximum latency and tolerated jitter (the maximum delay variation). ertPS is designed to support VoIP with silence suppression. No traffic is sent during the silent periods. ertPS service is similar to UGS in that the BS allocates the maximum sustained rate in active mode, but no bandwidth is allocated during the silent period. There is a need to have the BS poll the MS during the silent period to determine if the silent period has ended. The QoS parameters are the same as those in UGS. rtPS is for variable bit rate (VBR) real-time traffic such as MPEG compressed video. Unlike UGS, rtPS bandwidth requirements vary and so the BS needs to regularly poll each MS to determine what allocations needed to be made. The QoS parameters are similar to the UGS, but minimum reserved traffic rate and maximum sustained traffic rate need to be specified separately. For UGS and arts services, these two parameters are the same, if present. nrtPS is for non-real-time VBR traffic with no delay guarantees. Only the delay minimum rate is guaranteed. File Transfer Protocol (FTP) traffic is an example of applications using this service class. BE (most of data traffic falls into this category) guarantees neither delay nor throughput. The bandwidth will be granted to the MS if and only if there is a leftover bandwidth from other classes. In practice most implementations allow specifying minimum reserved traffic rate and maximum sustained traffic rate even for this class.

3. VIDEO PERFORMANCE METRICS

This section discusses some issues related to performance metrics of video transmission. Performance metrics can be classified as objective and subjective quality measures. Objective measures that observe packet transmissions include packet loss, packet delay, packet jitter, and traffic load throughput rates. Other objective metrics that attempt to quantify video quality perceptions include the ITU video quality metric (VQM) and peak signal to noise ratio (PSNR), which measure the codec's quality of reconstruction. Subjective video quality is a subjective characteristic of video quality. It is concerned with how video is perceived by a viewer and designates his or her opinion on a particular video sequence. The main idea of measuring subjective video quality is the same as in the

Table 3 . Path-loss models.

Propagation Model	Mathematical formulation	Description
Free space model	$P_{rx}(r) = P_{tx}G_{tx}G_{rx}/\left((4\pi)^2 r^2 L\right)$. P_{rx} and P_{tx} are received power in watts. respectively; G_{rx} and G_{tx} are the gain of the receiving and transmitting antennas, respectively; L is the system-loss factor.	It is a mathematical model hardly applicable without considering the fading effect due to multi-path propagation.
Erceg's suburban fixed model	$PL = H + 10\gamma \log_{10}(d/d_0) + X_f + X_h + s$. PL is the instantaneous attenuation in dB, H is the intercept and is given by free space path-loss at the desired frequency over a distance of d_0 = 100 m. ∂ is a Gaussian random variable over the population of macro cells within each terrain category. X_f and X_h are the correlation factors of the model for the operating frequency and for the MS antenna height, respectively	It is based on extensive experimental data collected at 1.9 GHz in 95 macro cells of suburban areas across the United States. Very large cell size, base stations with high transmission power and higher antenna height. Subscriber stations are of very low mobility
Outdoor-to-indoor and pedestrian path-loss environment	$PL = 40\log_{10} R + 30\log_{10} f + 49$. PL is the instantaneous attenuation in dB. R is the distance between the base station and the mobile station in kilometers and f is the carrier frequency	Small cell size, base stations with low antenna heights and low transmission power are located outdoors while pedestrian users are located on streets and inside buildings and residences
Vehicular environment	$PL = 40(1 - 4*10^{-3} * \Delta h_b)\log_{10} R - 18\log_{10} \Delta h_b - 21\log_{10} f + 80dB$. R is the distance between the base station and the mobile station. f is the carrier frequency and Δh_b is the base station antenna height in meters	Larger cells and higher transmitter power. All subscriber stations have a high mobility

Mean Opinion Score (MOS). QoS requirement is very important for deploying IPTV and VoD as real time services over WiMAX networks. In order to assess the performance of video transmission systems, a suite of relevant performance metrics was identified to appropriately benchmark the system. Video on demand (VoD) deployments over WiMAX is affected by time varying bandwidth, packet delays, and losses. Since users expect high service quality regardless of the underlying network infrastructure, a number of metrics were collectively used to measure the video content streaming performance to ensure compliance and user's quality of experience (QoE) [18]. The following objective measures, widely used in the video content performance analysis employed, are packet loss ratio (PLR), packet delay (PD), packet jitter, and minimum throughput. The performance parameters affecting video have been shown in Table 4 as in [19].

4. VIDEO STREAMING OVER WIMAX

To provide a proper IPTV service to end users, this section describes the architecture of IPTV over WiMAX. IPTV providers must have an appropriate IP network to guarantee QoS at the services level. Our IPTV architecture over WiMAX network split into five subsystems, as can be seen in Figure 1: Video content is the first subsystem of our model, in this subsystem the servers store video content of any type of movies and audio content source. Different video types store the video source as national TV broadcasters, local broadcasters, Internet TV operations and any other future video broadcast service. Our video content held in this part has been variable bit rate videos encoded. The video content is delivered to the WiMAX network through the long distance, high capacity content distribution core network. The core network distributes the video flows from the header to the distribution network of the service provider. The distribution network goes from the end of the core network to the aggregation router, where the access network starts. The access network lets the user connect to the service provider and allows access to the multimedia content. The first requirement of an access network is to have enough bandwidth to support multiple IPTV channels for each subscriber, while it allows other services (telephony and data). Finally, the customer network (mobile receivers) enables communication and information exchange between the computers and devices connected to the services offered by a service provider.

There are some of researches work have explored WiMAX in the context of real-time and stored video applications. An in-depth performance evaluation of mobile WiMAX is carried out using adaptive modulation and coding under the real-like simulation environment of OPNET presented in [6]. They have evaluated the per formance parameters of mobile WiMAX with respect to different modulation and coding schemes. Their performance has been evaluated in terms of average throughput, average data-dropped, the MOS value of voice application and the BW usage in terms of UL data burst usage when deployed VoIP on WiMAX Networks. They have been observed that using lower order modulation and coding schemes, the system provides better performance in terms of throughput, data dropped and MOS at the cost of higher BW usage.

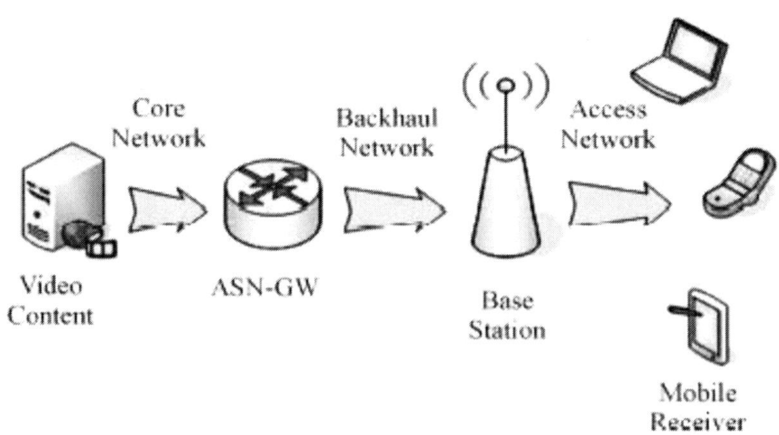

Figure 1. Mobile TV over WiMAX architecture.

Table 4. Performance parameters for deploying VoD.

Metrics	Mathematical formulation	Description	Acceptable
Packet loss ratio (PLR)	$PLR = \left(\frac{lost_{packet}}{lost_{packet} - received_{packet}} \right)$	**PLR** is the corrupted, lost, or excessively delayed packets divided by the total number of packets expected at the video client station.	10^{-3}
Packet End-to-End Delay (E2E) (ms)	$\mathbf{D_{E2E} = Q\,(d_{proc} + d_{queue} + d_{trans} + d_{prop})}$, where: **Q** is the number of network elements between the media server and mobile station . $\mathbf{d_{proc}}$ is the processing delay at a given network element. $\mathbf{d_{queue}}$ is the queuing delay at a given network element. $\mathbf{d_{trans}}$ is the transmission time of a packet on a given communication link between two network elements. $\mathbf{d_{prop}}$ is the propagation delay across a given network link	**Packet delay** is the average packet transit time between the media server and the video client station.	<400
Packet delay variation (PDV) or packet jitter (ms)	$\mathbf{j_{pkt} = t_{actual} - t_{expected}}$, where: $\mathbf{t_{actual}}$ is the actual packet reception time. $\mathbf{t_{expected}}$ is the expected packet reception time.	**Packet jitter** is defined as the variability in packet delay within a given media at video client station.	<50
Throughput (bps)	The **throughput** for variable bit rate (VBR) traffic loading is dynamic in nature and it is a function of the scene complexity and associated audio content. Variable bit rate (VBR) traffic loads is typically quoted as peak throughput ranges.	**Throughput** is defined as the traffic load that the media stream will add to the network. It can be measured in bits/sec.	[221-5311]

Challenges for delivering IPTV over WiMAX are discussed in [20]. These include the challenges for QoS requirements. Also, they describe the transmission of IPTV services over WiMAX technology, and the impact of different parameters in WiMAX network when deploying this service. An intelligent controller has been designed based on fuzzy logic to analyze QoS requirements for delivering IPTV over WiMAX. Also, an intelligent controller based on fuzzy logic in [21] is used to analyze three parameters: jitter, losses and delays that affect the QoS for delivering IPTV services. The aim is to define a maximum value of link utilization among links of the network.

An OPNET Modeler is used to design, characterize, and compare the performance of video streaming to WiMAX and ADSL. The simulation results indicate that, ADSL exhibits behavior approached the ideal values for the performance metrics while WiMAX demonstrates promising behavior within the bounds of the defined metrics. The work in [18] is extended work in [22] to include generation and integration of a streaming audio component, also enhances the protocol stack to include the real time protocol (RTP) layer. Network topology is redesigned to incorporate WiMAX mobility. Also, they include characterization of WiMAX media access control (MAC) and physical (PHY) layer. Simulation scenarios are used to observe the impact on the four performance metrics. Also, in [23] simulation is used to compare the performance metrics between ADSL and WiMAX by varying the attributes of network objects such as traffic load and by customizing the physical characteristics to vary BLER, packet loss, delay, jitter, and throughput. Simulation results demonstrate considerable packet loss. ADSL exhibits considerably better performance than the WiMAX client stations.

However, many of recent works explore the performance studies of WiMAX under different modulation and coding schemes. For example, Telagarapu et al. [24] analyzed the physical layer of WiMAX with different modulation techniques like BPSK, QPSK, QAM and comparison of QPSK modulation with and without Forward Error Correction methods. Singh et al. [25] provided an analysis of the Bit error rate Vs SNR and Block error rate Vs SNR curves using QPSK, 16 QAM and 64 QAM in an AWGN channel with parameters analyzed for different code rates. It was found that the greater number of symbols transmitted per block, the transmission quality decreases.

Islam et al. [26] evaluated WiMAX system under different combinations of digital modulation (BPSK, QPSK, 64-QAM and 16-QAM) and different communication channels AWGN and fading channels (Rayleigh and Rician), and the WiMAX system incorporates Reed-Solomon (RS) encoder with Convolutional encoder with 1/2 and 2/3 rated codes in FEC channel coding. Also the work in [27] evaluated and analyzed the bit error rate performance of WiMAX Physical layer with the implementation of different concatenated channel coding schemes under various digital modulations over realistic channel conditions. Simulation results demonstrate the out performance of the concatenated Cyclic Redundancy Check and Convolutional (CRC-CC) code when compared to Reed-Solomon and Convolutional (RS-CC) code and the proposed WiMAX system achieves good error rate performance under QAM modulation technique in AWGN, Rayleigh and Rician fading channels. Whereas Rehman et al. [28] used the Simulink tool to evaluate WiMAX system under different combinations of digital modulation and different communication channels AWGN and fading channels. Both Reed Solomon (RS) encoder with Convolutional encoder with 1/2 and 2/3 rated codes in FEC channel coding are incorporated in WiMAX system.

5. SIMULATION MODEL

The network topology for our test-bed network is given in Figure 2. The simulation model of our case study network deployed with 7-Hexagonal celled WiMAX, with multiple subscriber stations in the range of a base station. The base stations are connected to the core network by an IP backbone. The IP backbone is connected to the server backbone via an ASN-GW gateway to support the mobility in the WiMAX network. The ASN-GW gateway, IP backbone, and server backbone together represent the service provider company network. We have uses only one SS with varying speed.

This node is (mobile 5_1) move along the trajectories indicated by white color around cells. The green bidirectional dotted lines represent the generic routing encapsulation (GRE) tunnels. The common attributes used for network configuration are highlighted in Table5

Video streaming over wireless networks is a challenging task. This is due to the high bandwidth required and the delay sensitive nature of video more than most other types of application. As a result, a video trace of 2-hour MPEG-4 Matrix III movie is used in our simulation. This video coded by MPEG-4 part 2, is obtained from Arizona State University [29] with 352 × 288 frame resolution, and an encoding rate of 25 frames per seconds (fps). Table 6 shows the mean and peak rates for this video code, and reflects only video frames. Our proposed simulation model traces driven using video and audio traces from the Matrix III movie while these traces reflect the individual frame sizes rather than the actual frame data. So, this work also adds audio frames, which is 21.6 as in [18].

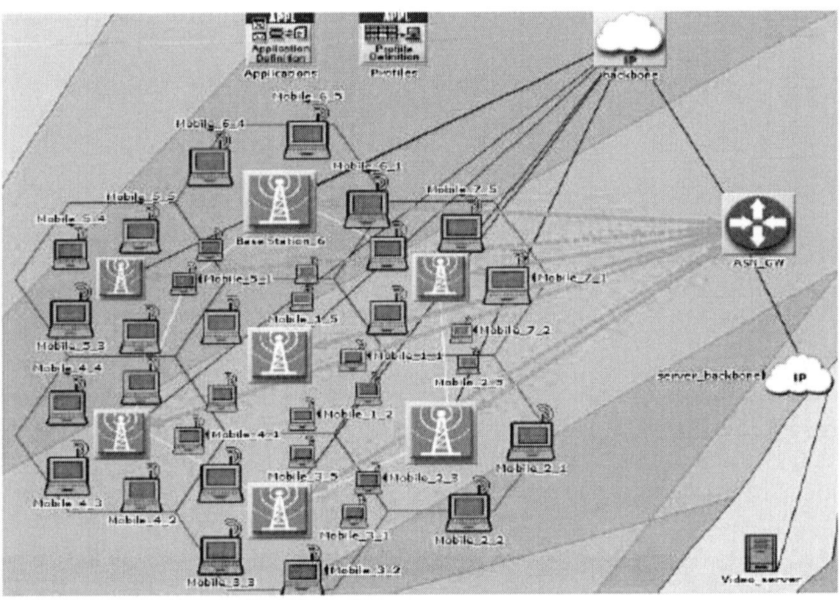

Figure 2. Network model of mobile TV over WiMAX.

Table 5. Network configuration details.

Network	Mobile WiMAX Network
Cell Radius	0.2 Km
No. of Base Stations	7
No. of Subscriber Stations	5
IP Backbone Model	IP32_cloud
Video Server Model	PPP_sever
Link Model (BS-Backbone)	PPP_DS3
Link Model (Backbone-server Backbone)	PPP_SONET_OC12
Physical Layer Model	OFDM 5 MHz
Traffic Type of Services	Streaming Video
Application	Real Video streaming
Scheduling	rtPS

Two independent instances of the video conferencing application are used to stream the separate and distinct video and audio components of the Matrix III movie. These two applications configured to work simultaneously stream in the profile configuration. The key parameters of this application configuration are the frame inter-arrival time and frame size. The incoming inter-arrival times are configured to the video and audio frame rates of 25 and 21.6, respectively. It should be noted that the outgoing inter-arrival time remains set to "none" in order to achieve unidirectional streaming from the video server. Furthermore, the frame size parameters are configured to explicitly script the video and audio traces.

Table 6 . Video codec traces characteristics.

Parameters	Matrix III
Codec	MPEG-4 Part 2
Frame Compression Ratio	47.682
Min Frame Size (Bytes)	8
Max Frame Size (Bytes)	36,450
Mean Frame Size (Bytes)	3189.068
Peak Frame Rate (Mbps)	7.290
Mean Frame Rates (Mbps)	0.637
Frame Rate (frames/seconds)	25

6. RESULTS AND DISCUSSION

The ninety nine scenarios were simulated and the results are collected and summarized in three cases depending on varying speed of the mobile node, varying path loss models, and for different types of service classes. For each case, the types of modulation and coding schemes are choosing one at a time to obtain one set of simulation results for the different performance measures of packet loss, packet delay, packet jitter, and traffic load throughput.

6.1. Case 1: Mobile Node with Different Speed

This subsection shows the simulation results of twenty seven scenarios which had been simulated under this case. This case is investigated how mobile TV speeds perform using different modulation and coding schemes offered by Mobile WiMAX. This evaluated the performance parameters, namely: packet jitter, packet E2E delay, data drop, and throughput of the mobile node, and showed which the appropriate speed is the best for Mobile TV. The speed of the mobile is provided in kmph while path-loss model and scheduling service are kept constant. Path-loss model is chosen as free space and service class as rtPS is considered.

The average packet jitter, and average E2E delay with fixed and adaptive modulation have been shown in Figures 3(a) and (b). Figure 3(a) shows the average variation of jitter for audio/video Mobile TV over Mobile WiMAX networks for different speed, video quality is best if the jitter is zero. Axis scale of Figure 3(a) is as log scale because the variation results are very small around zero. As shown in Figure 3(a), average audio/ video jitter is approximately zero for higher MCS and AMC under different speeds which is about 2.5753E−05 seconds, whereas other MCS as QPSK has a more average variation of jitter. From the results in Figure 3(a), it is observed that WiMAX using higher MCS (64 QAM 2/3, 64 QAM 3/4) and AMC (AMC-1, AMC-2) as a modulation technique shows better jitter compared with other MCS (QPSK, 16 QAM). Average End-to-End delay for different speeds under MCS shows in Figure 3(b), as we can see the average E2E delay of different speeds give lower packet E2E delay for audio/video Mobile TV under higher modulation and coding MCS (16 QAM, 64 QAM) and AMC compared with QPSK.

Intuitively, increasing data drop is affected by the speed of SS. However, when the speed of SS increases, the hand-off frequency increases, which results in increased data dropped and thereby decreased throughput. As can be viewed from Figure 4(a), the average data drop is significantly higher when SS moves with a greater speed as 150 kmph. The effect of data drop naturally decreases the average WiMAX throughput as shown in Figure 4(b). From Figure 4(a) it is observed that the data dropped is very low for 16 QAM 3/4, and 64 QAM 1/2 modulation schemes and varies with speed. We know that a higher order modulation scheme is more sensible to

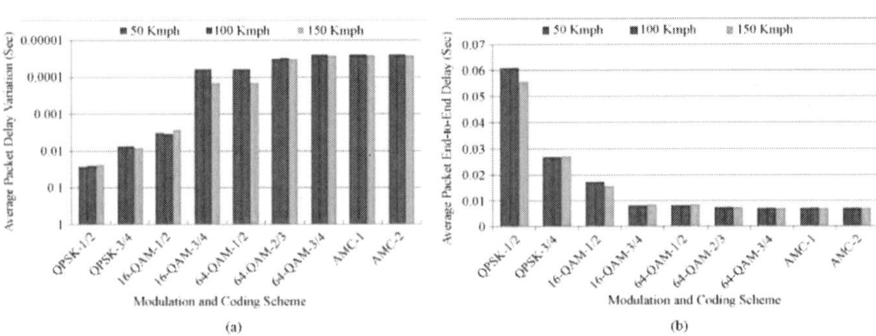

Figure 3. (a) Average video jitter; (b) Average packet End-to-End delay.

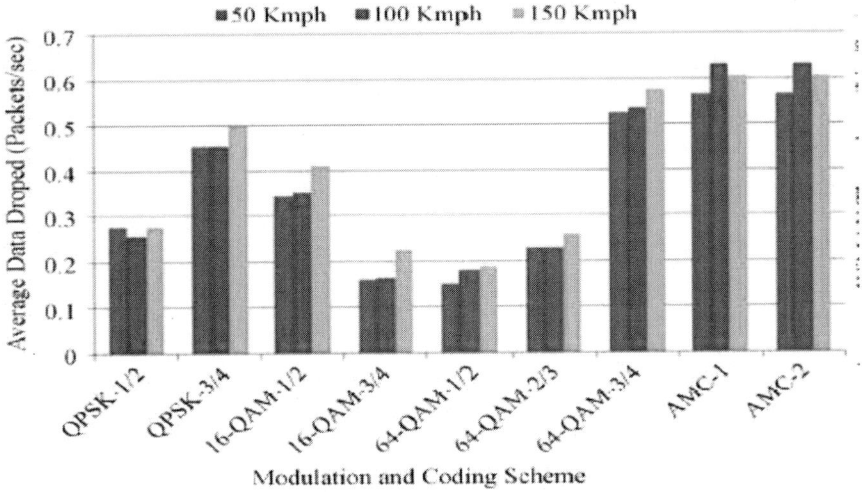

(a)

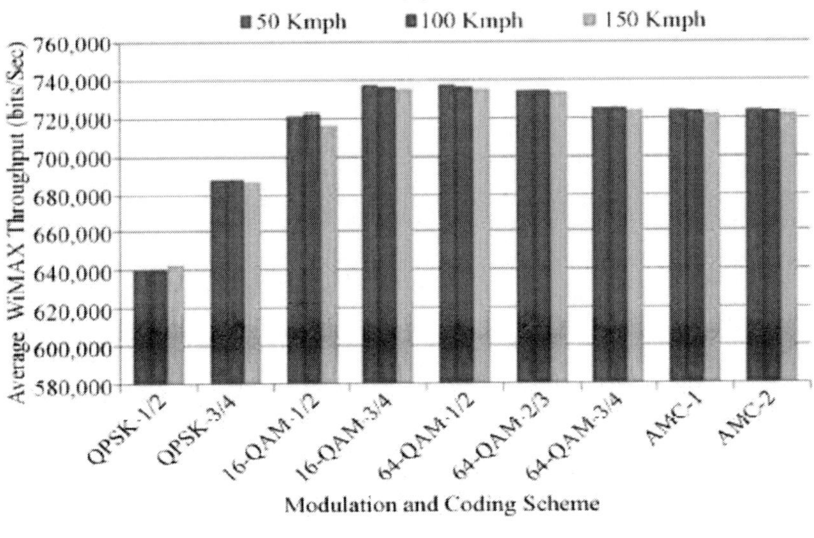

(b)

Figure 4. (a) Average packet data dropped for SS node; (b) Average WiMAX throughput for SS node.

SINR. As the SS is moving through the cell, it faces a different SINR value depending upon the distance from the BS and the propagation environment. With increasing distance, the SINR decreases and the higher order MCS gives more BLER than the lower order MCS for the same SINR value. For that the higher order 64 QAM 3/4 has more data dropped and AMC modulation schemes with varying speeds. Figure 5 shows the BLER for different MCS under 150 Kmph, yellow line in Figure 5 shows the 64 QAM 3/4 which has more BLER than other MCS. The average throughput is taken as a measure that will give the average of observed throughput right through the simulation. Thus, the higher order of MCS 16 QAM 3/4, 64 QAM 1/2, and 64 QAM 2/3 has the best performance (throughput) as observed in Figure 4(b). According to the results as in Figures 4(a) and (b), it is observed that WiMAX uses 64 QAM 1/2 and 16 QAM 3/4 as modulation techniques show better performance (high throughput) compared with other modulation techniques.

6.2. Case 2: Mobile Node with Different Path Loss

This subsection shows the simulation results of 36 scenarios, performance parameters of each scenario observed for various modulation and coding schemes with respect to various path-loss models. We have considered in this case keeping the speed of SS constant as 50 Kmph and scheduling service classes as rtPS.

In this study fixed radius WiMAX networks are considered for all the path loss. As in our knowledge, outdoor to indoor and pedestrian path-loss model is designed for small and micro cell WiMAX network. For outdoor to indoor and pedestrian propagation model, as SS node that moves away from BS will encounter a significant drop in SINR and as the higher order MCS (such as 64 QAM 3/4) requires high value of SINR to give a good throughput; higher order MCS will face a very large amount of data drop as revealed from Figure 6(a). However, outdoor to indoor and pedestrian has lower packet jitter and lower packet E2E delay as shown in Figures 7(a) and (b). Path-loss of free space is lowest; hence, the reduction of SINR with the distance from BS is less which leads to better throughput, lower packet jitter, lower packet E2E delay, and lower packet data dropped from all varies MCS as shown in Figures 6 and 7. Again, as the reduction of SINR with distance from BS is less, the SS has to change its modulation scheme less frequently which results in very high throughput for AMC as shown in Figure 6(b). Here, the suburban fixed model has been considered as hilly terrain with moderate to heavy tree density that representing rural environments

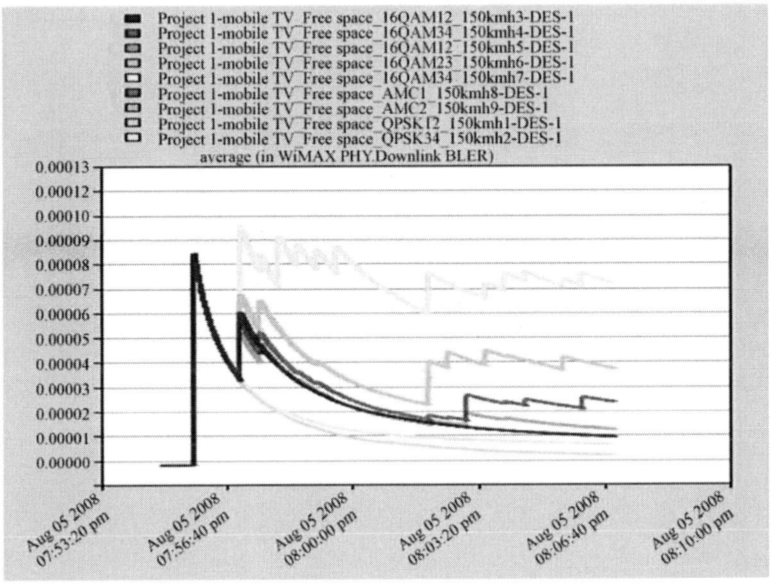

Figure 5. Average download BLER.

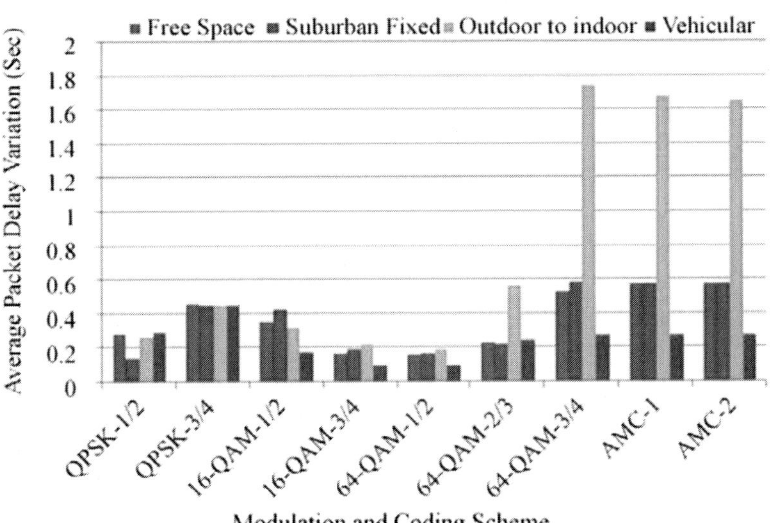

(a)

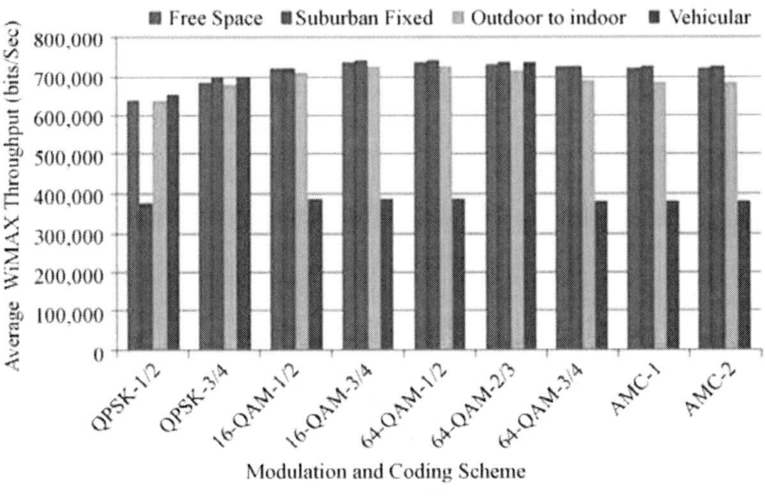

(b)

Figure 6. (a) Average packet data dropped for SS node; (b) Average WiMAX throughput for SS node.

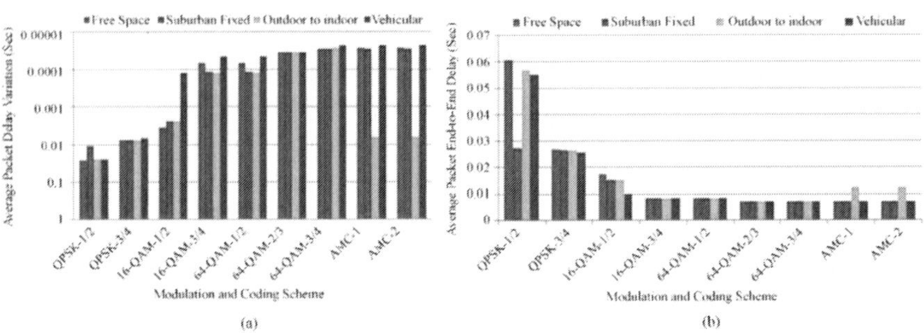

Figure 7. (a) Average video jitter; (b) Average packet End-to-End delay.

and has highest path-loss. Also, same terrain is considered here in a vehicle model. As outdoor to indoor and pedestrian propagation model experiences a very high packet drop compared with the others, it gives the lowest throughput compared with other propagation model as can be observed from Figure 6(b). For free space propagation model, 16 QAM, and 64 QAM has the same throughput under varied coding as in Figure 6(b).

6.3. Case 3: Mobile Node with Different Classes

This subsection exhibits the simulation results of 36 scenarios which has been simulated under this case; speed and path loss kept constant. The speed of SS is chosen as 50 kmph while path-loss model as free space is considered. Different service classes are used in this case under various modulations and coding scheme to obtain the performance metrics: packet delay variation, packet Endto-End delay, data drooped for a mobile node, and WiMAX throughput for the mobile node.

To the best of the authors, UGS and ertps were designed to support VoIP. However, UGS is used for CBR, whereas, A/V Mobile TV has VBR, it cannot use UGS in our simulation. So, UGS classes are not supported in our case study. On the other hand, ertps are designed to support real-time applications generating variable bit rate traffic, and are designed for VoIP service with activity detection. Results obtained shows that this class has worse performance in all metrics as shown in Figures 8 and 9 it has lower throughput, higher jitter, and higher packet E2E delay. Figures 8(a) and (b) show packet delay jitter and packet E2E delay for different service classes: rtPS, nrtPS, ertPS, and BE. rtps, nrtps, and BE has given the best performance, which reveals that for higher modulation and coding, all service classes give equal throughput, equal packet jitter, and equal packet E2E delay. In Figure 9(b), it can be observed that rtps classes' providers have greater throughput than other classes, nrtPS, and BE. The reason is that rtPS is designed for streaming Audio or Video.

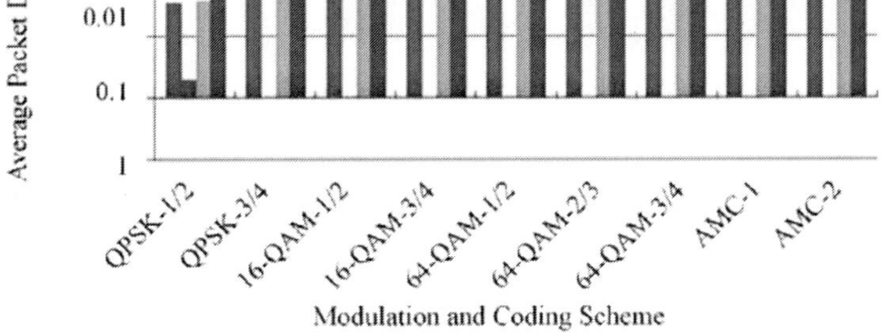

(a)

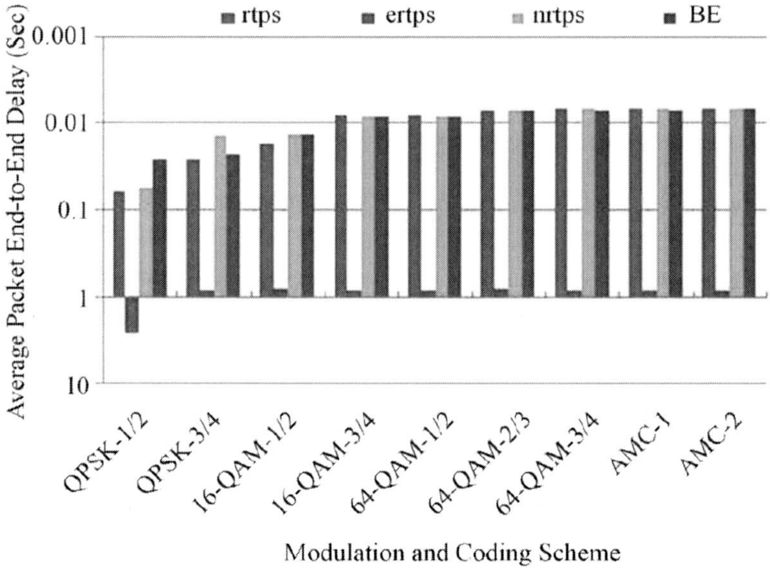

(b)

Figure 8. (a) Average video jitter; (b) Average packet End-to-End delay.

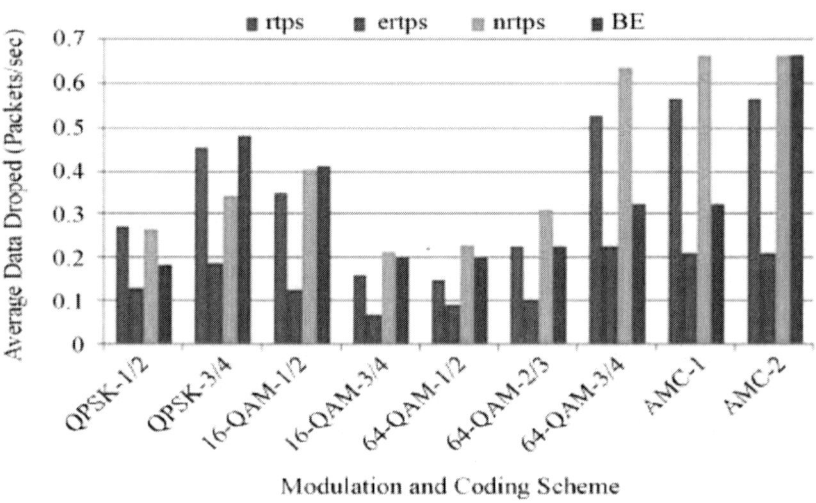

(a)

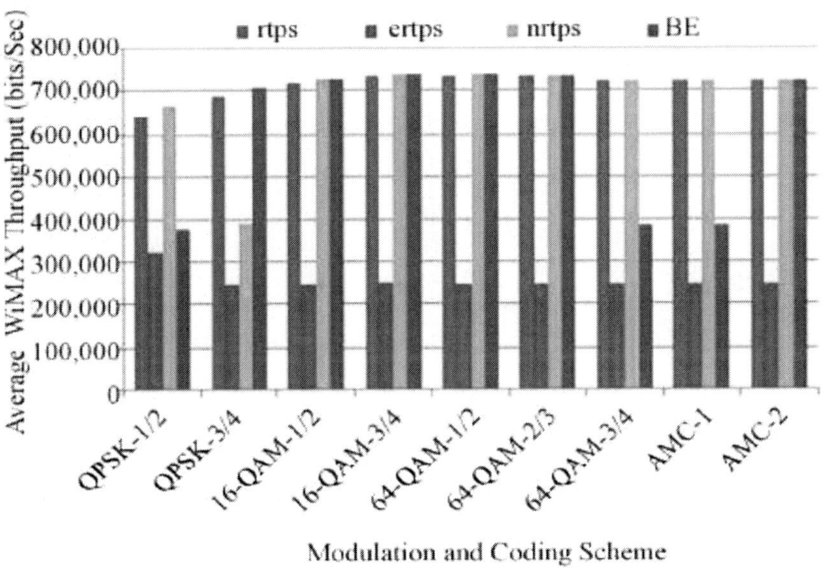

(b)

Figure 9. (a) Average packet data dropped for SS node; (b) Average WiMAX throughput for SS node.

7. CONCLUSION

This paper presents a comparative performance study of Mobile TV over mobile WiMAX network with respect to different modulation and coding schemes under different parameters including speed of mobile node, path-loss models, and MAC service classes. The performance has been evaluated in terms of average packet jitter, average packet E2E delay, average throughput, and average datadropped. It has been demonstrated, using OPNET simulation, that higher order modulation and coding schemes (namely, 16 QAM and 64 QAM) provide better performance. Also, it has been shown that AMC schemes (namely, AMC-1 and AMC-2) give relatively comparable performances to 64 QAM 3/4 scheme. In this work, simulation results show that the free space path loss is best for deploying A/V video application over different mobile node speeds, whereas "outdoor to indoor" is the worst case with the highest packet drop rate. Moreover, our simulation showed that rtPS scheduling service class is the most appropriate scheduling service for A/V video application.

ACKNOWLEDGEMENTS

The authors thank Professor Hussein Omar Qadi, President of Hodeidah University for his supporting to develop the academic research and researchers in the Hodeidah University.

REFERENCES

1. M. Chen, L. Zhou, T. Hara, Y. Xiao and V. Leung, "Advances in Multimedia Communications," International Journal of Communication Systems, Vol. 24, No. 10, 2011, pp. 1243-1245.
2. K. Ain, M. Tarafder, Sh. Khan and Md. Ali, "Path Loss Compensation Technique for WiMAX Technology Based Communication System," International Journal of Engineering Science and Technology, 2011.
3. I. Uilecan, C. Zhou and G. Atkin, "Framework for Delivering IPTV Services over WiMAX Wireless Networks," Proceedings of IEEE EIT, Chicago, May 2007, pp. 470-475.
4. I. Adhicandra, "Using AMC and HARQ to Optimize System Capacity and Application Delays in WiMAX Networks," Journal of Telecommunications, Vol. 2, No. 2, 2010, pp. 15-20.
5. Motorola "Adaptive Modulation and Coding (AMC)," Motorola, Stockholm, 2000.
6. S. Bhunia, I. Misra, S. Sanyal and A. Kundu, "Performance Study of Mobile WiMAX Network with Changing Scenarios under Different Modulation and Coding," International Journal of Communication Systems, Vol. 24, No. 8, 2011, pp. 1087-1104.
7. J. Andrews, A. Ghosh and R. Muhamed, " Fundamentals of WiMAX: Understanding Broadband Wireless Networking," Prentice Hall, New York, 2007.
8. Mobile WiMAX, "Mobile WiMAX—Part I: A Technical Overview and Performance Evaluation," WiMAX Forum, 2006.
9. D. Niyato, E. Hossain and J. Diamond, "IEEE802.16/ WiMAX-Based Broadband Wireless Access and Its Application for Telemedicine/E-Health Services," IEEE Wireless Communications Magazine, Vol. 14, No. 1, 2007, pp. 72-83.

10. WIMAX Forum, "WiMAX's Technology for LOS and NLOS Environments," https://sarwiki.informatik.hu-berlin.de/images/3/30/WiMax-LOS.pdf

11. OPNET Technologies, "Introduction to WiMAX Modeling for Network R&D and Planning," Proceedings of OPNETWORK, Washington DC, 2008.

12. I. Adhicandra, R. Garroppo and S. Giordano, "Optimizing System Capacity and Application Delays in WiMAX Networks," Proceedings of the 6th International Symposium on Wireless Communication Systems, Siena, 7-10 September 2009, pp. 540-544.

13. L. Nuaymi and E. Bretagne, "WiMAX-Technology for Broadband Wireless Access," Wiley, 2007.

14. J. Doble, "Introduction to Radio Propagation for Fixed and Mobile Communications," Artech House Inc., Norwood, 1996.

15. Ch. Dalela, "Propagation Path Loss Modeling for Deployed WiMAX Network," International Journal of Emerging Technology and Advanced Engineering, Vol. 2, 2012.

16. Y. Zhang, "WiMAX Network Planning and Optimization," CRC Press, New York, 2008.

17. IEEE802.16e, "IEEE Standard for Local and Metropolitan Area Networks, Part 16: Air Interface for Fixed and Mobile Broadband Wireless Access Systems," IEEE802.16e, 2005

18. W. Hrudey and Lj. Trajkovic, "Mobile WiMAX MAC and PHY Layer Optimization for IPTV" Journal of Mathematical and Computer Modeling, Vol. 53, No. 11-12, 2011, pp. 2119-2135.

19. IETF, "Y.1541 QoS Model for Networks Using Y.1541 QoS Classes Draft," 2009. http://tools.ietf.org/html/draft-ietf-nsis-y1541-qosm-07

20. A. Shehu, A. Maraj and R. Mitrushi, "Analysis of QoS Requirements for Delivering IPTV over WiMAX Technology," Proceedings of International Conference on Software, Telecommunications and Computer Networks (SoftCOM), 2010, pp. 380-385.

21. A. Shehu, A. Maraj and R. Mitrushi, "Studying of Different Parameters That Affect QoS in IPTV Systems," Proceedings of International Conference on Telecommunications and Information (WSEAS), 2010.

22. W. Hrudey and Lj. Trajkovic, "Streaming Video Content over IEEE 802.16/WiMAX Broadband Access," OPNETWORK, Washington DC, August 2008.

23. R. Gill, T. Farah and Lj. Trajkovic, "Comparison of WiMAX and ADSL Performance When Streaming Audio and Video Content," OPNETWORK, Washington DC, August 2011.

24. P. Telagarapu, G. Naidu and K. Chiranjeevi, "Analysis of Coding Techniques in WiMAX," International Journal of Computer Applications, Vol. 22, No. 3, 2011, pp. 19-26.

25. M. Singh, R. Uppal and J. Singh, "WiMAX with Different Modulation Techniques and Code Rates," International Journal of Information and Telecommunication Technology, Vol. 3, No. 1, pp. 31-34.

26. Md. Islam, R. Mondal and Md. Hasan, "Performance Evaluation of WiMAX Physical Layer under Adaptive Modulation Techniques and Communication Channels," International Journal of Computer Science and Information Security, Vol. 5, No. 1, 2009, pp. 111-114.

27. Md. Islam and T. Isalm, "Performance of WiMAX Physical Layer with Variations in Channel Coding and Digital Modulation under Realistic Channel Conditions," International Journal of Information Sciences and Techniques, Vol. 2, No. 4, 2012, pp. 39-47.

28. A. Rehman, T. Khan and S. Chaudhry, "Study of WiMAX Physical Layer under Adaptive Modulation Technique using Simulink," International Journal of Scientific Research Engineering & Technology, Vol. 1, No. 5, 2012, pp. 5-11.

29. Arizona State University, "Video Traces for Network Performance Evaluation". http://trace.eas.asu.edu/mpeg4/index.html

5

CHAPTER

SENDING SAFETY VIDEO OVER WIMAX IN VEHICLE COMMUNICATIONS

Jun Steed Huang [1], Oliver Yang [2,*] and Funmilyo Lawal [2]

[1] *Suqian College, Jiangsu University, Jiangsu 223800, China*

[2] *Computer Communication Network Research Lab, School of Electrical Engineering and Computer Science, University of Ottawa, 161 Louis Pasteur, Ottawa, ON K1N 6N5, Canada*

ABSTRACT

This paper reports on the design of an OPNET simulation platform to test the performance of sending real-time safety video over VANET (Vehicular Adhoc NETwork) using the WiMAX technology. To provide a more realistic environment for streaming real-time video, a video model was created based on the study of video traffic traces captured from a realistic vehicular camera, and different design considerations were taken into account. A practical controller over real-time streaming protocol is implemented to control data traffic congestion for future road safety development. Our driving video model was then integrated with the WiMAX OPNET model along with a mobility model based on real road maps. Using this simulation platform, different mobility cases have been studied and the performance evaluated in terms of end-to-end delay, jitter and visual experience.

Keywords: vehicular camera; WiMAX; Ad-hoc; video; modeling

1. INTRODUCTION

By virtue of its versatility and utility convenience, mobile wireless communication has now become a necessity. Over the years, different technologies have been developed to increase the data rate in wireless networks. Technologies like WiMAX (Worldwide Interoperability for Microwave Access) and 3G/4G devices can support 100 Mbps per data channel [1], and LTE (Long Term Evolution) can go as high as 200 Mbps. This has created room for different types of applications in different specialties. Users have come to desire not only the basic voice functionality of wireless devices, but also different forms of mobile multimedia communication. Video communication via a VANET (Vehicular Adhoc NETwork) is one such application.

Safety and environmental issues in managing existing road networks have become one of the greatest concerns of both the automotive industry and governmental bodies. A VANET consisting of clusters of vehicles communicating with each other and also with roadside infrastructure can provide a solution to these issues. Telematics for the telecommunications of safety information is one major tool used in this area. Progressively, sharing real-time driving video traffic among vehicles and roadside infrastructures would be the next natural step to enhance the safety feature. One motivation of this paper comes from the statistics of road traffic fatalities which shows that there were 44,192 accidental deaths in Canada from the years 2000 to 2004 [2]. California, with a population of 37 million people reports, one million vehicle accidents each year [3]. The increase in road traffic volume over the years has led to a concern among researchers to create an effective, efficient and safe automotive environment with the help of video communication. One example is the "smart car" in the state of Nevada which, as a first step towards this goal, uses video to assist hands-free driving.

Unlike other wireless environments that are mostly stationary or under slow motion, transmitting video signals in a VANET poses more challenges to resolve. Electromagnetic interferences such as those from engine electronics, from Additive White Gaussian Noise, or those due to weather conditions can affect the QoS (Quality of Service) as seen by a wireless user. Since the topology is constantly changing, a vehicle could move out of sight to cause an outage in data transmission. Obstacles along LOS (Line of Sight) communication are another major reason to cause an outage. Finally, packet loss due to data traffic congestion will definitely make things worse, and communication disruptions may become inevitable.

Understanding the characteristics of real-time video would allow the design of efficient real-time transmission/transportation protocols to address the above challenges and to allow us to carry out further study. There has been continuous and extensive work on VBR video from various specific video sources, (e.g., from video conferencing cameras, web cameras, broadcast TV or HDTV [4,5], and non real-time processing/storage of data taken from the cameras installed in a real public safety law enforcement car system [6]. However, images from a vehicular camera have higher activity around all edges than the middle portion. This is very different from a normal camera constructed for stationary scenarios, and can be addressed/utilized in the design of video communication. So far, we are not aware of any comprehensive study related to real-time driving video traffic originating in the actual vehicle camera, such as GV1500 from GenieView Inc. [7]. Furthermore, all these prior arts focus on the modeling of separate and independent video traffic before the wireless networks, and there is no work done on the video model of the traffic coming out of a wireless link, as will be needed in our later study of highway video through a VANET.

Higher layer protocols are usually required to support various video applications. TCP (Transport Control Protocol) [8] and UDP (User Datagram Protocol) [9] for real-time transportation of video have been shown to be ineffective. As is further reviewed in Section 2, TCP is best used for lossless data transfer as it ensures packet delivery by its support of packet retransmission. This poses a problem for real-time traffic, as there is no time for retransmission [10]. UDP is better for video traffic, but unfortunately still cannot guarantee real-time and loss performance. RTP (Real-Time Transport Protocol) [11] offers an alternative but has other shortcomings that need to be remedied, as is discussed later. Most of the aforementioned transport algorithms and the traffic control protocols below are built on top of TCP and have only been tested in wired networks. Quite often, packets in a wireless network are stored in a buffer waiting for proper bandwidth allocation before they can be delivered. In real-time communication, a poorer quality can be tolerated, but delay should be reduced to a minimum if not eliminated.

Since a large volume of video traffic can easily consume the limited wireless bandwidth, there is a need to throttle the video traffic for better bandwidth utilization and avoid congestion. Traffic/Congestion control can provide smooth throughput and high link utilization for real-time data. Of the two main types of traffic congestion, control in computer networking (reviewed in Section 2), implicit congestion control does not provide smooth throughput for real-time and low link utilization. Also, the control is based on packet loss-events which are not a big concern for real-time traffic as mentioned earlier. Explicit congestion control of traffic from the router node becomes impractical in a high speed when the CPU cannot catch up with the updating of events arising from changing driving speeds. Consequently, a different protocol (desirably with a low complexity) should be considered.

Finally, it is safer and more cost efficient to simulate possible solutions than actual field experimenting with highway driving. An effective VANET simulation platform/model has pertinent importance in research and industry to emulate a real world situation. One major challenge is the development of an effective platform that can bring all issues described earlier under one simulation model. A major issue is the integration of an effective mobility model that considers vehicle-to-vehicle interaction and vehicle-to-infrastructure interaction along with the full functionalities of a communication device capable of effective receiving, processing and transmitting capabilities. Although there has been work on creating a similar platform, e.g., [12,13,14], little work has been done on certain issues related to video streaming from vehicle camera such as the characterization of the data traffic (e.g., its unique packet size distribution) and the behavior under different driving conditions (e.g., highway vs. local).

In view of the above and others reviewed in the related work in the next section, we seek to create a robust platform for effective VANET simulation, one that can provide an environment to test, develop and deploy effective communication protocols that would enhance real-time communication of video information from vehicle camera. As forerunners, in this paper, we would like to:

(i) create a real-time camera video model under VANET conditions;
(ii) integrate a realistic VANET mobility model to test the testbed at various speeds;
(iii) integrate a practical algorithm to survive network congestion;
(iv) demonstrate the capability of such a platform.

To achieve the above objectives, we first adopt the protocol architecture of our simulation platform as shown in Figure 1. We need to adopt appropriate protocols for the lower layers consisting of the Data Link Layer and the Physical Layer, and the upper layers for the remainder. After a comparison of several viable technologies for the lower layers such as the pros and cons as reviewed in Section 2, we have chosen WiMAX (Worldwide Interoperability for Microwave Access) for its cost-effective approach that can provide a high data rate that can satisfy the needs of our vehicle camera users (low latency and high coverage) at high speed at an affordable cost. It is scalable because its uses OFDMA (Orthogonal Frequency Division Multiple Access) to allow flexible frequency re-use. It implements full MIMO (Multiple-Input and Multiple-Output) which is good for mobile applications by enhancing timely information delivery to save lives and improve quality of life. In view of the complexity problem in implementing an explicit congestion control mechanism in the upper layers to control streaming traffic from driving video in real-time, our present approach is not to "cure" congestion but to "survive" congestion. We propose to reduce the compression rate of the video packets once congestion is detected in the network. We use the RTCP (Real Time Control Protocol) to feed such information back to the sender.

Layer 4	Sub-layer 3	RTCP
	Sub-layer 2	RTP
	Sub-layer 1	Data Traffic Controller
Layer 3	Sub-layer 2	UDP
	Sub-layer 1	IP
Layer 2	Sub-layer 2	Data Link Port
	Sub-layer 1	Data Link Buffer
Layer 1	Sub-layer 2	Physical Port
	Sub-layer 1	Physical Transponder

Figure 1. Overall Architecture.

The next step is to find a simulation tool and a realistic vehicular mobility model for integration. Of the few available network simulation languages such as Qualnet and NS-2, we have chosen OPNET because it has different wireless models including WiMAX. It has a rich set of video streaming protocols, and it provides a good framework, which allows us to integrate our mathematical driving video model effectively so that we can explore different scenarios during network traffic congestion. After searching, we have also decided that VanetMobiSim [15,16] is the only VANET mobility model that would allow the output code generator to be modified in order to generate and output traces useable by other simulators. Consequently, we can integrate its traces into our existing OPNET model using the open-loop approach.

To support the experiments from our simulation platform, we need to use realistic driving video traffic streams. Since we do not have an additional ACE [17] license to support the feeding of life data into our OPNET simulator, we shall generate our traffic streams based on a model established from real video traffics transported over wireless networks. Not only do we want to use this model to generate traffic traces quickly, we would also use it to alter the traffic characteristic under different scenarios later on without having to recapture traffic traces from life experiments, which can be both costly and time consuming.

The contributions of this paper lie in the following:

(i) The design and implementation of a theoretical but more realistic driving video model than the traditional video model. To the best of our knowledge, there is no comprehensive study related to real-time driving video traffic originated from a vehicle camera, nor any work done on the modeling of the video traffic from the vehicular camera system through the wireless link;
(ii) The integration of a VANET mobility model with such a driving video model to provide a practical VANET simulation platform for the evaluation of video traffic under different scenarios;
(iii) The design and implementation of a practical traffic controller for driving video over RTP applications.

The rest of the paper is organized as follows. Section 2 discusses existing works related to our research, especially their deficiencies. Section 3 discusses the network operations and models used in our testbed. To illustrate the procedure of modeling the driving video traffic in VANET, Section 4 presents an example of the trace collection and stochastic analysis of the trace. We then discuss the OPNET simulation of different scenarios and evaluate their performance in Section 5.

2. RELATED WORKS

Below are works related to our research work on VANET testbeds, models for driving video, as well as the lower and upper layer protocols.

2.1. VANET Testbeds

There have been practical implementations of VANET testbeds. For example, a testbed was implemented to extract H.263 video streams through an ADSL (Asymmetric Digital Subscriber Line) gateway to an SQL server in the Internet [18]. A prototype testbed [19] using a handheld device with an Android operating system was presented for data-intensive application. While these approaches are effective, their hardware implementations are expensive and do not allow quick and easy changes to adapt to various industry scenarios and implementation of special safety applications. Video cameras make up a key component in VANET due to their ability to aid navigation safety and give road users and relevant authorities a precise and clear image of the traffic conditions when necessary. Intel Corporation has decided to lead the development of a video camera system on board of a car [20]. Others include the CogniVue system [21]. Since a special high-definition camera can pump into the network to degrade network performance and video quality, this has been a key concern in VANET design. Solving this issue will unlock the power of driving camera video applications.

In light of the outlined efforts, one of the major focuses of VANET research is the creation of an effective simulation platform that can integrate a network simulator with a realistic vehicular traffic simulation model. Unlike other network environments, VANET mobility has a peculiar and unique nature due to the randomness of human behavior. One classification groups vehicular traffic under four models [22]: Submicroscopic, Microscopic, Mesoscopic and Macroscopic. It is important to choose a realistic mobility model to correspond to each one. An open-loop integration approach and a closed-loop integration approach have been discussed for the integration of the network model with a VANET mobility model. Although there has been other work on creating a similar platform (e.g., [12,13,14]), to the best of our knowledge, there is hardly any work on the characterization (e.g., the packet size distribution) of video data streaming from the vehicle camera and the behavior under different data traffic conditions. The closest work is on UAV video [23], but this was done from an image process perspective. A broad view of the state-of-the-art mobility models adapted for VANETs was discussed [24], which also provided a detailed survey and comparison of mobility models. However, it only provides a model for data transmission and not for video transmission. Such a model needs to consider other factors (like the physical layer communication mechanism and control algorithms in the application layer for real-time video streaming) for effective VANET simulation. Note that a model or a tool may be good for one aspect/feature, but their integration is necessary to model an effective safety solution.

2.2. Modeling Driving Video Traffic

Unlike other forms of multimedia traffic (like data, voice and video), driving video traffic has a very dynamic stochastic nature. Different statistical studies [25] have already revealed that the inherent characteristics (not source nor codec specific) of Variable Bite Rate (VBR) video traffic is the LRD (Long Range Dependence) which has a significant impact on the queuing performance and characterization of video traffic [25,26]. Video traffic modeling over wireless has been studied extensively. There is work on the video traffic modeling in a wireless environment with emphasis on the streaming video over WiMAX networks [27]. The I, P and B frames have been modeled separately according to their types, and a fixed pattern is used to combine these frames together to form the GOP (Group of Picture) [27,28,29]. A video traffic model was obtained using a trace generated with a fixed frame inter-arrival time [30]. All these prior works have focused on the separate and independent video modeling before the traffic is fed into the wireless networks, and none has considered the video model after coming out of the wireless link. Such modeling is required in our highway video forwarding through a number of consecutive cars. Furthermore, we are not aware of any comprehensive study related to real-time driving video traffic originated from the actual vehicle camera, such as GV1500 made by GenieView Inc. [7]. Due to the very driving nature, such a video has an even higher statistical fluctuation that is very hard to control. More discussions can be found in further sections such as 1 and 4.3.

2.3. Lower-Layer Protocols for VANET

VANET applications can be supported through V2R (Vehicle-TO-Roadside) and V2V (Vehicle-to-Vehicle) communications. IEEE 802.11 (WiFi) [31], IEEE 802.16 (WiMAX) [32] and DSRC (Dedicated Short Range Communications) [33] technologies can be used in this model of communication.

With the DSRC standard, OBU (OnBoard Units) placed at each vehicle can send or receive data to or from roadside units (RSUs), as well as communicating directly with each other. There are thus many potential applications in urban areas (or in populated countries in Europe or Southeast Asia) such as forward collision warning, electronic parking payments, and approaching emergency vehicle warning [34]. Unfortunately, it is limited in its distance coverage and may not cover rural areas in countries like Canada and USA. Furthermore, DSRC is based on 802.11a and cannot handle high-speed mobility while maintaining an acceptable data rate. It also does not have the ability to hand over connections at high speed, which is essential for high-speed motion and cannot support high data rates. DSRC would probably be integrated with other technologies for use in the VANET if it is to be universally accepted. For example, we can use dynamic spectrum sharing for both DSRC and WiMax to improve the communication efficiency as well as the spectrum utilization to cover both the rural and urban areas [35].

WiMAX (Worldwide Interoperability for Microwave Access) [14] is a 4G equivalent technology that enables the delivery of last mile wireless broadband access. WiMAX technology [36] can eliminate the use of cables and can save costs when used in remote and rural areas. The technology is scalable and has a flexible frequency re-use scheme because it uses OFDMA (Orthogonal Frequency Division Multiple Access) technology. Unlike DSRC, WiMAX was designed to handle high data rates. It implements full MIMO (Multiple-Input and Multiple-Output) which is good for mobile and car applications by enhancing timely information delivery to save lives and improve quality of life. Many laptops in USA already support such interface even though its present price is still high. With the above observations on and comparison work with DSRC, one sees that WiMAX is the most cost effective approach by providing a data rate that can satisfy the needs of our vehicle camera users (low latency and high coverage) at high speed and at an affordable cost.

2.4. Higher-Layer Protocols for VANET

Real-time data has an arrival deadline to reach its destination, or else its information is no longer valid. TCP (Transport Control Protocol) [8] and UDP (User Datagram Protocol) [9] are protocols that have been used but were found to not be effective. TCP is best used for lossless data transfer as it ensures packet delivery via packet retransmission. This poses a problem for real-time traffic, as there is no time for retransmission [10]. UDP can be better for real-time video traffic, but unfortunately still cannot guarantee real-time nor loss performance because their flows cannot adapt their rates to congestion control mechanisms at the router [37]. They can provide the best-effort service only since they have no priority over TCP flows. Congestion will also lead to random jitter and eventually loss, as video traffic becomes prevalent in f VANET. Although in theory TCP or UDP should not be affected by the choice of physical transmission medium such as fiber optic, they often perform poorly in wireless networks [38] because their design and implementations were carefully optimized for wired networks. RTP (Real-time Transport Protocol) [11] is a protocol designed for end-to-end real-time transport of video traffic. However, it does not take many performance issues into account including channel outage, traffic congestion, the effect of overheads in uncast/multicast applications, and the end-to-end delay in a mobile environment.

In summary, the behavior of these protocols in the VANET environment are quite different as the wireless medium is prone to interference from forces of nature and manmade effects, and the situation is exacerbated when transmitter and/or receiver are moving.

Various works have been conducted on congestion control in wired environments [39], wireless environments [40,41] and VANET [42]. Congestion control protocols can also be classified into two main types [39]: window-based and rate-based. The window-based congestion control algorithms adjust the window-size according to the level of congestion detected in the network. Algorithms like RED (Random Early Detection) [43], its TCP variants (like Tahoe [44], NewReno [45]), and AIMD (Additive Increase and Multiplicative Decrease) [44] fall under this category. This type of control usually corresponds to the implicit congestion control as congestion is detected at the receiver through the observation of events such as packet loss. So many algorithms in this type of approach can cause fluctuations in the source sending rates, thus producing non-smooth throughput and low link utilization. These are evidently not effective for real-time video streaming [46] in practice.

Rate-based congestion control algorithms on the other hand, allow each source in the network to directly adjust its sending rate based on the congestion notification received, and usually from the router that knows the state of the network. TFRC (TCP Friendly Rate Control) [47] and API-RCP (Adaptive PI Rate Control Protocol) [39] falls under this category. API-RCP presents an effective way to calculate the source-sending rate based on the queue deviation, and is quite capable of producing steady queue length to meet the bandwidth requirement of a driving camera. The rate-based control usually makes use of explicit congestion control due to the nature of its operation (as a contrast to the implicit control discussed earlier). However, this type of control may not be effective when a vehicle is moving because constant updating of the router due to motion consumes more CPU resources. Eventually, the CPU cannot catch up with the events arising from high speeds.

Most of the algorithms mentioned previously are built on top of TCP and have only been tested in wired networks. In wireless networks, delay is caused mostly by wireless interference rather than network congestion. Quite often packets are stored in a buffer waiting for proper bandwidth allocation before they can be delivered. As mentioned earlier, in real-time traffic, buffering (which incurs delay) is not wanted. We can tolerate a poorer quality, but delay is unacceptable. Explicit congestion controls handle traffic more effectively by calculating the flow throughput at the router node, and then smoothing throughput and achieving high link utilization in real-time. However, they are not effective for practical purposes. This is because they are more tedious to implement; one has to take time to load the controller on every router, and constant updates are required for the router of the moving vehicles. This consumes more CPU resources, and may not catch up with the driving speed. We thus consider an automatic predictive source control to assist the explicit control. When P frames start piling up, we drop the even frames first and then the odd frames without waiting for control commands to come from the base station.

3. OPERATION AND MODELS OF THE VANET TESTBED

Figure 2 is a simple example of a VANET consisting of cars communicating their video information with each other and with the Internet (IP cloud) via the RSUs, which can be implemented as a WiMAX-enabled Base Station. The server is to forward or to archive the streaming video from the cars. Each car is acting as a WiMAX Mobile Station (MS) capable of receiving, processing and forwarding data to the destination and consequently to the backhaul network via the WiMAX Base Station.

In our VANET testbed, each vehicle has the ability to communicate with any neighboring vehicles. Depending on the nature of the messages, they will either stay within the VANET or exit the backhaul network via the RSU. For instance, tailgating, collision and braking (from preceding cars) are warning messages that can remain within the VANET network, while detailed regional weather forecast, notification of traffic jams and collision are usually exchanged between the backhaul network and the VANET. In many video applications we envision that video messages from the vehicular camera are usually required to be forwarded from the point of interest (such as traffic congestion, road block, and accidents) to a control center in the backhaul network to aid various parties (such as traffic personnel, emergency agents, insurance companies) to respond to the situations more effectively. We shall refer to these types of traffic as "safety driving video" or simply "safety video" later on.

To implement a testbed for the study safety driving video traffic among vehicles, we shall provide the operations at different layers as well as different models used in the testbed in the following two sections.

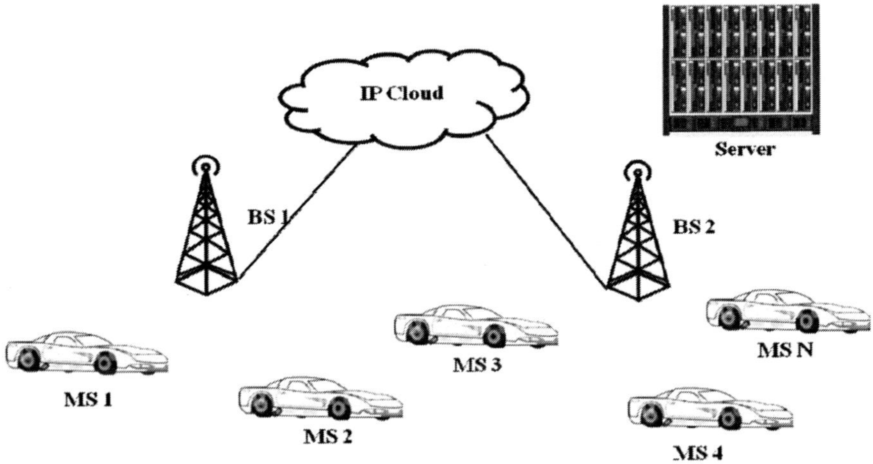

Figure 2. Vehicle camera network operation.

3.1. Protocol Layers and Operations

With reference to Figure 2, an RSU (BS 1 or BS 2) has the capability of handling up to N cars simultaneously. Each car/MS (Mobile Station) is associated with the closest RSU in the coverage area. Safety driving video packets are routed and given priority by recognizing their service class and scheduling type. For example, the silver service class and the rtPS (real-time Polling Service) scheduling [17] are assigned to the safety video. For each service class, there is a maximum sustainable traffic and reserved traffic rates (e.g., 384 kbps for the silver service class). Each video packet arriving from the higher layer at a MS is expected to be within a size limit. Otherwise, it will be segmented before being encapsulated into a PDU (Protocol Data Unit) with the appropriate header information, and then transmitted. For example, video packets longer than 1500 B must be segmented to fit the SDU (Service Data Unit) for the silver (safety video) service.

The driving video traffic is generated from the application layer when an MS wants to transmit. We use UDP (User Datagram Protocol) as the transport layer protocol to cater for the end-to-end delivery of packets. RTP (Real Time Protocol) is used at the application layer to combat the possible unreliability in UDP. RTP is a real-time streaming protocol designed for streaming audio and video. In conjunction, we use RTCP (Real-Time Transport Control Protocol) for controlling, signaling and quality service monitoring.

The unique nature of driving video traffic is due to its burstiness and persistence. The size of each frame generated is dependent on its previous frames and the driving speed, and the exact time a frame is generated or received also depends on previous frames and road conditions. These two correlated uncertainties contribute to the high variable bit rate characteristics of driving video communication. Coupled with the real-time and accuracy requirement of our safety applications, these factors pose a modeling challenge. Of the various video codec schemes that have been developed e.g., MPEG2/4, H.263/4, we have chosen the MPEG4 video compression for our vehicular cameras and studied its traffic model for our testbed. Our array-based MPEG encodes the input video into three types of compressed frames called the I, the Peven, and the Podd frames. I-frames are compressed using intra frame information only. Peven or Podd-frames are coded similarly but with motion compensation for even or odd lines respectively according to the previous I frame. Depending on the vehicle driving speed or congestion condition, Peven and Podd frames may be dropped alternatively, or altogether. In general, I–frames are 10 times bigger than P frames. After coding, the frames are arranged in a deterministic order, which is called GOP (Group Of Pictures). This pattern is not specified hence different coders may use different patterns for subsequent GOPs. A common feature, however, is that all video frames exhibit strong time dependent correlations among themselves. This gives rise to different queuing statistics and time-dependent rush hour congestion phenomenon. This is the subject of streaming traffic analysis and modeling later on, as well as the need for a delicate adaptive congestion controller design.

Each car has a WiMAX capability in transmitting (and receiving) at the physical and the MAC layers after receiving the safety video packets from (and delivering to) the network layer. At the MAC layer, the destination address is examined and the nearest RSU is chosen to handle the request. The packet is sent to that RSU, and further forwarded accordingly. In our investigation, the IP cloud in Figure 2 can be set to its default values and acts a router. The server is configured to accept packets generated by our simulator.

As suggested above, we shall use mobile WiMAX technology at the physical and MAC layers for our VANET. This is a revision of the fixed WiMAX IEEE 802.16e-2005 standard [48] and it provides functionalities such as Base Station handoffs, MIMO (Multiple Input Multiple output) transmit/receive diversity, and scalable FFT (Fast Fourier Transform) sizes. Table 1 lists some of the more relevant features. Although WiMAX can support high data rates providing up to 128 Mbps downlink and 56 Mbps uplink using its MIMO antenna techniques, we shall use SISO antenna technique, which supports up to 1 Mbps uplink and downlink. We use the WiMAX MAC architecture to define service flows that can be mapped into gradual IP sessions to enable end-to-end IP based QoS.

Table 1. IEEE 802.16e characteristics.

Sub-layers 2	Standard	IEEE 802.16e–2005
	Multiplexing	OFDMA
	FFT size	Scalable (512, 1024, 2048, *etc.*)
	Duplexing mode	TDD
Sub-layers 1	Modulation scheme	QPSK, 16-QAM, and 64-QAM
	Subcarrier spacing	10.94 kHz
	Signal bandwidths	5, 7, 8.75, and 10 MHz
	Spectrum	2.3, 2.5, and 3.5 GHz

Of the two-duplexing modes allowed by mobile WiMAX, we shall use the TDD (Time Division Duplexing) mode (the other is FDD (Frequency Division Duplexing) mode) as done in most deployments in order to accommodate the asymmetric VANET traffic. That is, our uplink traffic requires more bandwidth than the downlink, due to the uploading of safety (not downloading entertainment video that the other researchers are focused on) driving video traffic to the RSU. The WiMAX technology uses OFDMA frames for multiple users to receive data from the BS at the same time thereby increasing bandwidth utilization. The frames are structured into UL (Uplink) and DL (Downlink) subframe as shown in Figure 3 [49].

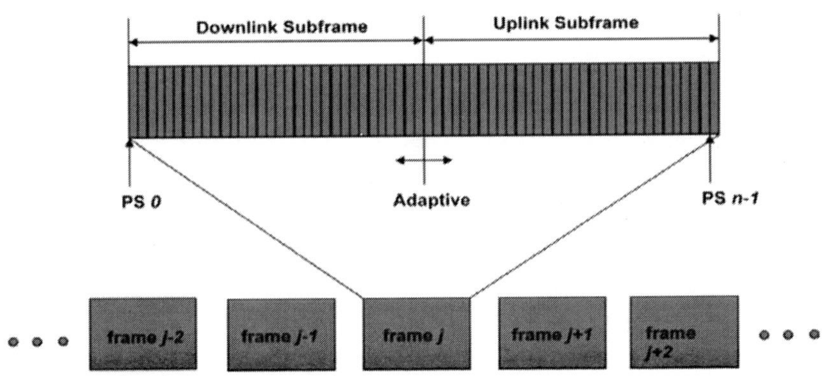

Figure 3. WiMAX frame structure.

All the MSes receive the frame control section of a DL subframe. The DL mapping field in the frame control section specifies the location and duration of DL data burst in the frame from a particular MS. Conversely; the UL mapping field specifies the location and number of time slots allotted in the frame for a particular MS to transmit its data on UL. All MSes listen to the control section in the DL frame, and receive or transmit data during the slots assigned to them by the BS [50]. As seen from Figure 3, this structure enables dynamic allocation of DL and UL resources to efficiently support asymmetric DL/UL traffic. A subchannel is allocated to each user there by reducing the channel interference in the frequency domain. OFDMA (Orthogonal Frequency Division Multiple Access) is the scheme used allowing multiple accesses to every user on our network.

3.2. Network Models

Figure 4 summarizes the various components of our Driving Video VANET OPNET model. It consists mainly of the video model and the VANET model. OPNET modeler then provides the platform for the communication model and allowed for the integration of the various components of the Driving Video VANET OPNET model and other considerations like the congestion controller. They are discussed in the following.

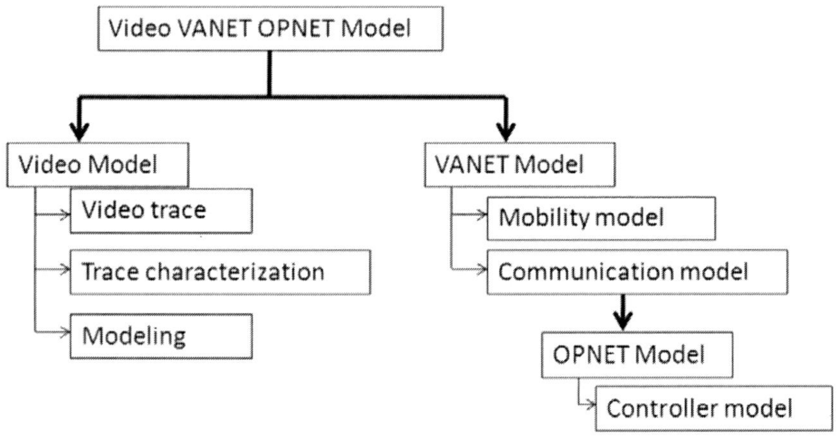

Figure 4. Driving video VANET OPNET Model structure.

3.2.1. VANET Model

The VANET model consists of the VANET mobility model for the movement of a vehicle, and a communication model for information exchange with neighboring MS and BS.

3.2.1.1. VANET Mobility Model

Of the four categories of VANET mobility models described previously, we adopted both the macroscopic and microscopic models to allow the modeling of vehicle-to-vehicle and vehicle-to-infrastructure interaction. VanetMobiSim is chosen to integrate these models in OPNET due to its flexibility to manipulate output files (by coding its output generator file to produce a format desired by our simulator). It also generates traces in a special Universal Trace Format [15] in the form of (time node_id pos_X pos_Y velocity acceleration) where time records the simulated time; node_id identifies the specific node; pos_X and pos_Y provide the X and Y coordinates of the specific node at the given time respectively; and the velocity and acceleration show respectively the velocity at which the node moves and its acceleration at that time. With these capabilities, VanetMobiSim can model traffic lights, stop signs, human mobility dynamics and safe inter-distance management. It allows the integration of road topologies from databases such as GDF [51] or TIGER [52] to user-defined topologies as well as random topologies. VanetMobiSim provides a flexible platform in which the user can configure the shortest path during a trip based on the criteria of Dijkstra, road speed and road density. The trip can either be generated by random source-destination or according to activities [15].

3.2.1.2. RSU and Vehicle Communication Model

The RSU and the vehicles are the major communication nodes in VANET. Our RSU is a WiMAX BS. Each RSU is non-application-sensitive so that a wide range of information can be sent and received. Each vehicle has WiMAX receiving and transmitting capabilities to enable vehicle to vehicle and vehicle to infrastructure (RSU in our case) interaction. Table 2 shows the basic essential characteristics of our model along with some typical settings.

Table 2. RSU parameters.

Parameter	Value	Unit
Physical Layer	IEEE 802.16e	NA
BS TX Power	5	W
Number of TX	SISO	NA
BS Antenna Gain	15	dBi
Minimum Power Density	−80	dBc
Maximum Power Density	−30	dBc
Link Bandwidth	20	MHz
Base Frequency	5.8	GHz
Physical Layer Profile	OFDM	NA

3.2.2. Driving Video Model

The driving video model is used to analyze and characterize the video trace of a live driving scene captured by a vehicle camera. The results are then used to generate models for use in our simulator for real-time driving video communication in a VANET environment. The implementation of this wireless driving video traffic model allows us to handle a wide range of burstiness in terms of packet inter-arrival and packet size arising from the unique nature of wireless video traffic during driving as mentioned above. A guideline to check the efficiency of a traffic model is, however, needed as to set a standard to be followed. Parsimony, analytic, relative accuracy, flexibility, implement ability and absolute accuracy are factors listed in [53] showing the engineering tradeoffs among different traffic models.

Using these factors on a scale of 1 to 3, 3 being the highest, Table 3 shows the settings of a few popular models including the FBM (Fractional Brownian Model) and the OPNET-emulated TCP traffic. By comparison, we adopt the Mini-Pareto model to be used later due to its overall score. In particular, we have taken a systematic approach in developing our mini-Pareto model. Driving video traffic traces were collected using the same camera used for a road test. In our testbed, we traced the frames from the receiver instead of the transmitter in order to correspond to video traveling from vehicle to vehicle, and through the RSU to the city's road monitoring center in a safety network. The traces were analyzed and stochastically represented and plugged into our simulation platform. A detailed discussion about these steps is presented in Section 4.

Table 3. Traffic model comparison.

Items	FBM	TCP Model	Mini-Pareto
Parsimony	3	1	2
Analytical	1	1	2
Flexibility	1	1	1
Implemental	2	1	3
Accuracy	2	3	2
Mobility	2	1	3

3.2.3. Traffic Controller Model

As discussed in Section 1 and Section 2.4, we only need to survive congestion to meet the QoS requirement of our real-time safety application scenario. This has given us the idea of reducing the complexity of the explicit control algorithm in order to cut down the CPU consumption when the vehicular camera is moving at higher speeds. Our careful study shows that reducing the sending bit rate is the key factor of success, and this can be achieved in two ways: dropping some frames (packets) or reducing quantization level per frame. There are different consequences and tradeoffs.

For driving video traffic, the drop pattern is very important to the quality of service as seen by user. Dropping just the even frames or just the odd frames is usually better than dropping both consecutively. The frequency of a drop can be a more critical factor than the percentage of drop. For instance, dropping 10 consecutive packets out of 100 packets is a 10% drop, which is the same as randomly dropping 1 out of 10 packets from a stream of 100 packets. However, the latter is more tolerable than the former in our system. Since there are many more P frames than I frames, we only drop P frames, first even then odd in our testbed. Thus we can use an automatic predictive source control to assist the explicit congestion control. That is, when P frames start piling up, we drop the P-frames by following a particular drop patter. For example, dropping the even frames first and then the odd frames without waiting for control commands to come from the base station. We can also set an upper limit of approximately 25% beyond which the video becomes less comfortable to watch (and may not satisfy customer expectations).

According to our industrial partners, changing the quantization level will be the next practical and cost-effective approach to reduce bit rate in terms of complexity and ease of deployment. We lower the quantization level when we use fewer coefficients from the coarse DCT (Discrete Cosine Transformation) algorithm, and thus reducing the packet size as well. Depending on the end-to-end delay, the receiver notifies the sender of the quantization level to use in encoding the subsequent video frames. This method will bring the total bit rate down to 50% of the original coding/sending rate. Beyond 50%, we need to drop the color components, thus leaving the video in its black-and-white mode.

In summary, the sending bit rate = frameRate*FrameSize which can be lowered by lowering one or both of the frame rate or frame size. There are tradeoffs between the two. Dropping packets (frames) will lower the frame rate while reducing the quantization level that will lower the resolution. The challenge in lowering the resolution is to keep the user experience/satisfaction high as long as possible, and various experiments have been conducted to find out the optimum point. Although our real GenieView camera has implemented in both methods in firmware to improve user experience, our OPNET model has chosen to drop the frames/packet only. This is reflected in Figure 5 and Figure 6 below where the called party collaborates with the calling party to determine the end-to-end delay based on how our controller uses one byte to notify the source to reduce its sending rate by dropping selected frames as discussed above. We also make use of RTP/RTCP at the application layer to provide an effective feedback mechanism.

Figure 5 depicts the video called party (receiver) procedure. Once the variables are initialized successfully, and a connection is established, data transmission can begin upon request. As soon as a frame is received, the frame count is incremented until it reaches the threshold value p. Once p is reached (e.g., p = 5 in our experiments), the frame count is reset to zero and the end-to-end delayassociated with the packet is checked. If the end-to-end delay is less than the threshold value t set by the end user on the receiving end (e.g., 100 ms), a default/designed rate of x2 is fed back to sender via the ACK packet. Else a lower rate x1 is returned instead. The x1 value can be a default value such as half of x2, or computed according to some formula)

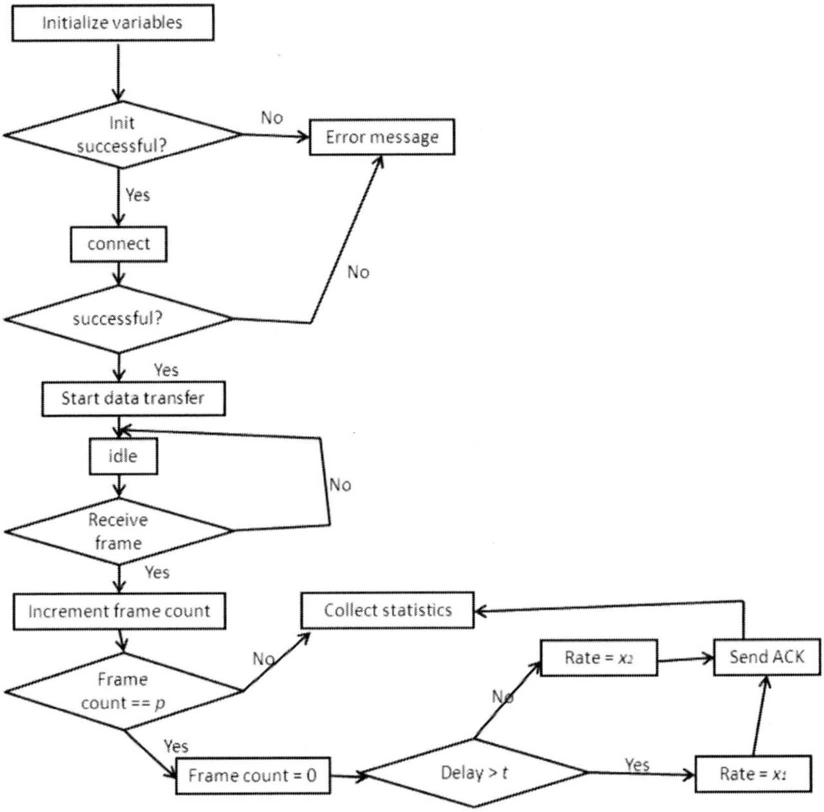

Figure 5. Video called party flow diagram with RTCP implementation.

Figure 6 shows the flow chart of the controller operation controller at the calling party manager (sender). Once the initialization process is successful, the video characteristics/properties are collected and a connection is established with the receiver. For every packet received, the algorithm checks if it is a control ACK packet. If not, the required statistics on packet type are collected and the packet is destroyed. If the packet is an ACK packet, the rate is retrieved and used to determine whether the sender should use the normal sending rate x2 or the reduced rate of x1. The parameters for the call establishment are also updated according to the new value obtained.

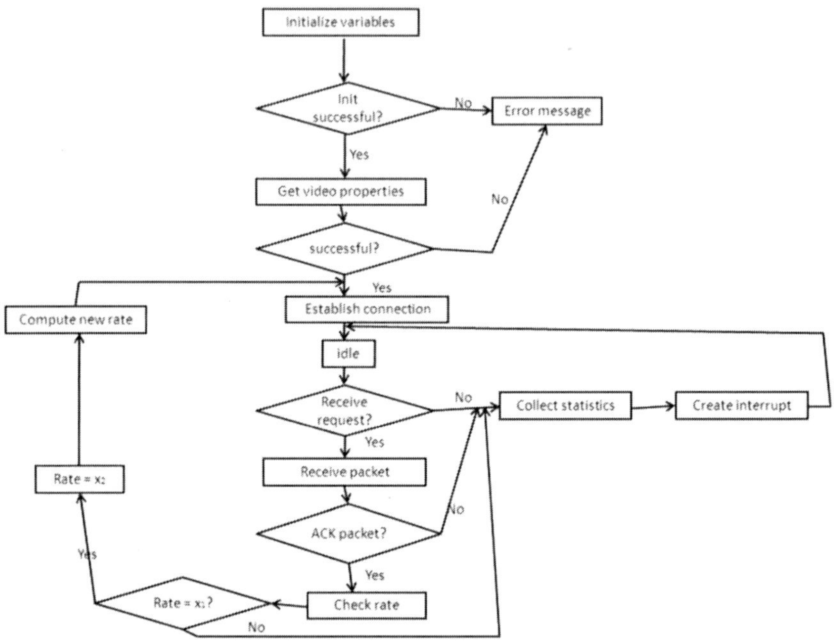

Figure 6. Video calling party flow diagram with traffic controller implementation.

4. EXPERIMENTAL SETUP

By collecting and studying video traces we want to come up with a model that can generate video traffic with characteristics as close to real life traffic as possible. This section discusses on the setup of our simulator testbed illustrates the trace collection process, its analysis and modeling. This will set the stage for traffic generation and performance evaluation in the next section.

4.1. Trace Collection

As shown in the setup of Figure 7, we use the monitor to play a series of driving video clips at 10 min each from driving action movies. Each video was captured by a PTS (Pan Tilt Zoom) camera taken from a vehicle. The video is encrypted and highly compressed at a ratio of 250:1 at the transmitter block. Using one processor for each section of the screen, the

96-array processors in the transmitter block can achieve fast independent image processing. The transmitter sends the highly compressed encrypted video images to the control center via a wireless link using the non-licensed 900 MHz FHSS radio-band. Please note that 900 MHz is more affordable to us presently than WiMax and DSRC. Since they are all wireless links, they will all be subjected to the same noise sources, such as EM interference. We only need to obtain similar type of interference and their statistics first to verify our platform.

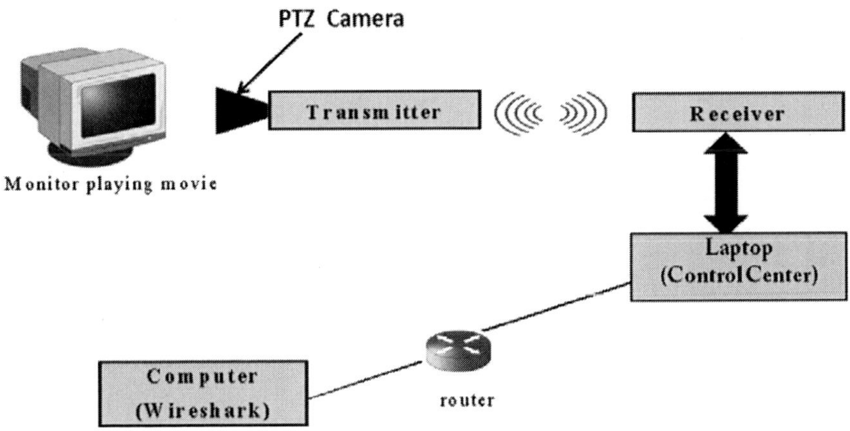

Figure 7. Setup for taking video traces.

The receiver block decodes and decompresses the received video frames and plays the image at the control center at about 20 fps. The control-center laptop and the receiver are connected using a RS232 serial port. The control center is then connected through a router to a computer hosting the packet trace-capturing tool—Wireshark™ [54]. Once the system is turned on, the computer with the Wireshark™ software is set to access the "capture" folder in the control center which would then stream the video via a router for trace capture.

A video trace (a record sequence of time and size of each arriving packet) captured by Wireshark is shown in Figure 8. Due to the loss and duplication that will change its original characteristics, this type of traffic trace is closer to those from our application in the real environments. We want to characterize such a life driving video trace using proven stochastic processes and integrate it into a model that has as few parameters as possible but which can produce traces to match the real measurements.

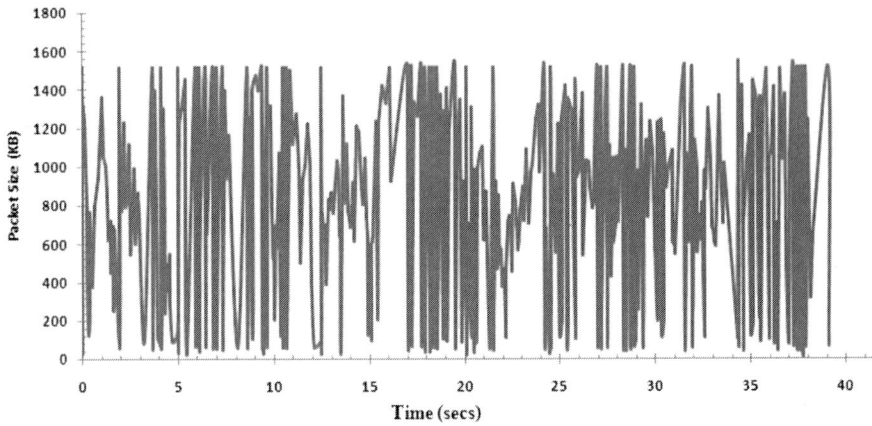

Figure 8. Live driving video trace.

In order to model video traces of different correlation/fractal characteristics arising from different activities, we collected traces playing three different video clips. The video clips were chosen based on their activity rates. Actions like highway chasing movies (with a lot of movement on the edge of the screen/camera and hence unique variation in frame sizes), stories in a parking lot (with less and slower movement) and dramas on the street (on the average driving speed) were chosen. We combine these clips to form a two-hour video trace (consisting of the time and size of each arriving frame), which is sufficient to characterize the fractal nature of the driving video traffic as well as the wireless channel effect.

4.2. Trace Analysis and Modeling

The next challenge is to analyze the trace and to create a driving video traffic model for traffic generation later. To verify the correlation of video frame and to obtain the modeling parameter, we obtain the frame-to-frame correlation of the video traffic. Figure 9 is an example using our video trace, which shows correlation is present up to 20 seconds apart.

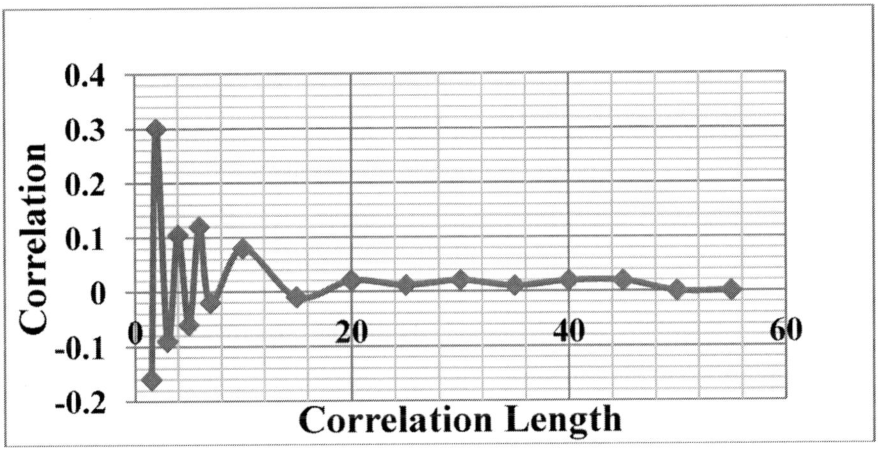

Figure 9. Correlation length of driving video.

We have used the Pareto distribution to characterize the general correlation of "action" video frame duration in wireless video streaming as well as the strong correlation between a cluster of byte loss and/or duplication caused by the air interface. The Pareto distribution is a long-tailed distribution with the following pdf (probability density function) g(x) for the consecutive I-frame duration, where the edge activity level is high. Note that x here is not the frame size but the group length rather.

$$g(x) = c\alpha^{c} / x^{c+1} \qquad \alpha \le x < \infty \tag{1}$$

where $\alpha > 0$ is the location parameter and $c > 0$ is the shape parameter. The location parameter value can be obtained from the mean correlation length E(x) of the trace where $E(x) = c\alpha/(\alpha - 1)$. On the other hand, the shape parameter has to be obtained via the Hurst parameter. That is,

$$H = \frac{1}{2}(3-c) \Rightarrow c=3\text{-}2H \tag{2}$$

Note that the Hurst parameter is a useful parameter to reflect the invariance according to the entropy conservation property [26]. Since our video is highly compressed with little loss of information, the entropy/uncertainty remains the same. Therefore, we can determine c from Equation (2).

An indirect but easy way to obtain the Hurst parameter value is via the slope of the IDC (Index of Dispersion for Count) curve as a function of the level of aggregation. Here IDC is defined to be normalized variance of aggregated variables [26], and the level of aggregation refers the logarithm of the number of variables base-10 (therefore level-0 has 1 variable, level-1 has 10 variables, level-2 has 100 variables and so on). Mathematically, the relationship is given by [26]

$$H = \frac{1}{2}(1 + slope) \tag{3}$$

The slope of the curve is usually measured from the first point to the last point on the graph. For example, Figure 10 gives the IDC of the video trace we use. From this figure, the slope is measured to be 0.33. Hence the Hurst parameter H is 0.67, and the shape parameter is c = 1.66.

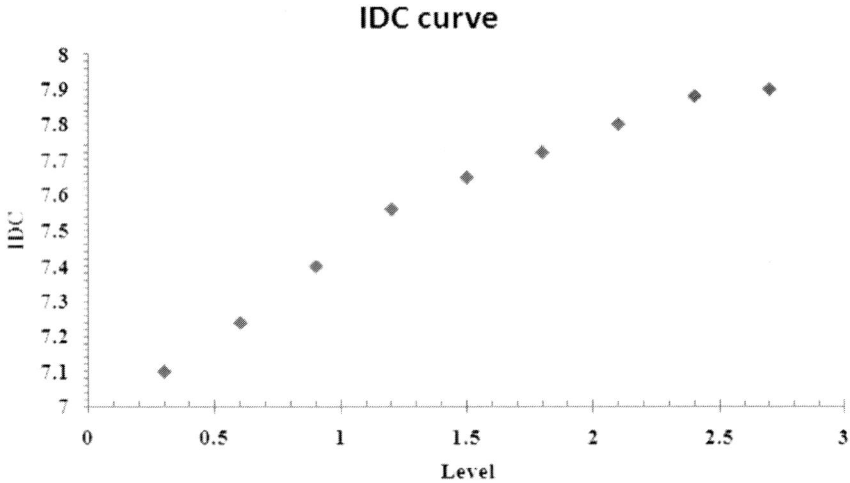

Figure 10. Entire trace IDC curve.

Once we obtain the pdf g(x) to characterize the activity duration (via the correlation of I-frames), we shall next determine the packet (frame) size distribution f(x), which is not Pareto in general. Furthermore, unlike the traditional video (e.g., video conference) with the moving part in the middle, our driving video has the moving part on the edge. Whereas the traditional video has a normal distribution in the middle for its packet size after 2D-DCT (Discrete Cosine Transform), the distribution for driving video is at the edge, which becomes very hard to model analytically to match the measured curve. After many trials, we have determined that an empirical model using nine mini-sources laid out in a grid can capture the distribution properly. The surrounding sources will capture the distribution of packets further away from the middle (edges and corners) which are usually moving and which produce large packets. The mini-sources closer to the center will capture the scenes in the middle which are relatively still and that produce smaller packets.

Figure 11 demonstrates the empirical analysis we performed on frame size (horizontal axis) of our traffic trace. For simplicity, here we are using the basic camera configuration, which produces one packet per frame. We use a tool called EasyFit™ [55] to obtain the pdf of the n = 9 video traces. The orange curve in the figure shows that the lognormal distribution is the best distribution to fit the given histogram data.

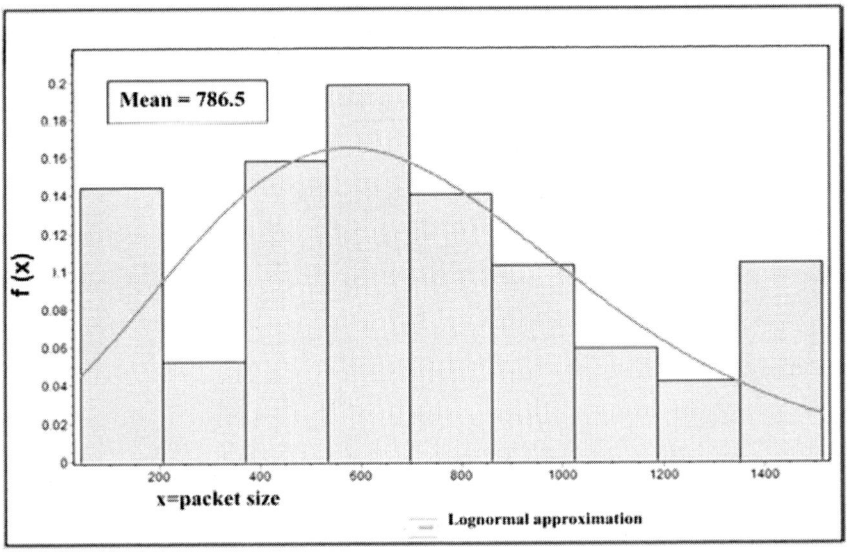

Figure 11. The pdf f(x) of video trace frames.

Finally, we want to determine the mean off-time of each source in order to model the inter-arrival time between each driving scene (such as trees or buildings on the road side). We have determined that the distribution is relatively memoryless, and therefore its pdf is exponential given by

$$a(x) = \begin{cases} \lambda e^{-\lambda x} & x > 0 \\ 0 & \text{otherwise} \end{cases} \tag{4}$$

where $1/\lambda$ is the mean inter-arrival time.

4.3. Driving Video Traffic Generation

Having obtained all the characterization parameters for traffic generation in previous sections, we are ready to emulate our wireless video traffic. Figure 12 shows the schematics of our wireless driving video traffic generation model from the superposition of n mini-sources [56]. Each mini-source represents a video traffic stream regulated by a switch to give a long-range dependency. We shall model the on-time switch by a Pareto distribution given in Equation (1) and the off-time by a Poisson process [an exponential distribution as given in Equation (4)].

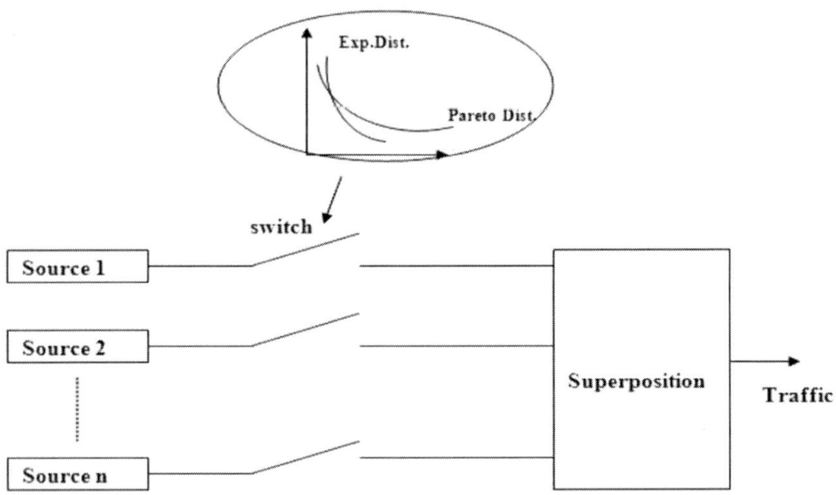

Figure 12. Model description for driving video traffic generation.

Table 4 shows an implementation of using nine mini-sources, each representing one of the 3 × 3 sectors of a camera screen. Here, frame duration is the transmission time of a frame (i.e., frame size divided by bit rate). Assume the camera is facing the front when driving on a road. The middle is usually relatively still when compared to the sides that see more activities. The left turns, right turns, and uphills or downhills are all different scenes that can lead to special packet size distributions. Big packet size is produced by a fast side view of the camera while the small size is produced by the almost still middle view. In the characterization of each mini-Pareto source in the table, we use the same shape parameter obtained from the Hurst parameter, but the location parameter is different for different packet size bin.

Table 4. Mini-sources with characterization values.

Bins	Mean Off-Time (ms)	Mean On-Time (ms)	Frame Duration (ms)	Mean Packet Size (Bytes)
mini1	7531	195	8	125
mini2	20603	453	18	290
mini3	6911	702	28	450
mini4	5515	960	38	615
mini5	7800	1216	32	779
mini6	10400	1472	39	943
mini7	18200	1723	46	1104
mini8	27300	1984	53	1272
mini9	10400	2232	59	1430

We have also tried other numbers such as n = 3 × 4 = 12 mini-sources to match the horizontal and vertical ratio of the camera. As verified by our experiments, the more mini-sources created, the higher the accuracy. However, the computational complexity also increases due to the more parameters needed to match the trace. This is translated into longer simulation time. So we have chosen n = 9 for our model for the best trade-off we found between parsimony and accuracy.

Figure 13 shows the video traffic generated mini-pareto video traffic, which can be compared to the live video traffic in Figure 8. Although not exactly the same, one can verify the simulated traffic captures the level of fluctuation of the driving video traffic.

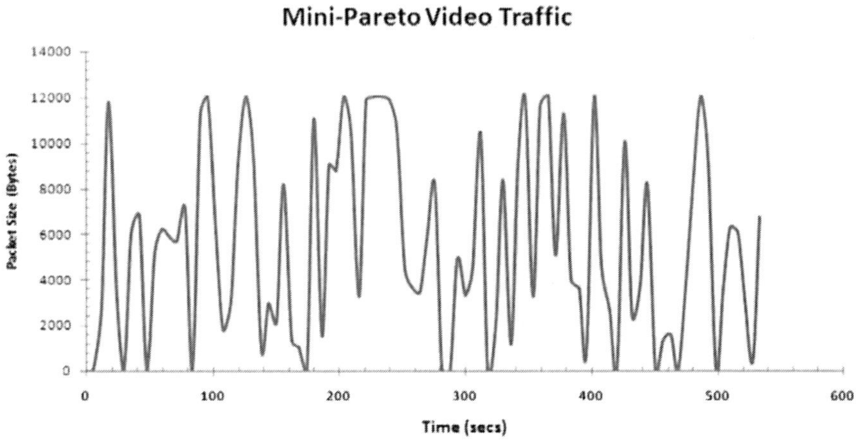

Figure 13. Mini-Pareto video traffic.

Table 5. Simulation parameters.

Parameter	Value	Note
Physical Layer	IEEE802.16e	WiMax
Data Rate	10	Mbps
BS T_X Power	5	W
MS Tx Power	1	W
Antenna Type	Omni-directional	Horizontal
BS Antenna Gain	15	dBi
MS Antenna Gain	9	dBi
Link Bandwidth	20	MHz
Modulation Scheme	16-QAM	Model
Path Loss Parameter	Vehicular Environment	ITU
Number of Vehicles	10	Cars
Mobility Model	VanetMobiSim	Open
Number of RSUs	2	Base
Simulated Time	3600	secs
Seeds	127	random
Terrain Dimensions	1300 × 1250	m^2

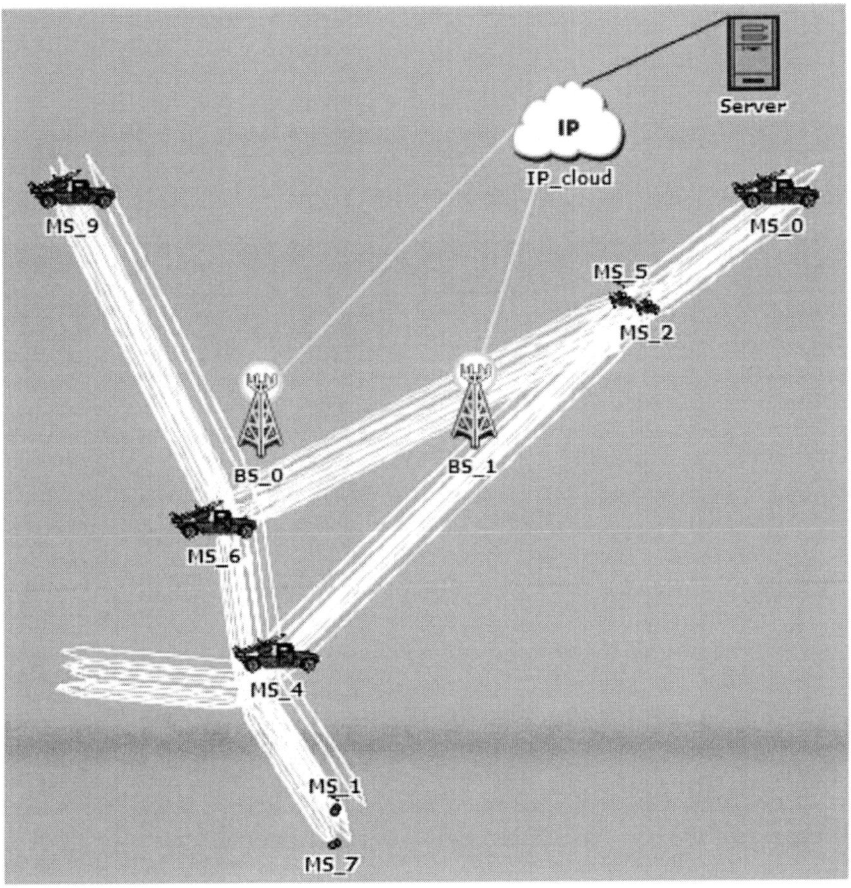

Figure 14. Highway scenario.

4.4. OPNET Simulation Setup

Table 5 gives the values of the general parameters we used in our simulations. The terrain has an area of 1300 × 1250 m. The relative position (x, y) on the terrain is used to integrate the VANET mobility model trajectories and obtain the initial positions of the vehicles. Vehicular environment for the path loss parameter is modeled according to the description in the "Radio Tx Technologies for IMT2000" white paper of the ITU [57]. The shadow fading standard deviation was set to 10 dB. The mobility pattern (trajectory) during the simulation is predefined using the VanetMobiSim discussed in Section 3.2.1.1.

We have investigated different scenarios such as highway and residential, and have studied the effect of varying traffic congestion (density), wireless interference and traffic speed limits. Due to space limitation, we shall present the highway scenario for this paper.

In the highway scenario shown in Figure 14, we consider cars moving along the highway with a minimum speed of 60 km/h and a maximum speed of 100 km/h. There is one traffic light, just before the cars enter the highway. The car trajectories are represented by the white lines. We use the free space loss model for our pathloss. Unless otherwise stated, the following are assumptions taken throughout the paper for our simulation study:

(1) WiMAX BS is a "stationary" node. This is required due to the limitation of our OPNET model and we need it to act as an intermediate node for packet forwarding to the destination;
(2) OFDMA is used by the RSU and there is always a slot available for each SS sending video traffic once the SS is within the communication range;
(3) No disruption in a communication channel once the node is allocated a dedicated channel;
(4) No obstacle along the LOS between the RSU and the mobile cars to allow the simulation to monitor the vehicles within the range of the RSU;
(5) Finite buffer size for each transmitter in order to be more realistic so that the trade-off between buffer size and end-to-end delay can be determined.

5. PERFORMANCE EVALUATION USING OPNET TESTBED

With the simulation platform established, we can now evaluate the performance of a practical controller through measurements of different performance measures using different network parameters such as network size and driving speeds, and different protocol parameters.

There are two network sizes we use to evaluate the scalability of our explicit congestion control algorithm: a small network consisting of 10 cars and 2 RSUs, and a larger network of 30 cars and 2 RSUs. Each car in these networks has a maximum sustainable traffic rate of 384 Kbps. With reference to Figure 2, the link between the RSU and the backhaul network was a DS3 link with a capacity of 44.736 Mbps. Section 4.4 has also provided the values of other parameters.

Each simulation took 3600 s, which is determined to be a good duration for the performance measures to converge. The following are the definitions of the performance measures we used in our investigation:

(1)End-to-End Delay (in seconds): This is defined as the time duration from the instant when a packet is generated at the source node (e.g., the safety camera on board a car) until the time it is received at the destination node. This performance measure presents a wholistic view of the time components in video capture, encoding and network delay before an end user visualizes the images. In real-time traffic, this parameter is very critical as it shows how "current" the information at the end user (e.g., a law enforcement officer) is. Since we track this at the application layer, the end-to-end delay also takes into account compression delay and all other delays incurred along the communication path and system;

(2)Jitter: Jitter is defined to be the variation of end-to-end delay between consecutive packets. If two consecutive packets leave the source node with time stamps t1 and t2 and arrive at the destination at time t3 and t4 after reassembly and playback, then the jitter will be represented by Equation (5).

$$\text{Jitter} = (t4 - t3) - (t2 - t1) \tag{5}$$

Negative jitter means that the packets where received in different time range i.e., (t4 – t3) < (t2 – t1);

(3)Visual MOS Rating: This rating is a subjective parameter that measures the quality of video received by the server, judging the end-user experience. This measure was borrowed from the same parameter used to measure the MOS (Mean Opinion Score) in voice. Using the definition given by the ITU-T recommendation G.107 [58], the score R is given by

$$R = Ro - Is - Id - Ie + A\,, \tag{6}$$

where Ro is the SNR (Signal to Noise Ratio); Is is the simultaneous impairment factor; Id represents impairment caused by delay; Ie is the equipment impairment factor as a result of low-bit codecs and packet loss and A is the advantage factor that compensates for the loss cause by other factors.

To customize this parameter for video evaluation, we have created a testbed case with all the "ideal" conditions. We allocate one base-station to one car so that there is more than sufficient bandwidth, hence keeping R as high as possible. We obtained a Visual MOS value of 3.6 that we should use as a reference to represent the best visual experience and 1 the worst visual experience, which is plot as the Y-axis in Figure 15.

(4)Buffer Overflow Percentage: This measure is the fraction of time during which arriving packets find the buffer at the Subscriber Station full and are not served. This measure implicitly reflects the percentage of time video packets lost due to buffer overflow.

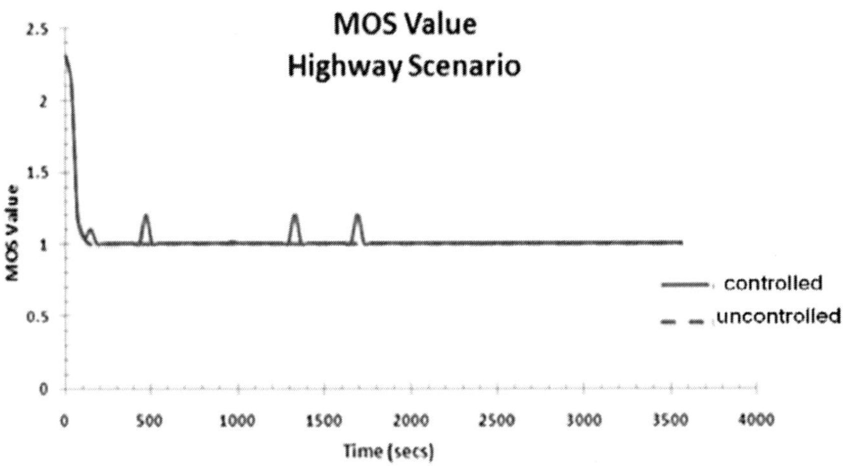

Figure 15. MOS value for a highway scenario.

5.1. Small Network

We only use 10 cars in this small network. After 10 s into the simulation, the WiMAX bandwidth is observed to vary from 125 kHz to 20 MHz (due to the adaptive modulation used as specified in Table 5). At the same time, we observe the video bandwidth also varies randomly with a peak of about 200 kbps and with an average of about 50 kbps at the Subscriber Station. Under this peak to average ratio of 6 dB, we set the controller threshold to 0.1 s in order to keep the controller active most of the time and to verify the operation of our congestion control algorithm. Below are more discussions of various performance measures.

5.1.1. Time Evolution Performance

Figure 16 shows the highly varying end-to-end delay, which can be attributed to the wireless interference and the high speed of the vehicular nodes in the highway. One can see that the controller is always active in reducing the end-to-end delay (blue solid curve) but not always to the minimal point. This is due to the coupling effect of the car speed and periodic feedback so that some packets go “unnoticed”, thereby losing it effectiveness.

MOS value for the highway scenario is shown in Figure 15. The score reflects the effect of fluctuations in end-to-end delay on the quality of video seen by the user. The controller tends to keep the visual experience at a constant rate, and improves when the end-to-end delay is reduced substantially.

Figure 17 shows the effect of the controller on the queue size. By controlling congestion in the network, our controller is able to reduce the occasions of buffer saturations and therefore the loss of video quality due to packet loss.

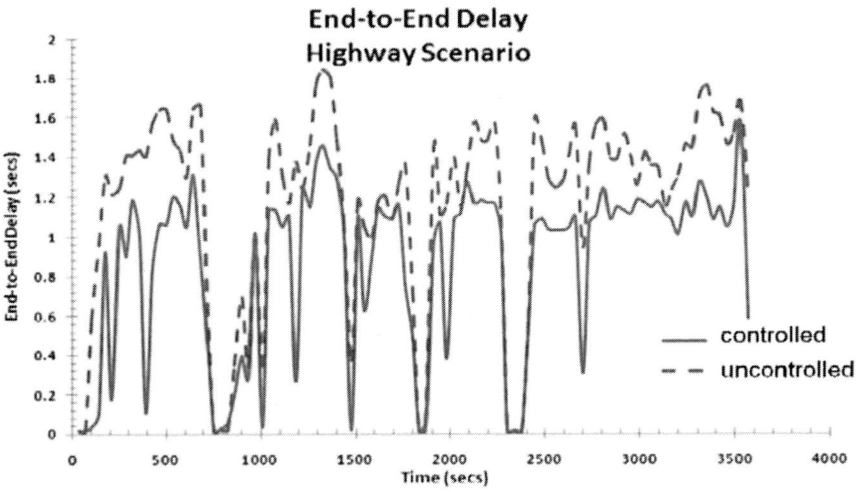

Figure 16. End-to-end delay for a highway scenario.

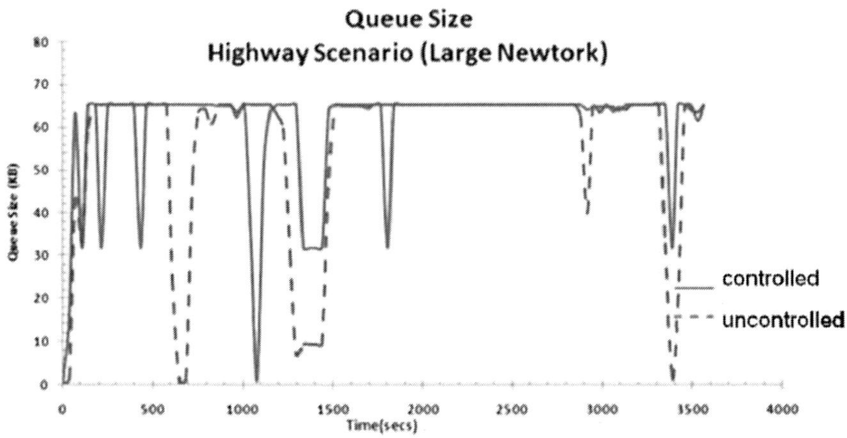

Figure 17. Instantaneous queue size for highway scenario.

The mean end-to-end delay performance for the highway scenario is shown in Figure 18. As expected, as the service rate increases, the end-to-end delay reduces. Increasing the service rate reduces the end-to-end delay and increasing the buffer size increases the end-to-end delay.

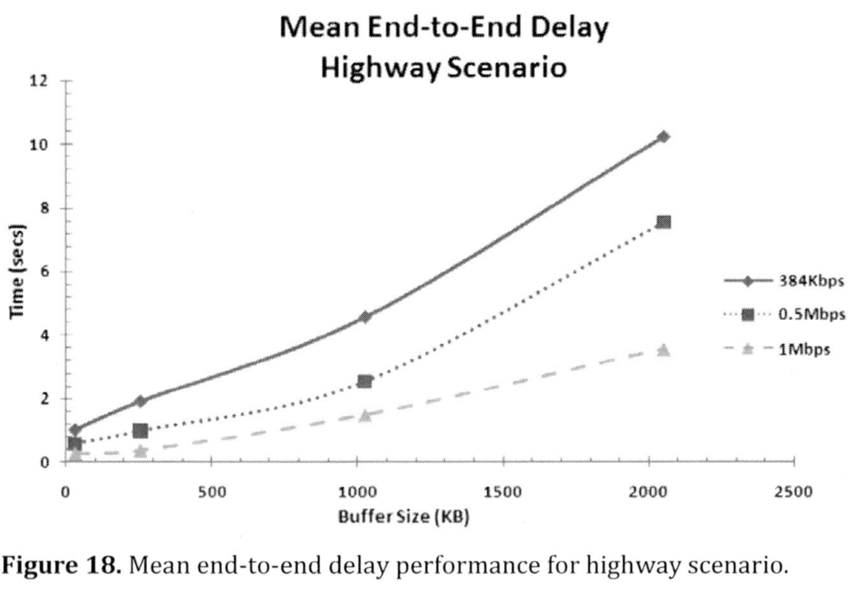

Figure 18. Mean end-to-end delay performance for highway scenario.

5.1.2. Buffer Overflow Percentage

The percentage of buffer saturation of the highway scenario is shown in Figure 19. As expected, the percentage of time for which the buffer is full decreases with respect to increasing buffer size. Note also the larger bandwidth also reduces the percentage as packets can be served faster.

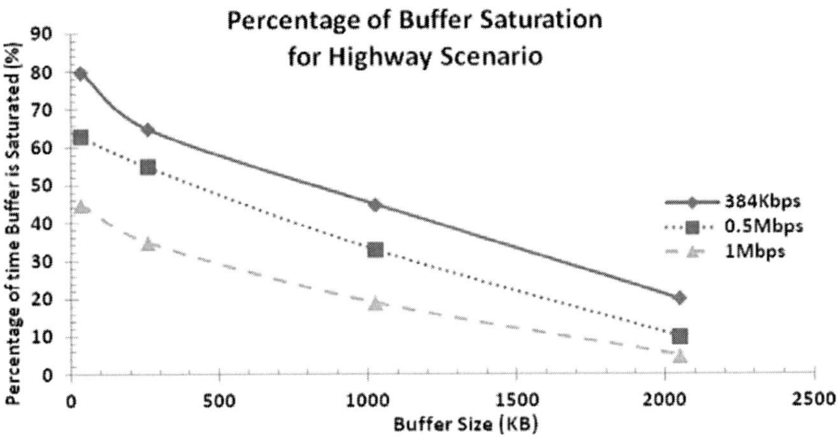

Figure 19. Percentage of buffer saturation performance for highway scenario.

5.2. Large Network

We next study a larger network with 30 cars. All other parameters and scenario remain the same as the small network unless otherwise stated.

5.2.1. Time Evolution Performance

Figure 20 shows the end-to-end delay for the highway scenario in the large network. The solid line shows the end-to-end delay with controller implemented and the dotted line shows the end-to-end delay without the controller. The behavior seen here is similar to what was seen in Figure 16 except that the end-to-end values are much higher in this case due to an increase in the network size (longer path length and queuing delay, etc.). The controller again helps to reduce the end-to-end delay in this case, although its effectiveness is influenced by other factors, especially the speed of the car.

Figure 21 shows the MOS value for the highway scenario for the large network. Similar to the small network, the effect of the controller on the quality of service as seen by the user is not obvious as the end-to-end delay and their fluctuations are still high in this scenario.

Figure 22 shows the maximum queue size is reached more in this case due to the increase in network size. This also contributes to the visual experience discussed earlier.

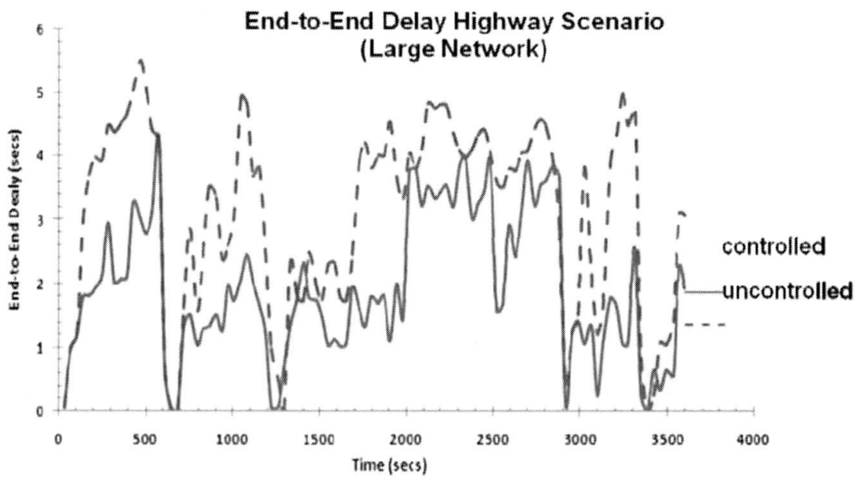

Figure 20. End-to-end delay for highway scenario (large network).

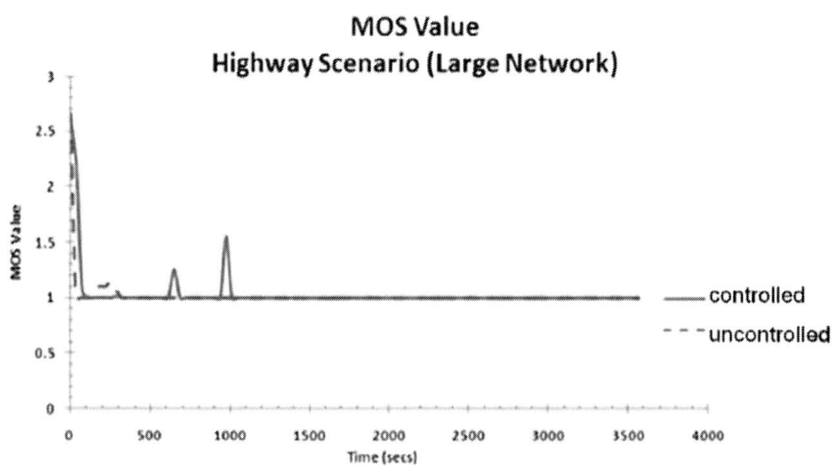

Figure 21. MOS value for highway scenario (large network).

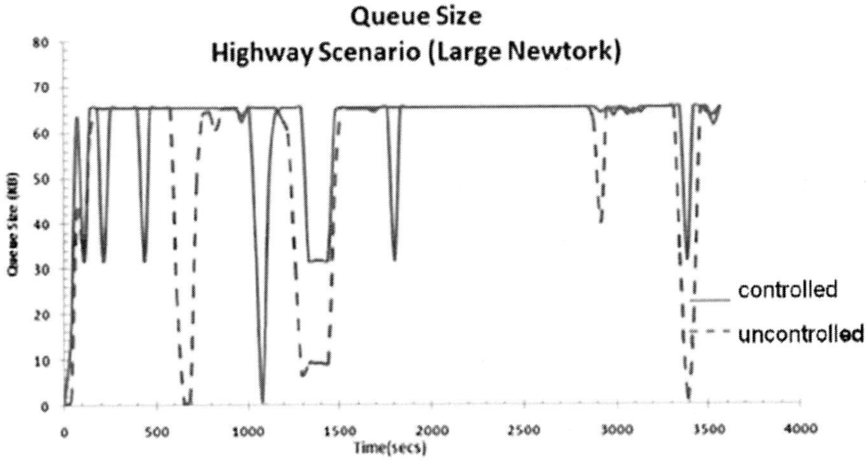

Figure 22. Instantaneous queue size for highway scenario (large network).

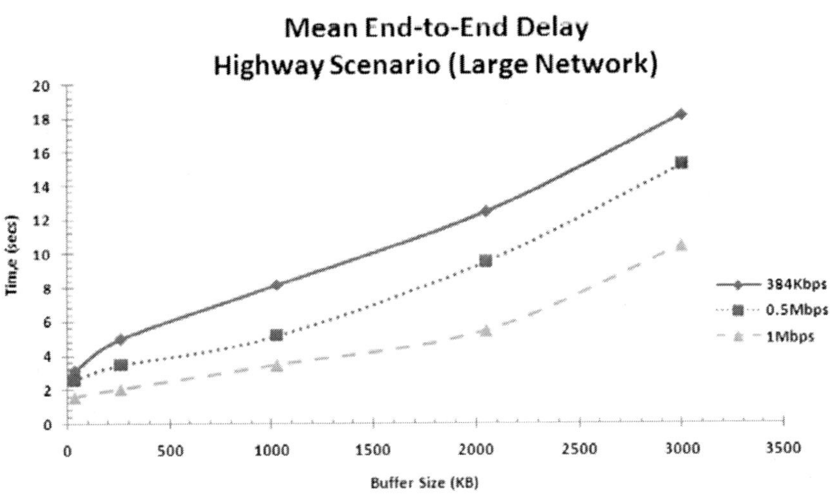

Figure 23. Mean end-to-end delay performance for highway scenario (large network).

5.2.2. Mean End-to-End Delay

Figure 23 shows the mean end-to-end delay performance for the highway scenario in the large network for varying buffer sizes and service rates. Observations are similar to those of Figure 18 and hence the same discussion applies, except the mean end-to-end delay is higher in this case due to the increase in network size.

5.2.3. Buffer Overflow Percentage

The percentage of buffer saturation performance curve for the highway scenario is shown in Figure 24 for the large network. It is seen that as the percentage of buffer overflow reduces as the buffer size increases and the service rate increases. This behavior is consistent with what is presented in Figure 19 except that the percentage of buffer saturation is higher in this case since the network size is increased.

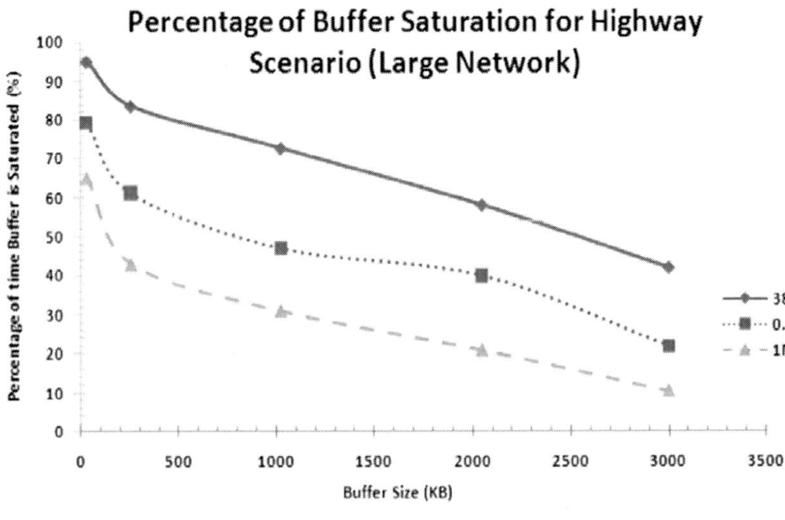

Figure 24. Percentage of buffer saturation performance for highway scenario (large network).

6. CONCLUSIONS

We have created and evaluated an effective driving video VANET simulation platform with which other researchers can develop and test various VANET applications. The platform was built using the OPNET simulation tool to integrate our theoretical driving video model with a VANET model. We have built a simple but practical traffic controller for driving video traffic over

RTP applications, and provided a complete tier of communication layers for proper performance analysis.

With the initial work done, there is still a lot of room for improvement and for exploring new model futures. Concurrent simulation using the simulation model and the mobility model would allow the wireless communication to affect the mobility of vehicles more directly so that closer reality results can be obtained. The traffic controller can be improved to keep the end-to-end delay closer to constant in order to reduce the quality impact on the viewers. Others include the comparison of our traffic controller under DSRC, and the validity of the Rician/Raleigh propagation model for moving vehicles.

ACKNOWLEDGMENTS

The research was supported by Auto21 Center of Excellence in Canada. Thanks go to Botao Zhu and Xiaoxiao Liu for reviewing and comments.

CONFLICTS OF INTEREST

The authors declare no conflict of interest.

REFERENCES

1. IEEE P802.16-REVd/D4, Air Interface for Fixed Broadband Wireless Access System. Available online: http://hunter.hosted.pl/tts/Strony_www_TTS_2004.2005/LMDS_Marcin.Gilewski_Lukasz.Waksmanski/airInterface.pdf (accessed on 17 July 2012).
2. Ramage-Morin, P.L. Motor Vehicle Accident Deaths, 1979 to 2004; Component of Statistics Canada Catalogue No. 82-003-X Health Reports. Statistics Canada: Ottawa, Ontario, Canada, 2008. Available online: http://www.statcan.gc.ca/pub/82-003-x/2008003/article/10648-eng.pdf (accessed on 17 July 2012).
3. Transport Canada. Canadian Motor Vehicle Traffic Collision Statistics: 2007. Available online: http://www.tc.gc.ca/eng/roadsafety/tp-tp3322-2007-1039.htm (accessed on 17 July 2012).

4. Heyman, D.; Lakshman, T. What are the implications of long-range dependence for VBR-video traffic engineering? IEEE/ACM Trans. Netw. **1996**, 4, 301–317.

5. Ryu, B.; Elwalid, A. The importance of long-range dependence of VBR video traffic in ATM traffic engineering: myths and realities. ACM SIGCOMM Comput. Commun. Rev. **1996**, 26, 3–14.

6. Li, X.; Huang, H.; Shu, W. VStore: Towards Cooperative Storage in Vehicular Sensor Networks for Mobile Surveillance. In Proceedings of the IEEE Wireless Communication and Networking Conference (WCNC), Budapest, Hungary, 5–8 April 2009; pp. 1–6.

7. Huang, J.; Qian, H. Video Encryption for Security Surveillance. In Proceedings of the 41st Annual IEEE International Carnahan Conference on Security Technology, Ottawa, Ontario, Canada, 8–11 October 2007.

8. Jacobson, V.; Braden, R.; Borman, D. TCP Extensions for High Performance. May 1992. Available online: http://www.ietf.org/rfc/rfc1323.txt (accessed on 17 July 2012).

9. Postel, J. User Datagram Protocol. 28 August 1980. Available online: http://www.ietf.org/rfc/rfc768.txt (accessed on 17 July 2012).

10. Balk, A.; Maggiorini, D.; Gerla, M.; Sanadidi, M.Y. Adaptive MPEG-4 Video Streaming with Bandwidth Estimation. In Proceedings of the 2nd International Workshop on Quality of Service in Multiservice IP Networks, Milano, Italy, 24 February 2003; pp. 525–538.

11. Schulzrinne, H.; Casner, S.; Frederick, R.; Jacobson, V. RTP: A Transport Protocol for Real-Time Applications. Available online: http://www.ietf.org/rfc/rfc3550.txt (accessed on 17 July 2012).

12. Liu, B.; Khorashadi, B.; Du, H.; Ghosal, D.; Chuah, C.-N.; Zhang, M. VGSim: An integrated networking and microscopic vehicular mobility simulation platform. IEEE Commun. Mag. **2009**, 47, 134–141.

13. Sommer, C.; Dressler, F. Progressing toward realistic mobility models in VANET simulations. IEEE Commun. Mag. **2008**, 46, 132–137.

14. Wegener, A.; Hellbruck, H.; Wewetzer, C. VANET Simulation Environment with Feedback Loop and its Application to Traffic Light Assistance. In Proceedings of the 2008 IEEE GLOBECOM Workshop, New Orleans, LA, USA, 30 November–4 December 2008; Volume 3, pp. 1–7.

15. Fiore, M.; Harri, J.; Filali, F.; Bonnet, C. Vehicular Mobility Simulation for VANETs. In Proceedings of the 40th Annual Simulation Symposium, ANSS 07. Norfolk, VA, USA, 26–28 March 2007; pp. 301–309.

16. D1.3.1—VanetMobiSim: The Vehicular Mobility Model Generator Tool for CARLINK. Available online: http://carlink.lcc.uma.es/doc/provis/D1.3.1-VanetMobiSim%20The%20vehicular%20mobility%20model%20generator%20tool%20for%20CARLINK.pdf (accessed on 17 July 2012).

17. OPNET. Analyze and Troubleshoot Application Performance from a Network Perspective. Available online: http://www.opnet.com/solutions/brochures/ACE_Analyst.pdf (accessed on 17 April 2010).

18. Tian, L.T. Research on Intelligent Transportation Systems in Taiwan. In Proceedings of the 27th Chinese Control Conference, 16–18 July 2008; pp. 18–23.

19. Laskowski, M.; Friesen, M.R.; McLeod, R.D.; Ferens, K. Rapid Prototyping Vehicle-to-Infrastructure Applications Using the Android™ Development Platform. In Proceedings of the 2009 9th International Conference on Intelligent Transport Systems Telecommunications, (ITST), Lille, France, 20–22 October 2009; pp. 273–278.

20. Intel. Intel CTO Rattner Outlines Research Challenges in Home, Car and Network Energy Consumption. Available online: http://www.intel.com/pressroom/archive/releases/2010/20100413comp.htm (accessed on 17 April 2010).

21. Lalonde, J.; Laganière, R.; Martel, L. Single-View Obstacle Detection for Smart Back-Up Camera Systems. In Proceedings of the 2012 Computer Vision and Pattern Recognition Workshops (CVPRW), Providence, RI, USA, 16–21 June 2012.

22. Hoogendoorn, S.P.; Bovy, P.H. State-of-the-art of vehicular traffic flow modelling. Proc. Inst. Mech. Eng. **2001**, 215, 283–303.

23. Zhang, S. Object Tracking in Unmanned Aerial Vehicle (UAV) Videos Using a Combined Approach. In Proceedings of the 2005 IEEE International Conference on Acoustics, Speech, and Signal Processing, Philadelphia, PA, USA, 18–23 March 2005.

24. Harri, J.; Filali, F.; Bonnet, C. Mobility models for vehicular ad hoc networks: A survey and taxonomy. IEEE Commun. Surv. Tutor. **2009**, 11, 19–41.

25. Beran, J.; Sherman, R.; Taqqu, M.S.; Willinger, W. Long-range dependence in variable-bit-rate video traffic. IEEE Trans. Commun. **1995**, 43, 1566–1579.

26. Hong, S.H.; Park, R.; Lee, C.B. Hurst parameter estimation of long-range dependent VBR MPEG Video traffic in ATM Networks. J. Vis. Commun. Image Represent. **2001**, 12, 44–65.

27. Haghani, E.; Parekh, S.; Calin, D.; Kim, E.; Ansari, N. A quality-driven cross-layer solution for MPEG video streaming over WiMAX networks. IEEE Trans. Multimed. **2009**, 11, 1140–1147.

28. Krunz, M.; Tripathi, S.K. On the characterization of VBR MPEG streams. ACM Sigmetrics Perform. Eval. Rev. **1997**, 25, 192–202.

29. Matrawy, A.; Huang, C.C. MPEG4 Traffic Modeling Using the Transform Expand Sample Methodology. In Proceedings of the 2002 Gaithersburg Proceedings 2002 IEEE 4th International Workshop on Networked Appliances, Gaithersburg, MD, USA, 15–16 January 2002; pp. 249–256.

30. Yang, S.; Daigle, J. A Source Model of Video Traffic Based on Full-Length VBR MPEG4 Video Traces. In Proceedings of the Global Telecommunications Conference, St. Louis, MO, USA, 28 November–2 December 2005; Volume 2, p. 5.

31. Luo, J.; Hubaux, J.-P. A Survey of Inter-Vehicle Communication; EPFL Technical 1618 Report IC/2004/24; École Polytechnique Fédérale de Lausanne: Lausanne, Switzerland, 2004.

32. Gray, D. Mobile WiMAX Part I: A Technical Overview and Performance Evaluation. v2.8. April 2006. Available online: http://www.wimaxforum.org/techno-Logy/downloads/ Mobile_WiMAX_Part1_Overview_and_Performance.pdf (access on 17 July 2007).

33. Jiang, D.; Delgrossi, L. IEEE 802.11p: Towards an International Standard for Wireless Access in Vehicular Environments. In Proceedings of IEEE Vehicular Technology Conference (VTC) Spring, Singapore, 11–14 May 2008; pp. 2036–2040.

34. Tufano, D.R.; Knee, H.E.; Spelt, P.F. In-Vehicle Signing Functions of an In-Vehicle Information System. Available online: http://www.osti.gov/bridge/product.biblio.jsp?query_id=3&page=0&osti_id=219502 (accessed on 17 July 2012).

35. Hossain, E.; Chow, G.; Leung, V.; MacLeod, R.D.; Misic, J.; Wong, V.W.S.; Yang, O. Vehicular Telematics over heterogeneous wireless networks: A survey. Comput. Commun. **2010**, 33, 775–793.

36. Ghosh, A.; Ghosh, M. Fundamentals of WiMAX: Understanding Broadband Wireless Networking; Prentice Hall: Upper Saddle River, NJ, USA, 2007.

37. Zukerman, M. Back to the future. IEEE Commun. Mag. **2009**, 47, 36–38.

38. Tanenbaum, A. Computer Networks; Prentice Hall: Upper Saddle River, NJ, USA, 1996.

39. Hong, Y.; Yang, O. Design of adaptive PI rate controller for best-effort in the internet based on phase margin. IEEE Trans. Parallel Distrib. Syst. **2007**, 18, 550–561.

40. Hac, A. Congestion control in a wireless network. Int. J. Netw. Manag. **1999**, 9, 185–192.

41. Rath, H.K.; Sahoo, A. Cross Layer Based Congestion Control in Wireless Networks. Available online: http://www.it.iitb.ac.in/research/techreport/reports/8.pdf (accessed on 17 January 2010).

42. Wischhof, L.; Rohling, H. Congestion Control in Vehicular Ad Hoc Networks. In Proceedings of the 2005 IEEE International Conference on Vehicular Electronics and Safety, Xi'an, China, 14–16 October 2005; pp. 58–63.

43. Floyd, F.; Jacobson, V. Random early detection gateways for congestion avoidance. IEEE/ACM Trans. Netw. **1993**, 1, 397–413.

44. Jacobson, V.; Karels, M.J. Congestion Avoidance and Control. In Proceedings of the SIGCOMM 88 Symposium on Communications Architectures and Protocols, Stanford, CA, USA, 16–18 August 1988; pp. 314–329.

45. Hassani, M.M.; Tavakolaie, B.H. An Analytical Model for Evaluating Utilization of TCP TAHOE Using Markovian Model. In Proceedings of the International Conference on Networking, Architecture, and Storage, NAS 2007, Guilin, China, 29–31 July 2007; pp. 257–258.

46. Huang, C.; Xu, L. SRC: Stable rate Control for Streaming Media. In Proceedings of the IEEE Global Telecommunications Conference GLOBECOM 03, San Francisco, CA, USA, 1–5 December 2003; pp. 4016–4021.

47. Hanley, M.; Floyd, S.; Padhye, J.; Widmer, J. TCP Friendly Rate Control (TFRC): Protocol Specification. Available online: http://www.ietf.org/rfc/rfc5348.txt (accessed on 17 July 2012).

48. Li, K. IEEE 802.16e-2005 Air Interface Overview; WiMAX Solutions Division Intel Mobility Group: Santa Clara, CA, USA, 2006.

49. IEEE WirelessMan 802.16, The 802.16 WirelessMAN MAC: It's Done, but What is it? Available online: http://ieee802.org /16/docs/01/80216-01_58rl/pdf (accessed on 17 July 2009).

50. Ahmadi, S. Introduction to Mobile Radio Access Technology: PHY and MAC Architecture. Available online: http://vivonets.ece .ucsb.edu/ahmadiUCSB_slides_Dec7.pdf (accessed on 12 December 2009).

51. Ertico. GDF—Geographic Data Files. Available online: http://www.eritco.com/en/links/links/gdf_-_geographic data_files.htm (accessed on 17 July 2010).

52. U.S. Census Bureau. Topologically Integrated Geographic Encoding and Referencing (TIGER) System. Available online: http://www. census.gov/geo/www/tiger (accessed on 17 February 2010).

53. Huang, J. Presentation for Generalizing 4IPP Traffic Model for IEEE 802.16.3. Available online: http://www.ieee802.org /16/tg3/contrib/802163p-00_58.pdf (accessed on 17 July 2012).

54. WireShark. Available online: http://www.wireshark.org (accessed on 17 July 2013).

55. Easyfit. Available online: http://www.mathwave.com /products/easyfit.html (accessed on 17 July 2009).

56. Gu, H.; Ji, Q. An Automated Face Reader for Fatigue Detection, Automatic Face and Gesture Recognition, 2004. In Proceedings of the Sixth IEEE International Conference, Seoul, Korea, 17–19 May 2004; pp. 111–116.

57. ITU-R, Guideline for Evaluation of Radio Transmission Technologies for IMT-2000. 1997. Available online: http://www.itu.int/dms_pubrec/itu-r/rec/m/R-REC-M.1225-0-199702-I!!PDF-E.pdf (accessed on 17 July 2012).

58. Mehmood, M.A.; Jadoon, T.M.; Sheikh, N.M. Assessment of VoIP Quality over Access Networks. In Proceedings of the First IEEE and IFIP International Conference in Central Asia on Internet, Bishkek, Kyrgyzstan, 26–29 September 2005; pp. 26–29.

6
CHAPTER

IMPROVING INTER-DOMAIN FAST HANDOVER USING MIH SERVICES IN MOBILE WIMAX

Ghaida A. Al-Suhail, Huda J. Al-Hammedy

Department of Computer Engineering, University of Basrah, Basrah, Iraq

ABSTRACT

Recently, the IP connectivity during the Mobile Node (MN) movement between Base Stations (BSs) belonging to different Internet Service Providers (ISPs) is still a key issue to be tackled. In this paper, therefore, we develop a new scheme to improve the performance of inter-domain fast handover over mobile WiMAX networks. The framework basically relies on the Fast Handover for Mobile IPv6 protocol (FMIPv6) when the Media Independent Information Services (MIIS) as defined in IEEE802.21 standard is applied to enable the Mobile Node in storing the information of the neighboring networks. A Fully Qualified Domain Name (FQDN) is also used to identify the IP address of the previous network operator and the MN during its movements. Since both MIIS and FQDN can support the node mobility between multiple domains, our proposed scheme can also be called P-FMIPv6. The numerical results show that the latency of IP connectivity of this proposed handover can be significantly reduced in addition to less service disruption time during handovers as compared to the existing FMIPv6 when IEEE802.16e network is considered.

1. INTRODUCTION

Providing seamless mobility support and maintaining the continuity of the service during the mobile node movement from one location to another is still an important issue for the mobile wireless communications. Mobile IPv6 (MIPv6) protocol [1] has proposed to support mobility in the internet between access networks, but the long handover latency in the MIPv6 degrades the perceived quality of service (QoS) essentially in real time services due to packet disruption and packet loss [2]. Therefore the Internet Engineer Task Force (IETF) has released other main protocol draft called Fast Handover for MobileIPv6 (FMIPv6) [3] which is designed to allow Mobile Node (MN) to anticipate its IP layer mobility by using Link layer triggers. Layer 2 triggers are required for anticipation and initiation handover [4]. They are delivered to the network layer modules as a notification event to report the changes in respect to the link and physical condition [4].

However, many researches have been presented for enhancing FMIPv6 and reducing the handover latency in it. In [5], an improvement to the handover delay is done by using an extension of the router discovery messages to initiate the handover. In [6], fast neighbor discovery and Duplicate Address Detection (DAD) scheme is presented by making each access point execute movement detection by unicast transmission of the stored router advertisement message and implemented a modified neighbor cache to handle configuration of the address under fast DAD. In [7], an enhancement for FMIPv6 handover based on link layer information is presented. One of the major things in FMIPv6 is not supporting global mobility, i.e. it doesn't support the discovery of the new network IP address when the mobile node moves between multiple domains. Moreover, during the handover initiation time, the MN could lose its connectivity to the Previous Access Router (PAR) due to the sudden disruption of the link. That's lead MN performs either a normal handover procedure in MIPv6 or the reactive FMIPv6 [8]. This may eventually cause long handover latencies which is not a desirable case. Therefore, in this paper a proposed fast handover scheme P-FMIPv6 is presented to support the MN movement between multiple domains.

The rest of the paper is organized as follows. Section 2 introduces a background about the existing FMIPv6 and Media Independent Handover (MIH) services. Section 3 describes the proposed P-FMIPv6 scheme. Section 4 introduces an analytical model to evaluate the latency performance of the proposed P-FMIPv6 scheme. The results evaluation is presented in Section 5, and finally the conclusions and some recommended future work are drawn in Section 6.

2. BACKGROUND

In this section, some related works are briefly addressed to support our proposed inter-domain Fast Handover (P-FMIPv6) scheme in this paper.

2.1. The Existing FMIPv6

Fast Handover for MIPv6 (FMIPv6) is an extension to MIPv6 for eliminating the handover latency. It depends on the link layer triggers to predict or respond rapidly to the handover events. It introduces a concept of Previous Access Router (PAR) and New Access Router (NAR), this extension allows MN to connect to NAR while still communicating with PAR. In Figure 1(a), once the MN is connected to NAR, it requests PAR to tunnel packets to NAR so that packets are not lost [7,9]. The MN registers its temporary location to its Home Agent (HA) and Correspondent Node (CN) through the Binding Update (BU) procedures. Figure 1(b) illustrates the basic handover procedure of FMIPv6 for intra-domain predictive mode [10]. In this mode, the same internet service provider is considered without the need for MIH services or IP change.

When FMIPv6 is applied to the IEEE802.16e based mobile WiMAX network; the mobile terminal scans to find the available target base stations. The scan results inform the available base station lists and their physical layer information such as strength of signal. The terminal selects one candidate base station and exchanges the router solicitation for proxy (RtSolPr) [3] and proxy router advertisement (PrRtAdv) [3] with current access router. When the terminal receives PrRtAdv message, it may configure its new IP address, referred as care-of address (CoA).

A fast binding update (FBU) [3] will be transmitted when the terminal indicated a handover. Sending of FBU activates the handover initiation (HI) [3] process and waiting fast binding acknowledgement FBAck [3] process. When the target access router receives HI, it confirms the procedure with duplicate address detection (DAD) process [3] and sends HACK message. Packet tunneling then is established, fast neighbor advertisement (FNA) [3] will be sent to inform the target router of the terminal's existence [3]. The definition of some terminologies used with the fast mobile IPv6 handover process is explained in Table 1.

2.2. Media Independent Handover (MIH)

IEEE 802.21 is a framework that enables seamless handover between networks, which is based on defined protocol stack involved in all the handover devices and provides the interaction among devices for optimizing handover decisions [11]. The MIH introduces a new logical entity called Media-Independent Handover Function (MIHF). The MIHF logically resides between the link layer and the network layer. It is worth mentioning that although the purpose of the working group is to

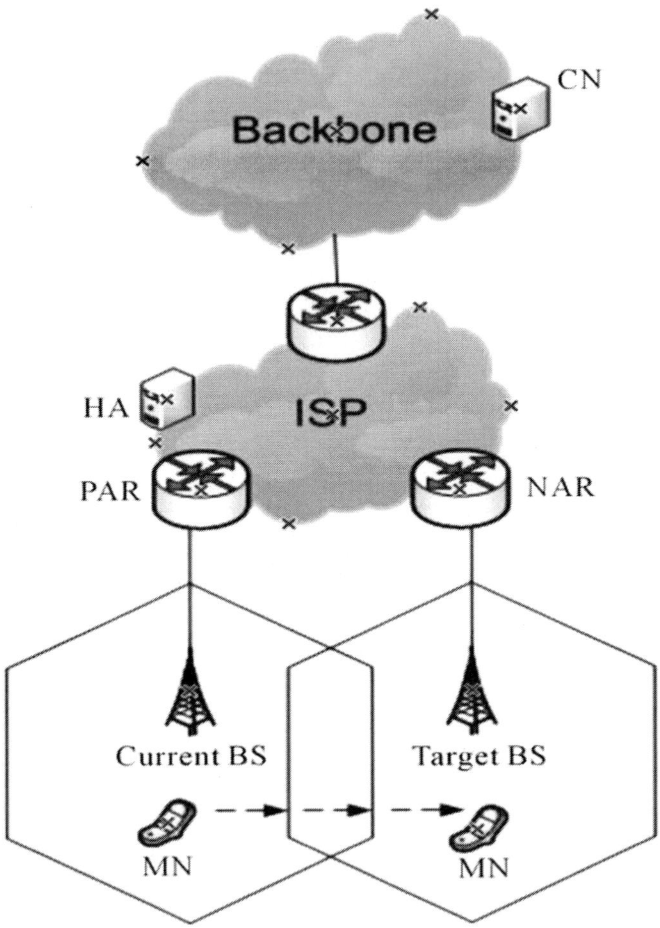

(a)

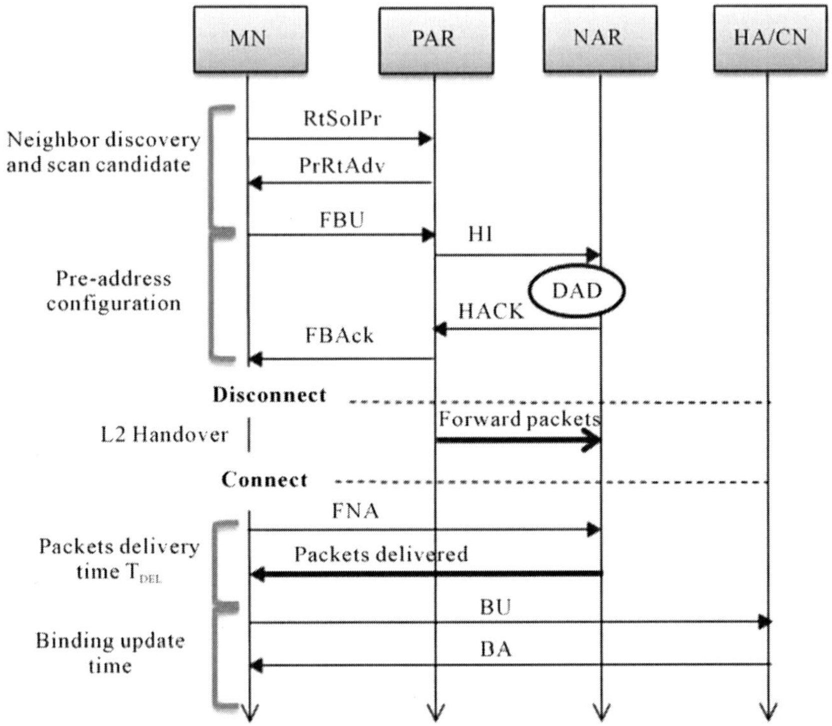

(b)

Figure 1. Inter-domain handover (a) Network model and (b) FMIPv6 (Predictive mode).

Table 1. The definition of some terminologies for FMIPv6.

Terminology	Definition
Mobile Node (MN)	A mobile node is a node that changes its location within the Internet topology.
Correspondent Node (CN)	Any node that communicates with the MN, it could be a server or another MN.
Home Agent (HA)	A router located on the home link that acts on behalf of the mobile node while away from the home link.
Previous Access Router (PAR)	The access router that the MN currently attached with.
New Access Router (NAR)	The access router that the MN attends to attach to.
Internet Service Provider (ISP)	A company which provides other companies or individuals with access to the Internet.
Base Station (BS)	A wireless station, providing services to the wired network for wireless nodes.

support seamless handover between heterogeneous networks; the recommendations can be applied to handover in homogeneous networks [12]. It provides, among others, abstracted services to entities residing at the network layer and above, called MIH Users (MIHUs).

MIHUs are anticipated to make handover and link-selection decisions based on their internal policies, and the information received from the MIHF. So, the primary role of the MIHF is to assist in handovers and handover decision making by providing all necessary information to the network selector or mobility management entities. The latter are responsible for handover decisions regardless of the entity position in the network. Figure 2 represents the 802.21 reference model. Three SAPs are included in the MIH reference model [13]:

- **MIH SAP:** This interface allows communication between the MIHF layer and the higher layer MIHF users.
- **MIH-LINK-SAP:** This is the interface between the MIHF layer and the lower layers of the protocol stack.
- **MIH-NET-SAP:** This interface supports the exchange of information between remote MIHF entities.

In IEEE 802.21, Service access point SAPs with associated primitives between the MIHF and MIHUs (MIH_ SAP) give MIHUs access to the following services that the MIHF provides [11-13]:

1) **The Media-Independent Event Service (MIES)** provides event reporting about, dynamic changes in link conditions, link status, and link quality. Events can be both local and remote. Remote events are obtained from a peer MIHF entity.

2) **The Media-Independent Command Service (MICS)** enables MIHUs to manage and control the parameters related to link behaviour and handovers. MICS provides a set of commands for accomplishing that. Commands can be both local and remote. The information obtained with MICS is dynamic.

3) **The Media-Independent Information Service (MIIS)** allows MIHUs to receive static information about the characteristics and services of the serving network and other available networks in range. This information can be used to assist in making a decision about which handover target to choose and to make preliminary preparations for a handover.

3. THE PROPOSED P-FMIPV6 HANDOVER

The FMIPv6 ensures a low-latency of the handover by anticipating the handover process before the MN loses connection with the current network. As explained in Figure 1, this goal is achieved by utilizing Layer 2 triggers. However, to reduce the overall handover latency in this scheme, we improve FMIPv6 predictive mode handover by developing new inter-domain handover scheme called P-FMIPv6 to reduce the L3 handover latency. In this scheme, MIIS is used to replace router discovery and to retrieve the information required about the current and surrounding networks for making the handover decision process.

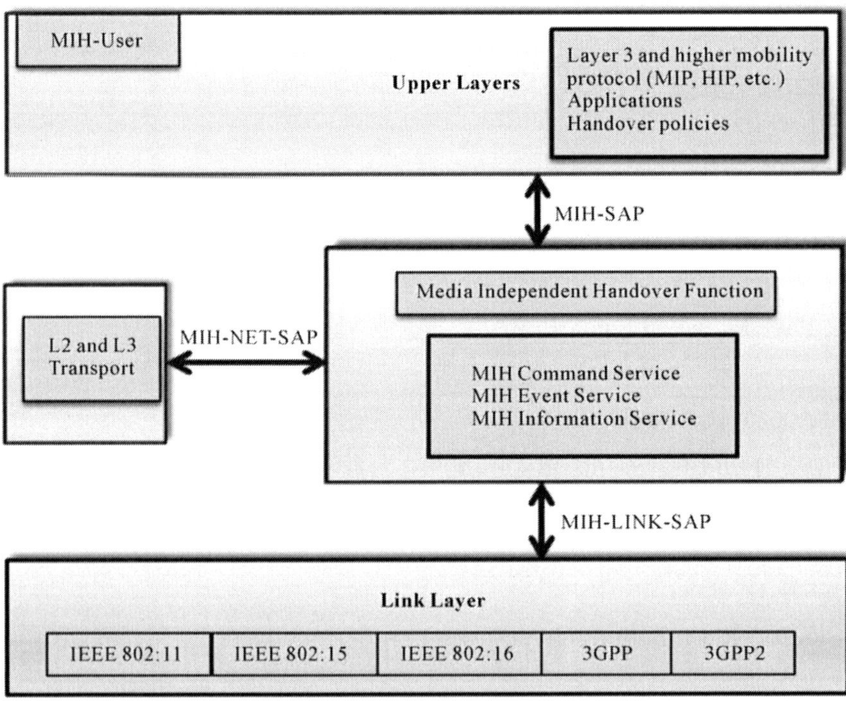

Figure 2. IEEE 802.21 reference model.

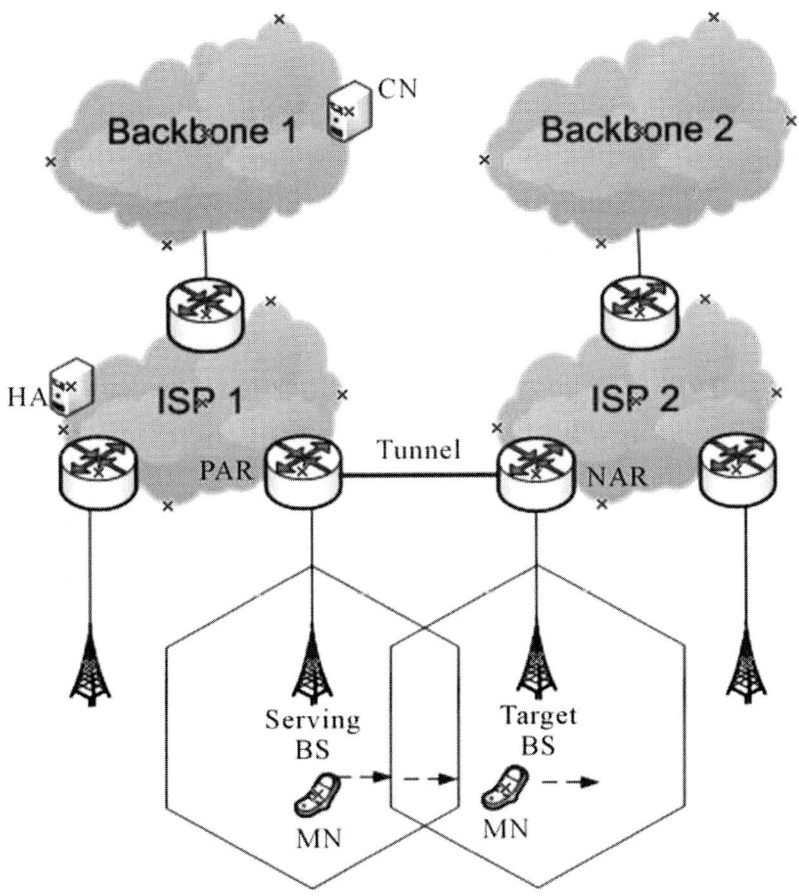

Figure 3. Network model supporting P-FMIPv6 scheme for inter-domain handover.

3.1. Network Model

We assume the MN is connected with the Serving Base Station (SBS) that belongs to the first domain, where each base station is integrated with exactly one Access Router AR. In this case every change of BS means a change of AR, as shown in Figure 3. The network model and each domain is served by an Information Server (IS) [2]. The IS could be co-located in the AR, and AR has access to the information to assist the handovers.

3.2. The MIH Services Required in P-FMIPv6

The FMIPv6 protocol in the MN is registered for Media Independent Event Service (MIES) notifications (i.e. L2 triggers) within its local stack. This will be done via MIH Event Registration service primitives that work in a request/response mode [6]. The events and commands services that will be registered by the FMIPv6 protocol are present in Tables 2 and 3, respectively. The MIH protocol defines the frame structure for exchanging messages between MIH function entities. These exchanged information via MIIS are structured in Information Elements (IEs) format coded using specific Type Length Values TLVs [6]. In the proposal subset of IEEE802.21, MIH services are utilized to enhance the handover process in FMIPv6. Tables 2 and 3 list the chosen MIH services and their corresponding primitives and parameters.

3.3. The P-FMIPv6 Procedure

To illustrate the proposed scheme, Figure 4 explains the details of timing signal flow during the handover process between PAR and NAR. In this scenario, the MN moves from the Serving Base Station (SBS) to the Target Base Station (TBS) that belongs to a different IP sub network in different domain. The MN must re-configure a new IP address and re-establish its IP connection. Hence, MN needs to perform the network layer handover besides the L2 handover. More specifically, the P-FMIPv6 solves the addressing problem during the MN movement between multiple domains by providing a mechanism for knowing if the NAR that the MN attends to attach to belongs to the same IP domain or to a different one.

We assume the existing of the information server in each domain in order to provide the router with the information of the domain. The IS could be thought of as the network knowledge reservoir which can be used to provide essential network related information, e.g., list of network providers, PoA MAC address, channel information, higher layer services etc., which may allow for enhanced network selection. In our network model the IS is a network entity which serves as a MIH Point of Service (PoS) but is located deeper in the access network.

Table 2. The MIH service primitives.

Primitive	Service	Parameters
Link Going Up	MIES	A signal from new link is found
MIH _Link_Going_Down	MIES	MN MAC Address, MAC Address of Current Point of Attachment (PoA)
MIH_Link_Down	MIES	MN MAC Address, MAC address of new PoA, Reason Code
MIH Link_Up	MIES	MN MAC Address, MAC address of new PoA, Link ID

Table 3. The description of MIH services for P-FMIPv6.

Trigger (Primitive)	Direction	Time of Triggering	Effect of Reception
Link_Going_Up	L2-L3 (MN)	This event is typically generated on the MN when a signal from the first PoA of an neighboring access network is received	Launching the detection of neighboring BSs procedures, and gathering the required information for handoff
Link_Going_Down	L2-L3 (PAR)	When the signal of the current link is likely to be down shortly	Indicates that the quality of a current link is decreasing and is now under an acceptable value
Link_Down	L2-L3 (PAR)	Forwarding the packets to the buffer in NAR	Ending the connection with the previous network
Link_Up	L2-L3 (NAR)	Completion of IEEE 802.16e Network entry procedure	The MN is now connected to the new network

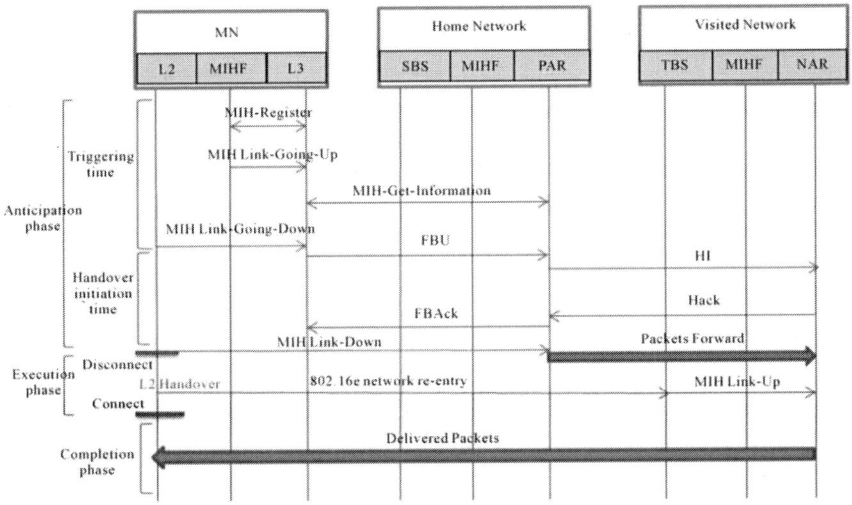

Figure 4. The proposed P-FMIPv6 signaling flow.

These servers need to be discovered, [14] describes the mechanism to discover the location of the IS via layer 3. In addition we will propose existing additional IE is the new network prefix IE transporting through the MIH messages to help in pre-configuration of the new IP address as explained in Figure 4.

To achieve a more clarity in describing the details of this proposed P-FMIPv6, the handover process is divided into three phases as follows. The first phase is called the anticipation phase; in which the MN anticipates its next (upcoming) connection before the disconnection with its current link. The second phase is the execution phase which through the MN performs L2 handover and finally the third phase is the completion phase, in which the MN completes its the handover process by delivering the buffered packets from the NAR to the MN.

The description of each phase is illustrated as follows:

1) **Anticipation Phase**: This phase is divided also into two interval processes: the triggering time (TT) part and the handover initiation time (HIT) part described as follows.

At the beginning of the triggering time (TT) process the MN registers for Media Independent Event Services MIES notifications (L2 triggers). And when the MN senses an increase in the signal strength of one of the neighboring network's BS, it will trigger MIH Link_ Going_Up event to the IP layer. Once the MN receives this trigger it will start its operation of scanning, detecting the candidates BS's of the neighboring network and gathering the important information that helps in the handover process. The required and necessary information is transporting through MIH messages (MIH Get_ Information_Request/MIH Get_Information Response) messages. The MN will retain this information stored in its neighboring binding cache, and the information is such as (lower layer information, L3 information of the neighboring access network).

Here, if the new network prefix IE between these Information Elements (IEs) is stored in the cache memory of the MN, the MN can pre-configure a new IP address automatically using the new network prefix and MN link Interface Identifier ID. The MN in this case will be provided with a pre-MN prefix to configure its address which will be a new IP address. By using this addressing, there will be no need to apply DAD procedure. Hence, the handover latency will be reduced and the chance of handover procedure to run in the Predictive Mode is largely increased since little or no time spent processing DAD. As a result, this procedure can support a much higher mobility speed [15].

After completing the triggering time (TT) process, the MIH Link_Going_Down (LGD) will be triggered to the IP layer as an indication for sending the FBU message and starting the handover initiation time (HIT). The LGD trigger is issued from MIH to IP layer in PAR and we specify this event as the start of handoff initiation phase. The FBU message with the new configured IP address is sent to the PAR. The previous IP address can be obtained by using the FQDN. This address configuration will be done early during the anticipation phase before occurring L2 handover and that will reduce the usual taken time for address configuration during the L3 handover procedure of FMIPv6. A Tunnel between the two ARs is established by exchanging HI/HAck messages between them and sending the FBAck message to the MN to validate the use of the new address as the new network IP address.

2) **Execution Phase:** When the MN receives the FBAck message from the PAR, its MAC layer triggers the MIH Link_Down event to the IP layer referring to the start of the Execution phase. The disruption with the previous link occurs in this phase. The packet forwarding to NAR starts right after the MIH Link_Down event. These packets are buffered in the NAR until the MN reaches in the new network. MIH Link_Down event is an indication to the disconnection with the old network and executing L2 handover, the L2 handover and the IEEE 802.16e network re-entry procedure is performed. Immediately after completing the network re-entry between MN and TBS, the TBS MAC layer triggers MIH Link_ Up event to the IP layer. This Link-Up event will replace the need for FNA message.

3) **Completion Phase:** In this phase the handover processes are completed by releasing the buffered packets from the NAR and delivers them to the MN without loss. Subsequently, the MN performs the registration with HA by sending and receiving a BU and a BA messages. The MN confirms the registration and removes the FMIPv6 tunnel.

As a result, Figure 5 can illustrate the flow chart of the P-FMIPv6 scheme; it describes the entire proposed handover scheme as follows:

a) The actual handoff procedure starts as soon as MIH Link_Going_Up event is arisen, so the MN will wait the arising of this event to start and perform the other procedures of the handover.

b) The MN exchanges MIH information messages to obtain the necessary information required for the handover process.

c) When the signal strength of the ongoing link is degraded, The MIH Link_Going_Down is triggered in response to this degradation of the current BS's signal strength.

d) The FBU will be provided with the pre-MN prefix, comparing the new network prefix with the previous one which can be obtained by using the FQDN of the previous network, if the new network prefix was different that means the NAR occurs in a different domain (different ISP) and that leads to the necessity of configuring a new IP address by combining the new prefix and the link interface identifier, and this case represents the scenario of the proposed P-FMIPv6 scheme, on the other hand if the two prefixes were the same that means the two AR's occur in the same domain and the comparison will depend on the subnet prefix, if it was different the MN will need to configure anew care of address instead of a new IP address to support its movement.

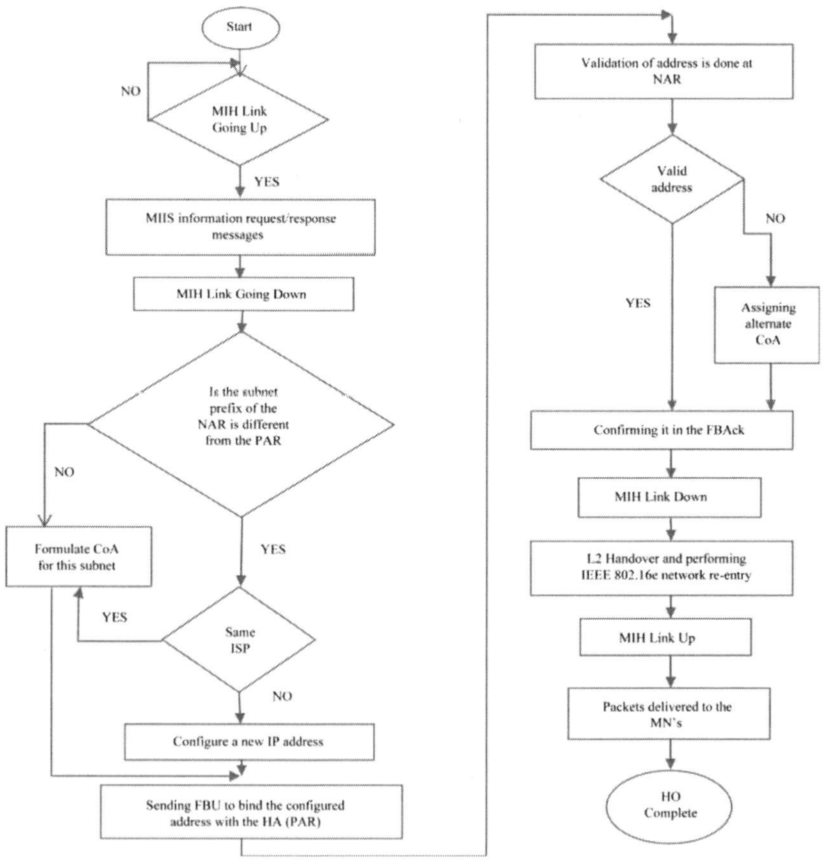

Figure 5. Flow chart of the P-FMIPv6 scheme.

e) Sending the FBU message to bind the configured address with its home agent (PAR). To establish the tunnel between the two AR's, PAR sends HI message to NAR in order to validate the new configured address and activate the tunnel between the two AR's, in response to this message NAR sends back HAck message to PAR to confirm this address validation, in case of the address was invalid a new alternate address will be assigned with the HAck message. Then PAR sends FBAck for address confirmation.

f) As soon as the FBAck message is received by the MN, tunneling the packets is being activated and forwarding the packets is starting from the current care of address to the new configured IP address of the new domain. Triggering MIH Link Down event is important for executing the handover properly.

g) The MN connection will be disrupted during L2 handover, and the IEEE802.16e network re-entry is performed. Later the Link_Up event is triggered from the MAC layer of NAR to the IP layer to advertise the connection with the new network, thus this event will replace FNA message. So there is no need to use this message any more.

h) Finally, the buffered packets are delivered to the MN indicating to the end and completing of the handover process.

4. ANALYTICAL MODEL

In this section, the analysis based on message triggers is done for examining the performance of the inter-domain P-FMIPv6 scheme to assess the total handover latency. Generally, the handover latency is defined as the time interval between last reception of data on the current Care of Address (CoA) and the moment when MN starts receiving packets at the new point of attachment [8,16]. Thus the handover latency can be evaluated by the sum of all particular delay intervals of L2 and L3 in the protocol stack where the mobile node experiences during the FMIPv6 and P-FMIPv6 handovers. In P-FMIPv6, it is found that the handover process of the wireless link layer (L2) has no a significant impact on the L3 handover process. Therefore, the delay caused by L2 mechanism can be discarded in this case because it does not have any impact on L3 handover mechanism. For this reason, L2 handover time interval will not be further examined and will be considered as a constant for P-FMIPv6 scheme.

To evaluate the total handover time of our P-FMIPv6 scheme, the typical network parameters defined in Table 4 are required. It is clearly noticed that there exist various types of delays introduced due to the handover

Table 4. The numerical parameters.

Parameter	Description	Values
T_{L2}	Latency of layer 2 handover	50 ms
T_{DAD}	Time delay for duplicate address detection	500 - 1000 ms
$T_{MN,BS}$	Transmission delay between MN and BS	20 ms
$T_{MN,AR}$	Transmission delay between MN and AR	20 ms
T_{NEG}	Negotiation time between MN and its BS	20 ms
$T_{PAR,NAR}$	Time delay between PAR and NAR	20 ms
T_{DEL}	Transmission time between NAR and MN delivery packets	40 ms
T_{FNA}	Time for the connection advertisement of the MN with the new network	20 ms
$T_{Re\text{-}Enrty}$	Time needed for network re-entry phase	50 ms

process in L2 and L3. In FMIPv6 scheme, the total handover time consists of the combination of the of transmission time between the MN and BS ($T_{MN,BS}$) during the scanning phase, the transmission time between the MN and AR during the transmission of Router Solicitation and Advertisement messages RtSolPr & PrRtAdv respectively and also the Binding messages ($4T_{MN,PAR}$). T_{NEG} indicates the negotiation time between the MN and BS. $T_{PAR,NAR}$ is the time required for establishing a tunnel between PAR and NAR, $T_{REENTRY}$ represent the time required for the re-entry phase of layer 2 handover and the time between MN and NAR for Fast Neighbor Advertisement (FNA) and forwarding packets messages ($2T_{MN,NAR}$).

In P-FMIPv6 handover, we can simply derive a delay formula which consists of different delay intervals caused by the binding messages between the MN and PAR ($2T_{MN,PAR}$), the delay occurred due to establishing the tunnel between the PAR and NAR ($T_{PAR,NAR}$), plus the re-entry time ($T_{Re\text{-}entry}$), and finally the time required to perform forwarding the packets from the NAR to MN($T_{MN,NAR}$). As a result, the total handover time for both handover schemes of RFC-5270 and the proposed scheme P-FMIPv6 are expressed as

$$D_{WiMAX}^{FMIPv6} = 2T_{MN,BS} + 4T_{MN,PAR} + T_{NEG} + 2T_{PAR,NAR} + T_{DAD} + T_{MN,BS} + T_{RE\text{-}ENTRY} + 2T_{MN,NAR} \quad (1)$$

$$D^{P\text{-}FMIPv6} = 2T_{MN,PAR} + 2T_{PAR,NAR} + T_{MN,BS} + T_{Re\text{-}entry} + T_{MN,NAR} \quad (2)$$

In addition, the service disruption time (SDT) of FMIPv6 and P-FMIPv6 can also be defined by Equations (3) and (4) respectively

$$SDT = T_{L2} + T_{FNA} + T_{DEL} \quad (3)$$

$$SDT^{P\text{-}FMIPv6} = T_{L2} + T_{DEL} \quad (4)$$

The service disruption time is known as the period of time when the MN cannot maintain active connection with the current BS to send and receive any data packets. In our proposed P-FMIPv6 the FNA is omitted by making use of MIES and thereby reducing the SDT which is an important parameter for assuring good QoS for realtime applications.

5. PERFORMANCE EVALUATION

To evaluate the P-FMIP protocol, the typical values of network parameters are defined in Table 4. Many research studies like [10,16-19] have basically used them under different handover procedures. By applying such numerical values in Equations (1)-(4), the following results are obtained as follows:

Figure 6 shows the handover latency of the P-FMIPv6 and the predictive RFC-5270 schemes. In P-FMIPv6, it is clearly shown that the proposed mechanism can remove the delay occurred by router discovery messages (RtSolPr and PrRtAdv) and FNA message, from the handover latency. We can also observe that due to the proposed P-FMIPv6 the handover latency will be reduced significantly by comparing it with the existing FMIPv6 scheme over 802.16e network [13]. The improvement in the handover latency roughly achieves 70% which is a very noticeable percentage value.

On the other hand, Figure 7 compares the service disruption time between the predictive FMIPv6 and the proposed P-FMIPv6 scheme. We assume the time for L2 handover (T_{L2}) is 50 ms and the time delay of FNA message T_{FNA} is 20 ms, and the delivery time delay (T_{DEL}) is about 40 ms. In the case of the proposed P-FMIPv6, it is found that we can obtain better performance than the original scheme.

In summary, the results show clearly that our proposed scheme features much shorter than the total handover delay and SDT as well compared to the original FMIPv6. Therefore, we can conclude that P-FMIPv6 can significantly improve the delay performance of Fast handover of MIPv6 for real time applications during the handover process.

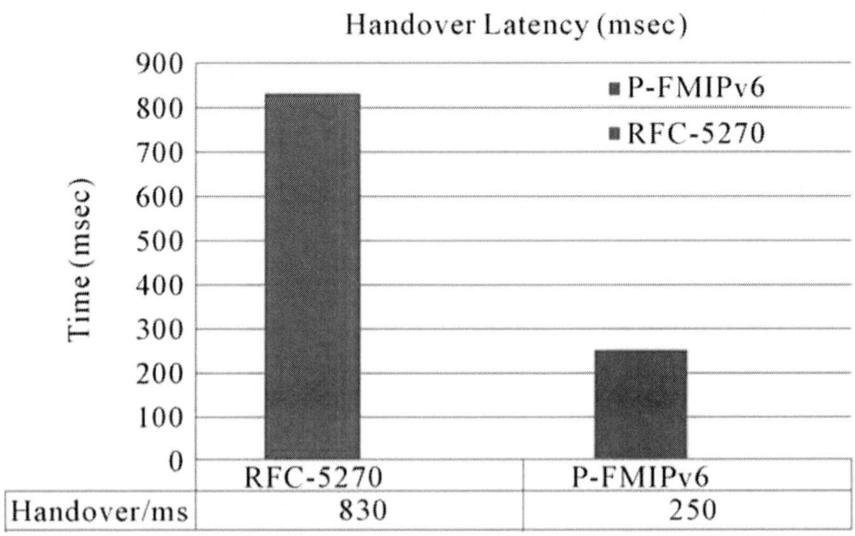

Figure 6. Handover latency of P-FMIPv6 vs RFC-5270.

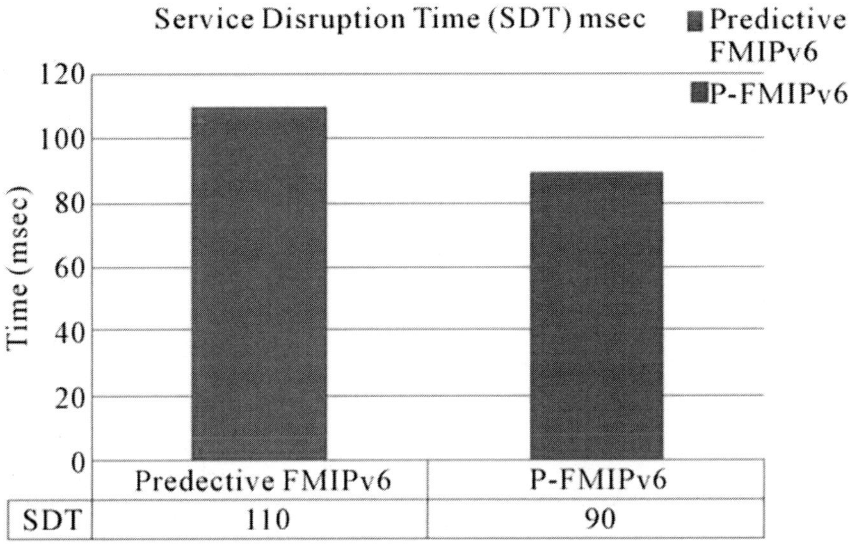

Figure 7. Service disruption time of P-FMIPv6 vs predictive FMIP.

6. CONCLUSION

In this paper, an inter-domain Proposed Fast Handover Scheme P-FMIPv6 is presented. The proposed scheme is based on MIIS information function and MIES, and the MN can perform the handover operation without the need for router discovery messages and FNA message. The analytical model of this P-FMIPv6 can introduce a significant reduction in the handover latency and a reasonable handover disruption time. The future work will be on examining the proposed P-FMIPv6 on intra-domain handover analysis and taking the cost evaluation in the consideration besides the delay of the handover process.

REFERENCES

1. D. Johnson, et al., "Mobility Support in IPv6," RFC 3775, IETF, 2004.
2. Y. Y. An, B. H. Yae, K. W. Lee, Y. Z. Cho and W. Y. Jung, "Reduction of Handover Latency Using MIH Services in MIPv6," Proceeding of AINA, Vienna, 18-20 April 2006.

3. R. Koodli, et al., "Fast Handovers for Mobile IPv6," RFC 4608, IETF, 2006.

4. N. Montavont and T. Noel, "Handover Management for Mobile Nodes in IPv6 Networks," IEEE Communications Magazine, Vol. 40, No. 8, 2002, pp. 38-43.

5. M. Alnas, I. Awan and R. D. W. Holton, "Performance Evaluation of Fast Handover in Mobile IPv6 Based on Link-Layer Information," Journal of Systems and Software, Vol. 83, No. 10, 2010, pp. 1644-1650.

6. C. Wu and P. Wang, "Improved Fast Handover Scheme for Hierarchical Mobile IPv6," 4th International Conference on Computer Science and Education, Wuhan, 25-28 July 2009.

7. B. Park, S. Lee and H. Latchman, "Performance Analysis of Enhanced-Mobile IPv6 with Fast Handover over Endto-End TCP," IEEE Communications Society Subject Matter Experts for Publication in the WCNC 2006 Proceedings, Las Vegas, 3-6 April 2006.

8. H. J. Jang, J. Jee, Y. H. Han, S. D. Park and J. Cha, "Mobile IPv6 Fast Handovers over IEEE 802.16e Networks," RFC 5270, IETF, 2008.

9. C.-H. Shih and Y.-C. Chen, "A FMIPv6 Based Handover Scheme for Real-Time Applications in Mobile WiMAX," Journal of Networks, Vol. 5, No. 8, 2010, pp. 929-936.

10. L.-L. Mo, "Research on Mobile IPv6 Technology and Handover Performance Optimization," American Journal of Engineering and Technology Research, Vol. 11, No. 9, 2011, pp. 941-946.

11. J.-H. Seol and J.-M. Chung, "IEEE 802.21 MIH Based Handover for Next Generation Mobile Communication System," 4th IEEE International Conference on Innovations in Information Technology, Dubai, 18-20 November 2007, pp. 431-435.

12. V. Gupta, "IEEE802.21 Standard and Metropolitan Area Networks: Media Independent Handover Services," Draft P802.21/D00.05., 2006.

13. M. Ali, P. Pillai, Y. F. Hu and K. Xu, "Load-Aware Radio Access Selection Heterogeneous Terrestrial Wireless Networks," International Journal of Computer Networks & Communications, Vol. 3, No. 6, 2011, p. 95.

14. T. Melia, G. Bajko, S. Das, N. Golmie and J. C. Zuniga, "IEEE 802.21 Mobility Services Framework Deign (MSFD)," IETF Draft, RFC 5677, 2009.

15. P.-Q. Huynh, P. Jangyodusk and M. Moh, "Supporting Video Streaming over WiMAX Networks by Enhanced FMIPv6-Based Handover," Proceedings of the 4th International Conference on Information Systems, Technology and Management, Bangkok, 10-12 March 2010.

16. Q. B. Mussabbir, W. Yao, Z. Niu and X. Fu, "Optimized FMIPv6 Using IEEE802.21 MIH Services in Vehicular Networks," IEEE Transactions on Vehicular Technology, Vol. 56, No. 6, 2007, pp. 3397-3407.

17. M. Skořepa and R. Klügl, "Analytical Method for L3 Handover Latency Evaluation," Advances in Communications, Computers, Systems, Circuits and Devices, World Scientific and Engineering Academy and Society Stevens Point, Wisconsin, 2010, pp. 342-347.

18. C. Makaya and S. Pierre, "An Analytical Framework for Performance Evaluation of IPv6-Based Mobility Management Protocols," IEEE Transactions on Wireless Communications, Vol. 7, No. 3, 2008, pp. 972-983.

19. M. Skorepal and R. Klugl, "Analytical Comparison of Mobile IPv6 Handover Schemes," Elektrorevue, Vol. 2, No. 2, 2011, pp. 22-26.

7

CHAPTER

NEW COMBINED WIMAX/DSRC INFRASTRUCTURE DESIGN FOR EFFICIENT VEHICULAR NETWORKING

Nabih Jaber1, Nicholas C Doyle2 and Kemal E Tepe1*

[1]WiCIP Laboratory, Department of Electrical and Computer Engineering, University of Windsor, 401 Sunset Avenue, Windsor, ON N9B 3P4, Canada.

[2]Mechatronic Systems Engineering, Simon Fraser University, Surrey, BC, Canada.

ABSTRACT

In this article, a new infrastructure of a combined Worldwide Interoperability for Microwave Access (WiMAX) and Dedicated Short-Range Communications (DSRC) link layer is proposed with the purpose of reducing simultaneous WiMAX connections. WiMAX offers wide area connectivity of vehicles to ground-based base stations, while DSRC offers relatively shorter communication that allows for vehicles in proximity of each other to communicate directly. The proposed design uses the fact that WiMAX amendments support the concept of a WiMAX relay node, and substitutes the WiMAX relay nodes with nodes that are capable of both WiMAX and DSRC communications. This change allows for the number of WiMAX connections to be concentrated while supporting more subscribing users via WiMAX tunnelled over DSRC relay. The focus of this design is on the use case of providing broadband Internet access to a large number of DSRC capable vehicles in a WiMAX served region. The design uses DSRC as a WiMAX tunnel, but with changes to the WiMAX protocol, specifically network entry and handover processes are redesigned to have different behaviour only when operating over DSRC. Network entry over DSRC modifications are described and illustrated with comparison to existing WiMAX standards. Handover process facilitated over both WiMAX and DSRC layers are described, illustrated and are also standard compliant. Unified modeling language is used to assist with the explanations of the components to improve understanding of the design in relation to existing WiMAX standards. In addition to standard WiMAX operability, the design can also support WiMAX data subscription using a software-defined WiMAX radio via DSRC relay connectivity. This proposed design improves WiMAX communication by reducing the number of WiMAX connections between vehicles. We plotted the throughput of various cluster sizes of WiMAX only mobile relay, versus our proposed DSRC-enabled WiMAX mobile relay in order to show the efficiency benefit of our design. We also provide a simulated curve of percentage improvement efficiency for varying amount of active users. We show that as the total number of users in the system increases, our proposed system significantly improves the overall system efficiency, especially in heavily congested traffic.

Keywords: Broadband communication Dedicated short-range communications Metropolitan area networks Mobile communication Multiaccess communication IEEE 80211 Standards IEEE 80216 Standards Vehicular and wireless technologies WiMAX, Wireless communication Pseudo-linear

INTRODUCTION

Ubiquitous wireless Internet seems to be the trend towards the future in our social media, video streaming, and Voice-over-IP (VoIP) calling culture. The advent of high-speed data broadband networking such as Worldwide Interoperability for Microwave Access (WiMAX), Long-Term Evolution, or other mobile communication standards do not appear to be a "one solution fits all" scheme. Even as the next generation of handheld wireless broadband solutions are being released, they are being released in a way that does not have full broadband coverage, but rather small pockets of high-speed data network within larger regions of lower speed data networking. These small pockets of high-speed network are not ideal for serving rich media applications to vehicles in vehicular ad hoc networking, due to frequent movements in and out of range of these high-speed areas. The metropolitan area WiMAX network is intended for blanketing a wide area with high-speed network capability, and correspondingly serves a relatively large number of users. We have previously shown in[1] that this high connection number seems to be Mobile WiMAX's weakness. We also show that by using an intermediate technology such as Dedicated Short-Range Communications (DSRC), concentrations of DSRC users can be served by a smaller number of WiMAX connections. At the same time, DSRC on its own does not lend itself well to serving high-speed broadband Internet without significant roadside infrastructure.

Mobile WiMAX[2] is the current revision of WiMAX, and is based on the 802.16-2009 revision[3] to the IEEE 802.16-2004 release of Wireless Wide Area Network. This system's physical layer (PHY) operates on frequencies between 2 and 11 GHz for non-line-of-sight applications and a Scalable Orthogonal Frequency Division Multiple Access (S-OFDMA) air interface. One important feature about the S-OFDMA system used in Mobile WiMAX is the easy scalability of the system to adapt to varying bandwidth configurations (1.25–20 MHz) in a cell, while keeping other system parameters such as frame size or sub-channel size constant. Adaptive modulation and coding allows each individual connection to adjust to changing channel conditions and provide an optimum balance between link throughput and robustness. Due to the fact that this scheme lends itself to the possibility of a large number of simultaneous streams, this increases the complexity of finding optimum coding and modulation due to more simultaneous transmissions. Therefore, the reduction of the number of simultaneous streams for the purpose of optimizing WiMAX transmissions is a significant motivation for our proposed system.

The IEEE 802.11p-2010[4] amendment to the IEEE802.11-2007 standard provided support for highly mobile ad hoc vehicular communications and introduced important connection optimizations that dramatically reduced the amount of time it took to make a connection between one vehicle and another, and formed the physical layer for the DSRC standard. Additionally, the increase in symbol duration facilitated a reduction in errors at the PHY normally due to severe inter-symbol interference (ISI) under high relative velocities. This is an important problem to address when working with mobile nodes at high relative velocities, which is the case for vehicular networking. This improvement allows for data rates up to 27 Mbps per 10 MHz channel to be transferred from one vehicle to another. There are a total of seven 10-MHz channels defined in the current DSRC standard, of which only three are reserved for safety and control. This large bandwidth of DSRC can be multicast and shared between several users at once within the same channel, but do not necessarily have Internet connection without nearby roadside service providing the connection. With our system, the Internet source can be a vehicle with both WiMAX and DSRC antenna systems, acting as both client for its own Internet requirements, and server to other DSRC connected vehicles.

The rest of this article is organized as follows: The following section provides the related study discussion. Section "Proposed combined WiMAX/DSRC system" contains the proposed combined WiMAX/DSRC system general description including resource usage mapping, a detailed efficiency model for downlink and uplink of WiMAX and DSRC channels, WiMAX connection concentration, DSRC Clustering procedure, and proposed combined WiMAX/DSRC layout; Section "New combined WiMAX/DSRC link layer infrastructure design" presents the new combined WiMAX/DSRC link layer infrastructure design from a deployment diagram perspective from subscriber station (SS) to cluster head relay (CHR) to base station (BS), and describes the connection environment, relationship between node types, and goes into detail with respect to existing WiMAX standards to describe the interoperability of our design. In addition, the same section presents an expanded view of the deployment diagram, with specific descriptions of the new elements of our architecture that allows interoperation between WiMAX and DSRC, specifically the WiMAX network entry over DSRC process and WiMAX/DSRC handover processes as compared to standard WiMAX handover operations; Section "Simulation results" provides the simulation results; Finally, the last section offers concluding remarks.

RELATED STUDY

All of the literature on IEEE 802.16j[5] amendment reviewed thus far concentrated on in-band relaying. As discussed in[6], the WiMAX subframes are further subdivided to provide the bandwidth for communications between the relay nodes and the subscriber nodes served by them. However, this is still subdividing the WiMAX bandwidth over multi-hop connections. The analysis done in[7] used the non-mobile or Fixed WiMAX standard with an OFDM frame structure. Mobile capable WiMAX described in the Mobile WiMAX standard[2] uses a somewhat different frame structure in order to preserve the robustness of the connection in highly mobile situations. Rather than OFDM, Mobile WiMAX uses an S-OFDMA frame structure[2]. Instead of dedicating all subcarriers to a single user, as is the case in OFDM, S-OFDMA defines bandwidth allocations both in terms of time and in terms of a subset of the available subcarriers.

DSRC is a technology for vehicle-to-vehicle communications that has seen much research. The DSRC standard only provides support for point-to-point communication. However, much research has investigated overlaying ad hoc routing network protocols on the physical layer. One scheme for ad hoc routing is clustering. Mobile nodes of physical proximity form groups, or clusters, with all traffic from within the clusters to other clusters being routed through a gateway node termed the cluster head node (CHR)[8]. In their proposed system, as nodes move, the clusters are independently controlled and dynamically reconfigured. Their architecture provides spatial reuse of the bandwidth due to node clustering, in each cluster, the sharing of the bandwidth is designed in a controlled fashion. Finally, and most importantly, their cluster algorithm is robust in the face of topological changes caused by node motion. Forming stable clusters of mobile nodes for ad hoc routing is an active research area with algorithms such as the Genetic Algorithm[9], a contention-based clustering algorithm[10] and the Distributed Clustering Algorithm[11], designed and tested to provide higher performance for generic mobile ad hoc networks (MANETs). The authors of[9] proposed a new clustering algorithm in MANET. Their resulting clustering algorithm provides a stable and generic cluster structure for the upper layer protocols. Hence, by applying the concepts of clustering to DSRC, which has been found in research to be realistic in heavily congested situations, the cluster head is in a position to concentrate the signals from all the cluster members. Authors of[12] investigated the concept of combining cellular backbone infrastructure with ad hoc relaying known as the iCAR project. The authors' iCAR system increases the conventional system's capacity cost effectively, and show that with a limited number of ad hoc relaying stations and some increase in the signalling overhead (as well as hardware complexity), the call blocking/dropping probability in a congested cell and the overall system can be reduced.

Authors of[13] investigated relatively stable clusters of moving airplanes. The authors referred to these systems as being "pseudo-linear", as they formed a relatively linear routing network, and had relatively constrained movement[14]. This study was subsequently granted a patent[15]. We propose extending this technique and lay the groundwork for applying this extension to congested highways and major roads for WIMAX and DSRC integrated system. For the described system, an optimized clustering and cluster head selection protocol will be investigated in detail through studies of real-world congested vehicle scenarios, using a test platform such as that described by Roccetti et al.[16]. Hence, in this manuscript, all the nodes are configured to have both DSRC and WiMAX system. To maximize network stability (and minimize the overhead of node hand-overs), the CHR selection protocol will have to factor in vehicle parameters such as relative speed and vehicle destination. Compared to the previous work, this clustering protocol will be specific for vehicles traveling on a road using a wide-area backbone network (WiMAX). This will simplify routing and allow an emphasis on speed and stable latency, as opposed to simple connectivity in a purely ad hoc system. Since the main contribution of this paper is to introduce a new and complete link layer infrastructure design for the WiMAX/DSRC system, network entry and handover processes are designed to have different behavior only when WiMAX operates over DSRC. The system described in this article assumes that all vehicles are equipped with both DSRC and WiMAX radios. These two systems for mobile nodes are available today, and as such the proposed system is capable of using either network.

The coexistence of WiMAX and DSRC in a heterogeneous environment has been discussed in[17], while this article is the first paper to design a new link layer protocol for the combined WiMAX and DSRC communication systems. There are many potential benefits to using this combined system rather than either WiMAX or DSRC on its own. Authors of[18] have found that using roadside stations for Internet access is impractical without a heavy investment in infrastructure. The improvement to the WiMAX system capacity minimizes the installation of costly WiMAX BS. On the other hand, one of the major problems with a DSRC-only system is the latency issues inherent with ad hoc routing systems. As the number of hops between the user and a network gateway is dynamic and the links can be fragile, it is difficult to provide bounds on the latency seen by the system. This can be a problem with Internet applications that are sensitive to latency, including voice communications and streaming media. A combined system also offers the ability to revert to WiMAX-only service, eliminating the sparse node problems seen in an ad hoc network where nodes are not dense enough to allow for continuous routing. This flexibility allows for rapid adaptation to changing system conditions such as exiting a busy highway onto a non-busy side road, without requiring the disruptive reestablishment of application-layer connections. Clustering and ad hoc routing becomes more effective as the road conditions become more congested, which are the conditions where traditional WiMAX routing is demonstrated to become ineffective. As such, the proposed system is far more robust to a wide variety of operating conditions, and provides bounds on the number of hops in the network, providing much more predictable latency conditions. Determining the conditions under which to run flat routing and when to cluster, as well as the size of the clusters and which vehicles to include in clusters, will be part of the cluster selection protocol through studies of real-world congested vehicle scenarios, using a test platform such as that described by Roccetti et al.[16]. In essence, this article extends the WiMAX protocol across DSRC for the purpose of providing broadband Internet access to DSRC connected clients, using a new link layer design.

PROPOSED COMBINED WIMAX/DSRC SYSTEM

Figure1 shows the structure of a Mobile WiMAX frame and illustrates some of the sources of overhead. System overhead is produced by fields in the frame not being used to transmit user data. The most important are the DL_MAP and UL_MAP fields, specifying the user bandwidth allocations for the frame in the downlink and uplink directions. As the number of active connections increase, so does the number of entries in each of these fields. Other system components, including the preamble and the ranging slot (in the uplink subframes), further reduce the system efficiency. Further system overhead comes from unused allocations. It is conceivable that not all of a user's allocated bandwidth will be used. This may be due to a user not having sufficient data to transmit, or when the user's data do not fill the entire allocation. This inefficiency adds up over multiple connections. A conservative estimate would be to assume that half of the last allocated slot is empty, and therefore contributing to the overhead. If the traffic is burstier in nature (such as VoIP), the losses due to this unused allocation could be much greater.

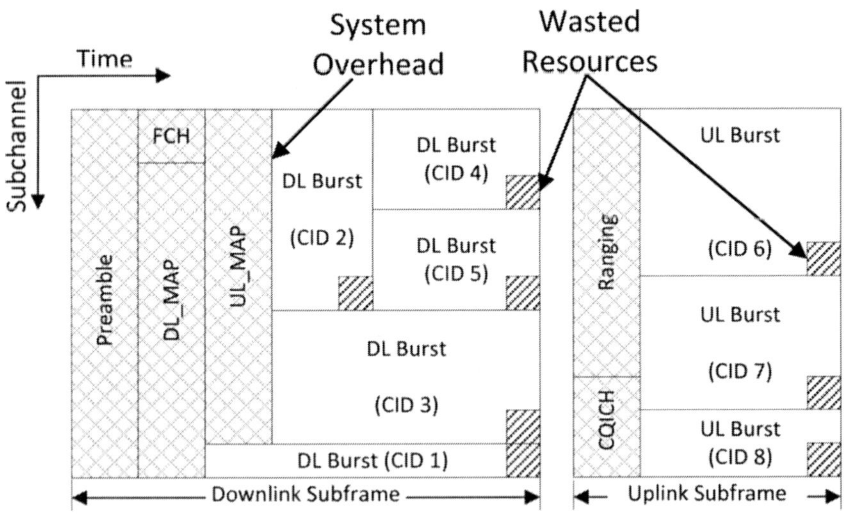

Figure 1 MOBILE WIMAX FRAME.

The analysis method used in this section to calculate the system overhead present in a Mobile WiMAX system is based on the works done in[11], but it is substantially modified to accommodate the differences in frame structure. In S-OFDMA, the available bandwidth is broken into a series of orthogonal (non-interfering) subcarrier frequencies, called subcarriers. Some of the subcarriers are dedicated to data transmission, while others are used for tasks such as pilots and guard bands. Groups of subcarriers are put together to form a channel, the smallest bandwidth allocation for a user.

The number of available channels (and thus users) is limited by the available channel bandwidth. S-OFDMA allows for flexible allocation of bandwidth depending on licensed bandwidth and allocation schemes. The four main bandwidth sizes described in IEEE 802.16 are 1.25, 5, 10, and 20 MHz[3]. S-OFDMA allows for a variety of bandwidth configurations, while all other system parameters (frame length, symbol length, etc.) remain the same. This was a design decision made to simplify the hardware in mobile devices, which were not anticipated to be very powerful. Mobile WiMAX uses Time Division Duplexing, dividing the frame into downlink and uplink subframes.

The smallest allocation that can be made in the OFDMA scheme is called a slot, which defines the allocation in bandwidth and time. In the downlink direction, it consists of one channel and two symbols in time. In the uplink direction, it is one channel and three symbols in time. Table1 shows the number of data carriers available for each bandwidth configuration.

Table 1 WIMAX CHANNELS BY AVAILABLE BANDWIDTH

Channel bandwidth (MHz)	FFT size (bits)	Channels (down/up)	N_{SD}—data carriers per symbol (down/up)
1.25	128	3/4	72/64
5	512	15/17	360/272
10	1024	30/35	720/560
20	2048	60/70	1440/1120

The number of channels in the uplink and downlink directions is defined in the IEEE 802.16 standard (8.4.6.1.2.2 for downlink, 8.4.6.2 for the uplink). A channel consists of a number of data carriers, which are a set of frequencies that can send encoded user data.

CONCENTRATION AND WIMAX

Concentration of connections is one of the items addressed by the IEEE 802.16j amendment[5]. As defined in[6, 7], the IEEE 802.16j amendment[5] introduces the relay station (*RS*) node. This node has some of the functionality of a BS and provides access to the WiMAX network for a number of users. While operating under the control of a single BS node, this node appears to end users as an independent *BS*. A new class of connection is defined between the *RS* nodes and the *BS* called a tunnel. These connections allow concentration of the connections for all SS nodes being serviced by the RS, transmitting this traffic across a single connection. This reduces the number of active connections in the system, improving system efficiency while servicing the same number of users.

All of the literature on IEEE 802.16j amendment[5] reviewed thus far concentrated on in-band relaying. As discussed in[6], the WiMAX subframes are further subdivided to provide the bandwidth for communications between the relay nodes and the subscriber nodes served by them. However, this is still subdividing a scarce resource (the WiMAX bandwidth). Furthermore, as the RSs are using the same frequencies with one another, interference starts becoming a problem (particularly if the RSs are moving and close to each other). In addition, particularly if non-transparent relaying is used, the RSs produce an additional set of system header messages. The WiMAX link can be identified as the scarcest bandwidth link in the system, and the one that must be optimized as much as possible.

Although not investigated in literature on multi-hop WiMAX thus far, the IEEE 802.16j standard also offers support for out-of-band relaying—that is, relaying using a separate set of frequencies. This will be the basis of the proposed network structure. By using a separate set of frequencies for relaying, the full bandwidth of the cell remains un-fragmented. Rather than use WiMAX for short-range relaying, this article proposes to use another technology. This will be better suited for short-range communications, allowing for better frequency reuse. This is a technology designed (and tested) specifically for vehicle-to-vehicle communications. As will be explained in the following section, this also allows for ad hoc clustering, which makes sense for mobile vehicle networking.

DSRC system

In 1999, the U.S Federal Communication Commission (FCC) allocated a 75-MHz spectrum at 5.9 GHz for DSRC system for services that involve vehicle-to-vehicle and vehicle-to-roadside communications[19]. The DSRC spectrum is structured into seven 10-MHz wide channels. One of these is a control channel and restricted to safety communications, two of these are reserved for critical safety of life and high power public safety, and the remaining four are service channels available for both safety and non-safety usage[19]. DSRC is a popular ad hoc protocol and is based on the popular IEEE 802.11 Wireless Local Access Network (WLAN) standards[20]. The main purpose of its establishment is for improving road safety, however, it is now considered for commercial purposes. The DSRC system is a short to medium range communication system, i.e., the distance range between the transmitter and the receiver should be around 1000 m. The IEEE 802.11p Wireless Access in Vehicular Environments (WAVE) amendment[4] makes WLAN suitable for vehicle communications[21].

DSRC is also known as the IEEE 802.11p[4] Wireless Access in Vehicular Environment (WAVE)[19], and was based on the IEEE 802.11a standard. The main difference between these two standards is the increase of the symbol duration in DSRC that resulted from the 10 MHz reduction in the channel spacing compared to the IEEE 802.11a standard, with 20-MHz channel spacing. IEEE 802.11a is one of the standards used in WLAN, and was designed for time-invariant channels suitable for stationary indoor environments with low delay spread; hence it makes sense to extend the symbol duration in DSRC, which was justified in[22]. The DSRC PHY employs the well-known orthogonal frequency division multiplexing (OFDM) PHY technique[23], which is popular for its ability to mitigate ISI through the use of sufficient guard interval.

DSRC PHY is defined by a partitioning of a signal across many low-symbol-rate orthogonal subcarriers. The message is encoded with a convolution encoder for forward error correction. The convolution encoder uses a generator [133_8, 177_8, and a constraint length of 7. The message is then interleaved. This interleaved message is now digitally modulated using one of four Gray-coded constellations: binary phase-shift keying, quadrature phase-shift keying (QPSK), 16-point quadrature amplitude modulation (16QAM), and 64QAM[4]. The OFDM packet is split into 64 subcarriers: 4 pilots, 48 data carriers, and 12 for padding. The symbol duration is 8.0 μs with 1.6 μs guard interval and signal bandwidth of 10 MHz. The subcarrier frequency is 0.15625 MHz (10 MHz/64). For phase tracking and frequency estimation purposes, the four pilot subcarriers are spaced apart at 1.875 MHz. These symbols are then multiplexed into a 64-point (inverse fast Fourier transform or IFFT) using the OFDM modulation scheme. The resulting symbol is transformed into an 80-point length vector after the addition of the guard interval, and then serially transmitted over the channel. The data rate is calculated as $R_d = \frac{R_c \log_2(m) N_{SC}}{T_S}$, where N_{sc}

is the number of subcarriers, R_c is the code rate, T_s is the symbol period, and m is the number of constellation points in the digital modulation scheme being used. Hence, the data rates depend on the code rates and modulation schemes for DSRC, which result in the following: 3, 4.5, 6, 9, 12, 18, 24, and 27 (Mbps).

At the receiver end, the reverse of the transmitter takes place. Hence, after converting the received signal to parallel signals and removing the guard interval, the signal is transformed back into a frequency-domain signal using the FFT algorithm. Then two training symbols are used to obtain the estimated channel response for the first OFDM symbol over the first two received symbols. The conventional DSRC system uses the same channel response estimated for the first OFDM symbol throughout the entire packet. The received data symbols of the packet are then compensated by the estimated channel response. Finally, the compensated data are deinterleaved and then Viterbi decoding takes place to recover the "message".

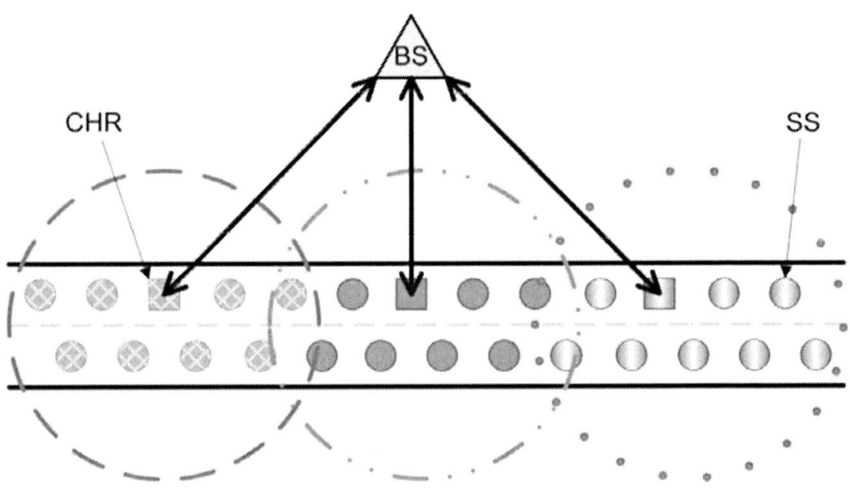

Figure 2 PROPOSED DSRC/WIMAX SYSTEM STRUCTURE.

Combined WiMAX/DSRC approach

The proposed system is to use ad hoc vehicular networking based on DSRC technology to form clusters of vehicular nodes, with the cluster head functioning as an IEEE 802.16j relay node in order to concentrate connections across the wide-area WiMAX link to the *BS*. The highly congested system conditions where WiMAX overhead becomes an issue is also the situation where clustering of DSRC nodes becomes most effective. Figure2 shows the suggested system layout. The smaller circles represent the short-range DSRC clusters, with one node identified as the cluster head (*CHR*). The different patterns represent the different clusters, using different frequencies. The identified *CHR* is then used to communicate over WiMAX to the *BS*. WiMAX is used to cover a relatively large geographic area with network coverage.

There are many potential benefits to using this combined system rather than either WiMAX or DSRC on its own. The improvement to the WiMAX system capacity minimizes the installation of costly WiMAX BS. In addition, we have shown in Figure1 that the structure of a Mobile WiMAX frame, and have also shown in Figure3 that the system efficiency is reduced with increased number of connections or users due to the fact that not all of a user's allocated bandwidth is used. This inefficiency adds up over multiple connections. Also, the proposed system solves one of the major problems with a DSRC-only system, which is the latency issue inherent with ad hoc routing systems. However, the use of the separate set of DSRC frequencies negates the need to divide the scarce WiMAX bandwidth for relaying, which will in turn solve the reduction of the system efficiency with increased number of connections problem seen in WiMAX systems. The benefit of frequency reuse in DSRC compared to the frequency reuse of a WiMAX-based relay system is also clearly seen due to the fact that a WiMAX system will have to reserve bandwidth for the relays over the entire cell size (possibly kilometres), so the bandwidth gets used up pretty quickly. A DSRC cluster, on the other hand, is going to cover a very small area. As such, especially in light of our pseudo-linear structure, we will be able to reuse DSRC frequencies regularly. For example, clusters separated by a few hundred metres can use the same frequencies and not interfere with each other, while the WiMAX system will allow these clusters to communicate with infrastructure over a wide coverage area.

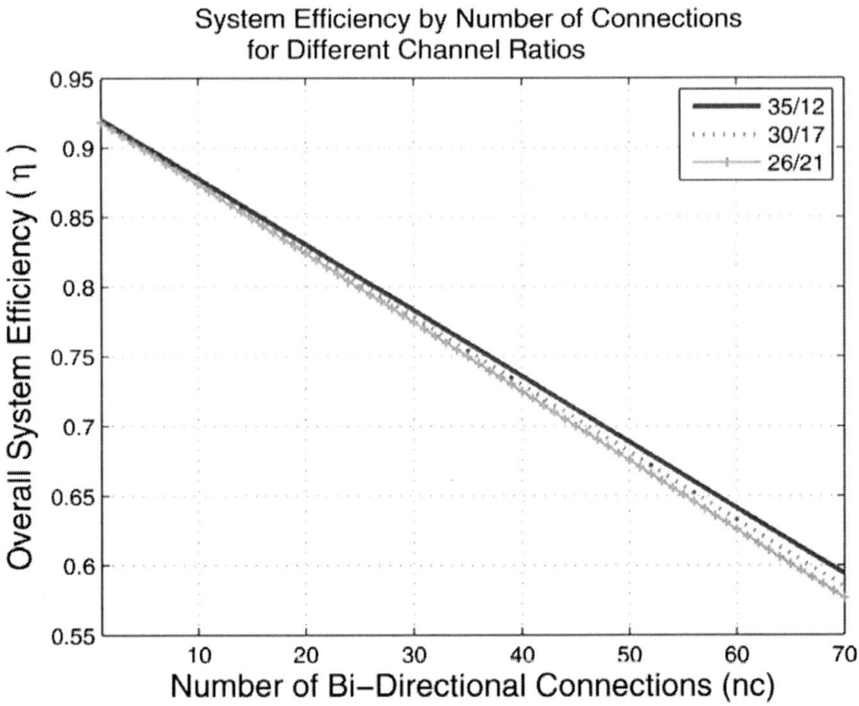

Figure 3 SYSTEM EFFICIENCY BY NUMBER OF CONNECTIONS.

Finally, this system will help encourage the adoption of DSRC, which is desired for its safety and Intelligent Transportation System (ITS) applications. The combined network also offers ITS systems an alternative to expensive[18] DSRC roadside equipment for data collection, electing instead to use the combined DSRC/WiMAX network to report ITS information such as traffic conditions[24]. An example structure of a combined WiMAX/DSRC system is shown in Figure2. The CHR nodes relay information with each other via Mobile WiMAX network, while the SS nodes relay information via each other via DSRC network.

In the proposed system, all subscriber nodes connected to the WiMAX network through the DSRC *CHR* are full members of the WiMAX network. In other words, *CHR* is not simply acting as an Internet access proxy. In this way, all *SS* nodes are registered onto the WiMAX network and are individually addressed by the system. By doing this, all nodes are better authenticated and audited by the system. This system is also more robust in mobility by taking advantage of WiMAX's built-in handover support. Finally, by registering all users on the WiMAX network, it further eases the switch between a direct WiMAX connection and a combined WiMAX/DSRC connection without resetting active connections, and allows for easier handovers between RS nodes. The switch between a direct WiMAX connection and a WiMAX/DSRC connection does not break the session that the user has with the *BS*.

Detailed efficiency model derivation

This article includes the first comprehensive and detailed mathematical derivation for the system's efficiency model. Efficiency is defined by the ratio of Medium Access Control (MAC) data transmitted to the total data transmitted. We start with calculating the number of data carriers (N_{SD}) in the downlink and uplink directions (and thus the data throughput capacity).

Calculating N_{SD} for the downlink subframes

The channels for downlink communications are defined by structures known as clusters. A cluster consists of 14 subcarriers spread over two symbols. Over the two symbols, this represents 24 data carriers and 4 pilot symbols. A channel consists of two clusters, for a total of 48 data carriers over 2 symbol times. An illustration of the OFDMA cluster structure is provided in Figure4.

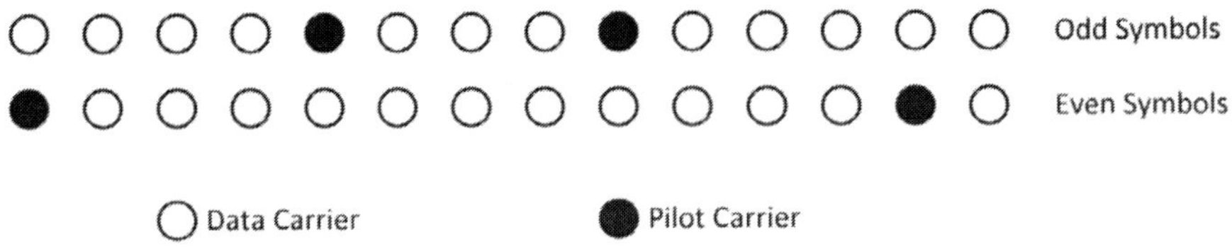

Figure 4 OFDMA CLUSTER STRUCTURE (DOWNLINK).

Using a 10-MHz channel as an example, the number of data carriers per symbol (N_{SD}) for the downlink subframes can be calculated to be:

$$\frac{30_{\text{subchannels/symbol}} \times 24_{\text{data carriers/cluster}} \times 2_{\text{clusters/subchannel}}}{2_{\text{symbols/cluster}}} = 720_{\text{data carriers/symbol}} \qquad (1)$$

Calculating N$_{SD}$ for the uplink subframes

In recognition of the less robust nature of a mobile client, the uplink communications uses a different allocation structure. As shown in Figure5, the available carriers are broken into tiles made up of four subcarriers and three symbol times. More of the carriers are dedicated to pilot signals for increased robustness. A channel consists of six of these tiles, for a total of 48 data carriers over 3 symbol times.

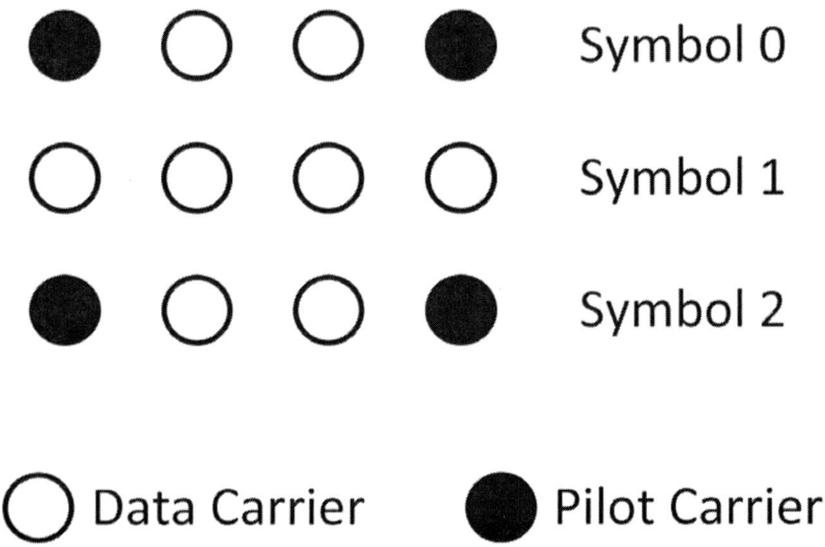

Figure 5 OFDMA TILE STRUCTURE (UPLINK).

This can consist of six consecutive tiles, or a more dispersed allocation for robustness against subcarrier specific interference. Furthermore, the allocated tiles can be changed between frames for more robustness.

Again, using a 10-MHz channel example, the number of data carriers per symbol (N_{SD}) for the uplink subframes can be calculated to be:

$$\frac{35_{subchannels} \cdot 6_{titles/subchannel} \cdot 8_{data\ \ carriers/title}}{3_{symbols/title}}$$
$$= 560_{data\ \ carriers/symbol} \qquad (2)$$

According to the IEEE 802.16 and Mobile WiMAX specifications for the OFDMA modulation scheme, the following modulation/coding rates are defined. Table2 is adapted from the one in[25], but updated to the different block size used in OFDMA.

Table 2 CODED AND UN-CODED BLOCK SIZE BY MODULATION

PHY mode (m)	Modulation and coding rate	Coding rate Rc	$\log_2 M$	Un-coded channel block size (bits) (BpS_m)	Coded channel block size (bits)
1	QPSK 1/2	1/2	2	48	96
2	QPSK 3/4	3/4	2	72	96
3	16-QAM 1/2	1/2	4	96	192
4	16-QAM 3/4	3/4	4	144	192
5	64-QAM 2/3	2/3	6	192	288
6	64-QAM 3/4	3/4	6	216	288
7	64-QAM 5/6	5/6	6	240	288

The selection of modulation and coding rate is a factor of the channel condition, with more complex modulation and lower coding rates improving the channel throughput, but decreasing the resistance against noise and interference. The modulation and coding rate will be dynamically adjusted to provide the best throughput for the channel condition, aided by the fast feedback mechanism defined in Mobile WiMAX. The physical mode (m) is an indication of which combination of modulation and coding rate was selected. The coding rate indicates the number of bits which will contain redundant information for channel robustness. The number of bits encoded per subcarrier is represented by $\log_2 M$. The term block refers to a group of 48 data carriers, which is the basic unit of data transmission over either two (for downlink) or three (for uplink) symbols. The coded channel block size is the number of bits that will be encoded in a block. The un-coded channel block size (BpS_m) is the number of user data bits that will be encoded in the block and is defined as:

$$BpS_m = N_{sc} \cdot R_C \cdot \log_2\ M, \tag{3}$$

where N_{sc} is the number of data carriers per channel, R_C is the code rate, and M represents the number of constellation points in the digital modulation scheme used.

The overall efficiency of the system can be defined by the ratio of MAC data transmitted over the air interface to the total data transmitted[1, 25]:

$$\eta = \frac{\theta_{Net/MAC}}{\theta_{Phy/symbol}} \tag{4}$$

where

$$\theta_{\mathrm{Net/MAC}} = \frac{\sum \mathrm{Payload}}{T_{\mathrm{frame}}} \tag{5}$$

and

$$\theta_{Phy/symbol} = \frac{N_{SD} \cdot R_C \cdot \log_2 M}{T_{symbol}} \tag{6}$$

The payload is the total number of data bits transmitted, and T_{symbol} is the frame duration (5 ms for Mobile WiMAX), while T_{symbol} is the symbol duration (102.9 μs for Mobile WiMAX including guard band).

Mobile WiMAX incorporates different effective symbol times and number of data carriers for the downlink and uplink subframes, as well as a variable ratio of symbol allocation. Therefore, the physical transmission rate can be further defined as:

$$\theta_{\mathrm{Phy/symbol}} = \frac{[(N_{\mathrm{DL\ Symbols}}) \cdot (N_{\mathrm{DL\ SD}} \cdot R_C \cdot \log_2 M)] + [(N_{\mathrm{UL\ Symbols}}) \cdot (N_{\mathrm{UL\ SD}} \cdot R_C \cdot \log_2 M)]}{(N_{\mathrm{DL\ Symbols}} + N_{\mathrm{UL\ Symbols}}) T_{\mathrm{symbol}}} \tag{7}$$

where $N_{\mathrm{DL\ Symbols}}$ and $N_{\mathrm{UL\ Symbols}}$ are the total symbols dedicated to the downlink and uplink and $N_{\mathrm{DL\ SD}}$ and $N_{\mathrm{UL\ SD}}$ are the number of data subcarriers in the downlink and uplink, respectively. This calculation does not take into account the overhead of the preamble, as well as the guarding between the transmit and receive portions of the frame, as they do not contribute to MAC level system overhead.

To calculate the payload, total slots available for data transmission are calculated as follows:

$$N_{Slots\ \ DL} = \left[\left(\left(N_{DL\ \ Symbols} - 1\right)/2\right) \cdot N_{DL\,SC}\right] \tag{8}$$

and

$$N_{Slots\ \ UL} = \left[\left(N_{UL\ \ Symbols}/3\right) \cdot N_{UL\ \ SC}\right] \tag{9}$$

where $N_{DL\ SC}$ and $N_{UL\ SC}$ are the number of channels in the downlink and the uplink, respectively. One symbol is removed from the downlink subframes to account for the preamble.

The number of slots available for payload (user data) is defined as:

$$N_{Payload} = N_{Slots} - \sum Overhead \tag{10}$$

This can be further be defined for the downlink and uplink subframes through the specification of the system overhead components specific to each direction as:

$$\begin{aligned} N_{\text{Payload DL}} = N_{\text{Slots DL}} - N_{\text{FCH}} - N_{\text{DL}-\text{MAP}} \\ - N_{\text{UL}-\text{MAP}} - N_{\text{DCD}} - N_{\text{UCD}} \end{aligned} \tag{11}$$

and

$$\begin{aligned} N_{Payload\ \ UL} = N_{Slots\ \ UL} - N_{Ranging} - N_{ACK} \\ - N_{CQICH} - N_{Padding} \end{aligned} \tag{12}$$

The system overhead takes the form of the preamble, the Frame Control Header (*FCH*), the *DL - MAP*, the *UL - MAP* on the downlink, and contention channels for ranging, bandwidth request, and fast feedback on the uplink. The preamble is one symbol long and allows for synchronization of all clients.

The *FCH* contains channel configuration information, including *MAP* message length, sub-channels, coding scheme, etc. The *DL - MAP* and *UL - MAP* messages provide the per-user bandwidth allocations in the downlink and uplink directions, respectively. The *FCH* consumes six slots, while the *DL - MAP* and *UL - MAP* are dependent on the number of connections. Each consists of a standard MAC header and ends with a CRC, for a total of 10 bytes. Furthermore, each message consists of the standard *DL - MAP* and *UL*

- *MAP* message, with an information element (*IE*) for each connection. The length of the *DL - MAP* and *UL - MAP* messages (in bytes) is defined as:

$$\begin{aligned} L_{\text{DL-MAP}} &= L_{\text{Header+CRC}} + L_{\text{DL-Map Header}} \\ &\quad + L_{\text{PHY Sync}} + L_{\text{DL-MAP-IE}} + L_{\text{Padding}} \\ &= 80 + 72 + 32 + n_c \cdot 60 + 4 \\ &= 188 + n_c \cdot 60 \end{aligned} \tag{13}$$

Similarly for the uplink,

$$\begin{aligned} L_{\text{UL-MAP}} &= L_{\text{Header+CRC}} + L_{\text{UL-Map Header}} + L_{\text{UL-MAP-IE}} \\ &\quad + L_{\text{Padding}} \\ &= 80 + 64 + n_c \cdot 48 = 144 + n_c \cdot 48 \end{aligned} \tag{14}$$

where n_c is defined to be the number of bidirectional connections active in the system. While connections are allocated on a directional basis, a bidirectional model makes sense for the kind of traffic being investigated in this article. The upload and download maps are modulated using 1/2 QPSK, the most robust modulation available.

The remaining variables from Equations (11) and (12) are defined as follows:

$$N_{DL-MAP} = \frac{L_{DL-MAP}}{BpS_1} \ \ and \ \ N_{UL-MAP} = \frac{L_{UL-MAP}}{BpS_1} \tag{15}$$

The Mobile WiMAX specification specifies a 6-slot channel for ranging and bandwidth requests. A 6-slot channel is used for channel quality information (*CQI*) requests, part of the fast feedback mechanism used for mobile systems to adapt modulation based on changing channel conditions. Hence from Equation (12):

$$N_{Ranging} = 6 \ \ and \ \ N_{CQICH} = 6 \tag{16}$$

Finally, overhead is introduced by each connection, as the *MAC PDU* will not perfectly align to the *MAC SDU* and padding will be added. It is reasonable to assume that this adds up, over time, to half of a slot in each direction per bidirectional connection. This is making the conservative estimation that nodes have information to transmit all the time. If the transmissions are burstier in nature, much more bandwidth will be lost in the form of unused allocations.

Using the conservative estimate, the overhead due to this unused bandwidth is found to be:

$$N_{Padding} = n_c \cdot 0.5 \tag{17}$$

With this information, the MAC data bit rate in the downlink and uplink directions are calculated to be:

$$\begin{aligned} N_{Payload\ DL} &= N_{Slots\ DL} \\ &\quad - \left[\left(\frac{6 + 188 + n_c \cdot 60 + 144 + n_c \cdot 48}{BpS_1} \right) \right. \\ &\quad \left. + n_c \cdot 0.5 \right] \\ &= N_{Slots\ DL} - \left[\left(\frac{388 + n_c \cdot 108}{BpS_1} \right) + n_c \cdot 0.5 \right] \end{aligned} \tag{18}$$

and

$$\begin{aligned} N_{Payload\ UL} &= N_{Slots\ UL} - (6 + 6 + n_c \cdot 0.5) \\ &= N_{Slots\ UL} - (12 + n_c \cdot 0.5) \end{aligned} \tag{19}$$

Finally, the MAC data transmission rate is calculated to be:

$$\theta_{Net/Mac} = \left[\frac{(N_{Payload\ DL} + N_{Payload\ UL}) \cdot BpS_m}{T_{Frame}} \right]. \tag{20}$$

The formulas derived above allowed testing of different configurations of the system, including the number of active connections, the ratio of uplink symbols to downlink symbols and the modulation and coding used for the data being sent. We have shown in[1] that the system efficiency decreased from 92% with a single active bi-directional connection to 58–59% with 70 active bi-directional connections. These results are also shown in Figure3. The channel ratio was also adjusted, using the WiMAX specification downlink:uplink ratio limits of 35:12 and 26:21 and a median value of 30:17. It can be seen that as more resources are dedicated to the uplink, a slight increase in the amount of overhead is produced. In addition, the results show that the system efficiency suffered when the systems were congested. Hence, these results found an overhead problem with the OFDMA system in Mobile WiMAX, which was similar to that found with Fixed WiMAX in[7].

This inefficiency is stemmed from the bandwidth allocation information included within each frame, coupled by unused parts of allocated bandwidth due to nodes not being able to fill their allocated bandwidth, and is multiplied by the number of link allocations within the system. By reducing the number of individual connections within the system, more of the bandwidth can be used for actual data transmission.

The following sections discuss methods to approach this issue. The concept of concentration of connections, introduced by the IEEE 802.16j amendment[5], is discussed as a possible solution. Network clustering and vehicular ad hoc networks are then investigated. Finally, the two techniques are combined and proposed as a solution to the issue of system overhead in WiMAX systems.

New combined WiMAX/DSRC link layer infrastructure design

System structure description and design of proposed system

Figure6 shows the overall structure of the proposed system. Three node types are identified, as well as the communication medium between one another. The three nodes SS, CHR, and BS provide the combined WiMAX/DSRC system.

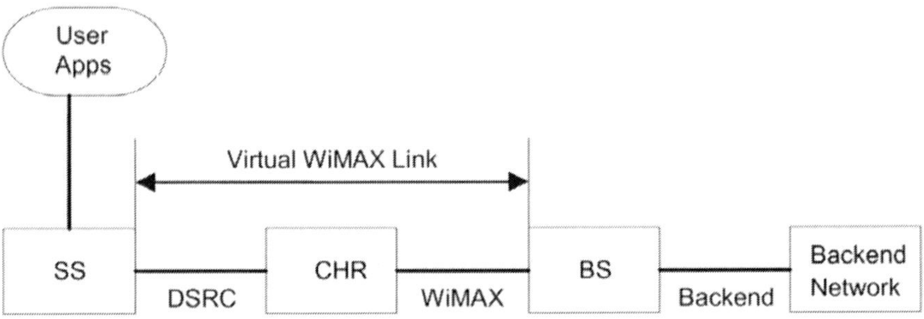

Figure 6 SYSTEM STRUCTURE.

The SS node represents the component containing the end user's applications in the system. In the context of a WiMAX system, the SS node is an agent in a mobile vehicle accessing Internet services through the network. In our proposed architecture, the SS node is a member of a DSRC network and communicates through the CHR cluster-head node. The CHR node is responsible for relaying information between the DSRC and the WiMAX networks. This configuration allows the CHR node to function as a pseudo BS to SS nodes operating under it. Data from these SS users is routed to the region's BS across a tunnelled WiMAX link. The BS node performs the functions of the BS node outlined in the IEEE 802.16j standard specification[5]. Within a WiMAX coverage region (WiMAX cell), the BS node is responsible for allocation of resources and handling authentication and handovers.

Overview of proposed system

The SS in our combined WiMAX/DSRC design is a mobile node equipped with both a DSRC transmitter/receiver and a WiMAX transmitter/receiver. However, instead of the WiMAX transmitter/receiver transmitting into the channel, it transmits and receives the WiMAX transmissions through DSRC transmissions and a cluster head/relay (CHR) station. The WiMAX network entry process has the required operations of Ranging, Registration, Security, and Service Provisioning WiMAX transmissions. The SS nodes can complete these operations by encapsulating them in DSRC packets for transmission through DSRC PHY. An SS can be elected to become a CHR station if the current CHR comes out of range, or a new CHR can be chosen from the group of in-range SS. When a reply message is received by the DSRC receiver, it must be processed by encapsulation and the reply processed appropriately. If no other SS nodes are in DSRC range, the SS node will become its own CHR station and communicate directly with WiMAX BS. This is particularly important because purely ad hoc systems suffer when there is not enough node density to ensure constant routing. Therefore, in situations where node density does not allow for clustering, SS nodes can therefore use the WiMAX radio to connect with the BS node directly. Through the use of the protocols described in this article, it will be possible for nodes to switch between being served in a DSRC cluster and being served through a direct connection and still maintain their network connections.

The CHR is a mobile node like the SS, but this node serves as a relay for multiple SS nodes to pass WiMAX transmissions that has been tunnelled through DSRC to roadside WiMAX BS. CHR nodes are elected from clusters of vehicles, and handovers can be made to ensure that SS nodes maintain connection with WiMAX BS, as will discussed later in this article. The CHR node essentially acts like a WiMAX multi-hop system, but with the subscriber side of the WiMAX relay being tunnelled through DSRC to an SS. This leads to some certain optimizations, that is, due to the fact that the SSs do not need to transmit over the WiMAX channel, the connection information setup is more efficient and is replaced on the DSRC side with an SYNC message that tells the SS stations how busy the WiMAX channel is.

This is so that the SS nodes know when and how fast to transmit to the CHR stations. It is recommended that each mobile node have two MAC addresses to readily distinguish between traffic intended for the node itself while operating in SS mode, and traffic intended for other nodes connected to it while operating in CHR mode. Some WiMAX tunnelled traffic is pre-processed at the CHR, especially during a handover session.

WiMAX BS are spread out across a coverage area. Each BS can serve and receive WiMAX data traffic from/to multiple mobile nodes. Additionally, a BS should directly be or indirectly connected to Backend Network, as seen in Figure6, to be able to provide Internet services to WiMAX connected subscribers. Our detailed proposed backend network or Internet Access Gateway (IAG) design and its functionality is presented in[26].

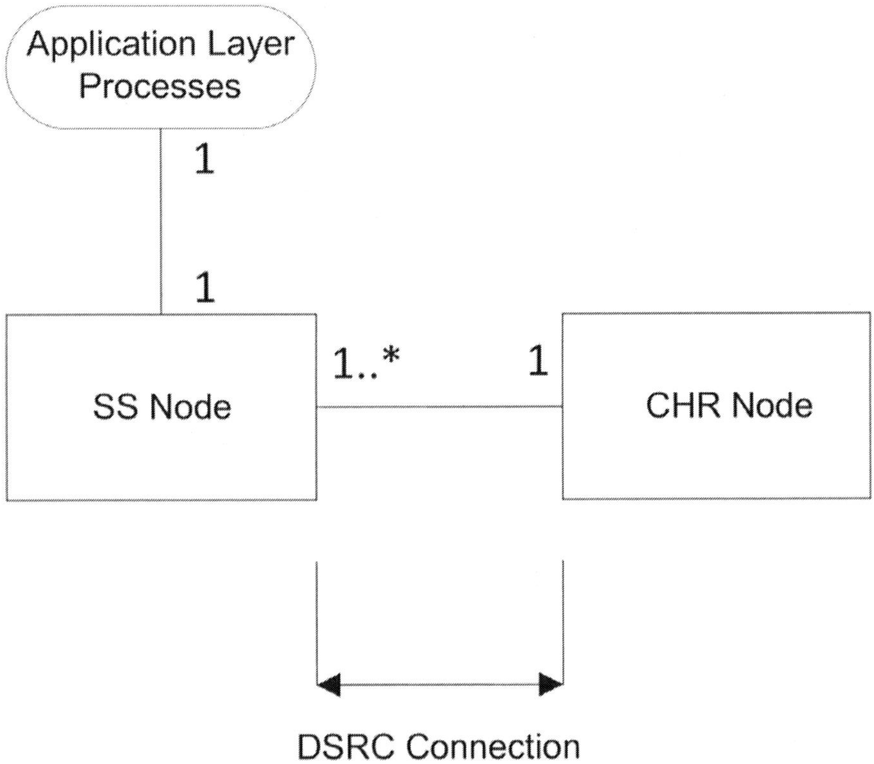

Figure 7 OVERALL SS NODE STRUCTURE.

Figure7 shows the relationship between SS and CHR nodes, without showing the BS node. The notation 1.* represents relative multiplicity of more than one. The SS node uses a single network link to communicate with the cluster head (the CHR node) in the DSRC network across a DSRC link. The details of this link are described in Section "Overview of proposed system". The network entry process between the SS node and the CHR and BS nodes is described in Section "Architecture interoperation and behavior". Figure7 also shows the CHR node which is based on the RS described in the IEEE 802.16j specification[5]. However, rather than relying on in-band or out-of-band WiMAX for relaying the messages, the CHR node uses DSRC as the relaying medium. The CHR and the SS nodes are identical in construction, with the CHR node simply being an SS node that was elected to being a relay over the underlying DSRC protocol. Similarly, as shown in Figure8, a CHR node (on the DSRC interface) can accept connections from multiple SS users, while on the WiMAX interface (BS side), a CHR node establishes an uplink connection with the BS node.

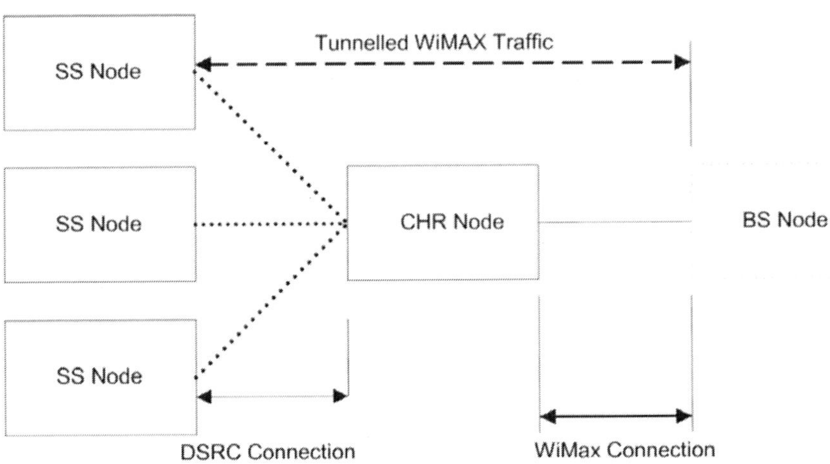

Figure 8 OVERALL CHR NODE STRUCTURE.

Due to the CHR node's dual PHY interface capabilities, two distinct communications occur. First, it is connected directly across the WiMAX network to the servicing BS node in the WiMAX cell. This link is described in detail in Section "Overview of proposed system", and is used to communicate both configuration networks for the CHR node, as well as for tunnelling traffic to and from the SS nodes served by the CHR node. The connection establishment process between the CHR node and its serving BS node is described in Section "Establishing CHR-BS connection over WiMAX PHY". Second, each individual SS node within its cluster establishes a connection with the CHR node over the DSRC link.

The overall BS node structure is shown in Figure9. It is responsible for the allocation of resources in the WiMAX network, authentication and accounting of users, and coordinating network handovers between BS. The BS node is also responsible for forwarding application layer data to the responsible application providers. Within the context of the proposed system, this includes forwarding Internet access data to the agent responsible for coordinating the network access. As described in the IEEE 802.16j specification[5], the BS node manages connections from RS nodes, as well as direct connections from SS nodes. As the primary controller for the WiMAX cell, it also accepts relayed connections from SS nodes. The BS node is also connected to the backend network set up by the WiMAX service provider. This network is used for many purposes, such as coordination between BS nodes for handovers and to authenticate users against a master authentication server. Another purpose for this connection is to provide services to the user, such as Internet access. The services that are involved in the BS nodes connecting with the fourth node, the Backend, or the IAG are described in our backend network proposal in[26].

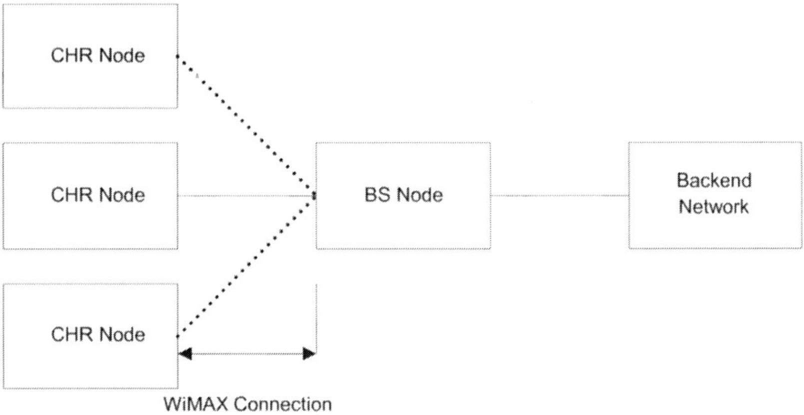

Figure 9 OVERALL BS NODE STRUCTURE.

The communication between the BS node and connected CHR nodes is described in the next section, which is important for understanding the appropriate handling of this communication when it comes to extending WiMAX services to DSRC connected SS nodes.

WiMAX communication details

As defined in the IEEE 802.16 specification[3], there are two main classes of communications: management messages and user data. The management messages are used to establish communications and update the link as circumstances change. The user data messages are used to transmit application layer data on behalf of the clients using the system. Within the context of the proposed system, this represents Layer 3 Network protocol[26] packets and Internet traffic. The IEEE 802.16j specification[5] adds another class of messages for traffic between the RS and the BS called relay messages. Relay messages allow an RS to group together traffic to and from the BS by SS nodes served by the RS, and transmit them using a single connection, represented by a connection identifier (CID), making more efficient use of the WiMAX bandwidth.

Both management and user data PDUs use a common MAC header. As defined in the IEEE 802.16 specification, this header is shown in Figure10. The details on the different fields in this header are shown in Table3. The different message types are identified by the CIDs used in the message. During the initial connection process, nodes will be assigned several CIDs. Some (the basic and primary management CID—PMCID) are used for management messages. RS nodes get assigned CIDs for tunnelling, while SS nodes will request CIDs for service flows. The details on management messages can be found in the IEEE 802.16 specifications[5]. The main purpose of management messages are to manage the user sessions on the WiMAX network. This includes tasks such as network entry, ranging (initial and periodic), bandwidth requests, and handover requests. Specific CIDs are assigned to each user within a WiMAX cell for management purposes. These include the assigned basic CID (BCID) and the assigned PMCID, as well as a default CID (usually 0) used during the original network entry. The structure of management messages is shown in Figure11. Some of the management messages used heavily during a typical WiMAX session (and in the proposed system) include the ones listed in Table4.

LSB

<table>
<tr><td>HT
(1)</td><td>EC
(1)</td><td>Type
(6)</td><td>ESF
(1)</td><td>CI
(1)</td><td>EKS
(2)</td><td>Rsv
(1)</td><td>LEN
(MSB)
(3)</td></tr>
<tr><td colspan="3">LEN (LSB)
(8)</td><td colspan="5">CID (MSB)
(8)</td></tr>
<tr><td colspan="3">CID (LSB)
(8)</td><td colspan="5">HCS
(8)</td></tr>
</table>

Figure 10 GENERIC WIMAX MAC HEADER.

Table 3 GENERIC MAC HEADER FIELDS

Name	Length (bits)	Description
CI	1	CRC Indicator
		1 = CRC is appended to the end of the PDU
		0 = No CRC is used
CID	16	CID
EC	1	Encryption Control
		1 = Payload is encrypted
		0 = Payload is not encrypted
EKS	2	Encryption key sequence
ESF	1	Extended subheader
		1 = Using extended subheader
		0 = Extended subheader absent
HCS	8	Header check sequence
HT	1	Header type (use 0 for generic header)
		1 = Bandwidth request header
		0 = Generic MAC header
LEN	11	Length
		Length of the MAC PDU in bytes (including header and CRC, if used)
Rsv	1	Reserved for future use. Set to 0
Type	6	Flags to indicate the type of payload attached

LSB

Generic MAC Header (See Figure 10)	**Management Message Type** (1 byte)	**Management Message Payload** (Variable Size)	**CRC (If Used)** (4 bytes)

Figure 11 WIMAX MANAGEMENT MESSAGE FORMAT.

Table 4 WIMAX MANAGEMENT MESSAGE TYPES

Message name	Type	Spec definition	Description
RNG-REQ	4	6.3.2.3.5	A message from an SS/RS node requesting ranging information
RNG-RSP	5	6.3.2.3.6	The reply from the BS/RS with ranging information and basic/PMCID parameters
REG-REQ	6	6.3.2.3.7	A request from an SS/RS node to finalize registration on the WiMAX network
REG-RSP	7	6.3.2.3.8	A reply from the BS node confirming or denying registration. Tunnel CIDs for RS nodes are provided at this point as well
PKM-REQ/PKM-RSP	9/10	6.3.2.3.9	Key sharing for authentication and securing of the connection
DSA-REQ	11	6.3.2.3.10	A request from an SS node to open a service flow with the BS node
DSA-RSP	12	6.3.2.3.11	The BS reply. If successful, includes the CID for the service flow
DSA-ACK	13	6.3.2.3.12	Acknowledgement by SS of the new service flow
MOB_BSHO-REQ	56	6.3.2.3.52	A request from the BS node to an SS node to initiate a handover process
MOB_MSHO-REQ	57	6.3.2.3.53	A request from the SS node to initiate a handover process
MOB_BSHO-RSP	58	6.3.2.3.54	A reply from the BS node confirming initiation of the handover process
MOB_HO-IND	59	6.3.2.3.55	Indication from the SS node that the handover process is complete

User data messages are transmitted across the WiMAX system in structures known as service data units (SDUs). This data come from applications operating at a higher protocol level, such as network layer traffic. The Layer 3 Tunnelling Protocol frame is an example of this kind of traffic. Frames containing SDUs are structured like the one in Figure12. User data are transmitted across service flows. IEEE 802.16 defines these as "a MAC transport service that provides unidirectional transport of packets either to uplink packets transmitted by the SS or to downlink packets transmitted by the BS". During the connection establishment process with the BS node, the SS node will be assigned a Service Flow CID associated with this service flow. This flow will be linked to a network service and both the SS and the BS nodes will forward it properly. In the case of the proposed system, the SS and BS nodes will transmit data from the upper layer network protocol using this service flow. The BS node will forward these SDUs on to the IAG for processing. The SS node will forward service flow traffic to the upper layer process, which will be responsible for providing Internet access for user applications. This process will be described in more detail in Section "Simulation results".

LSB

Generic MAC Header (See Figure 10)	**Service Flow Data** (Variable Size)	**CRC (If Used)** (4 bytes)

Figure 12 WIMAX USER DATA MESSAGE.

Relay frames are special frames defined in the IEEE 802.16j amendment. They allow an RS to group protocol data units (PDUs), which are WiMAX frames in this context, from multiple users into a single connection between the RS and the BS. This reduces the number of active connections in the system, which in turn lowers the system overhead produced by controlling the individual connections. Likewise, the BS will group PDUs destined for users served by the RS into this single connection. According to the 16j specification, the relay frames take the format of Figure13. The relay header itself is shown in Figure14. The fields in the WiMAX Relay Header are described in more detail in Table5.

Relay Header	User PDU #1	User PDU #2	...

Figure 13 WIMAX RELAY FRAME STRUCTURE.

<table>
<tr><td>HT
(1)</td><td>Rsv
(1)</td><td>RMI
(1)</td><td>Rsv
(7)</td><td>Priority
(3)</td><td>LEN (MSB)
(3)</td></tr>
<tr><td colspan="3">LEN (LSB)
(8)</td><td colspan="3">CID (MSB)
(8)</td></tr>
<tr><td colspan="3">CID (LSB)
(8)</td><td colspan="3">HCS
(8)</td></tr>
</table>

Figure 14 WIMAX RELAY HEADER.

The virtual WiMAX link is a key component to the proposed system. The system allows out-of-band relaying of a WiMAX signal across another technology, in this case DSRC link, which is more appropriate technology for vehicles. The virtual WiMAX link will attempt to implement as much of the WiMAX protocol as possible. This allows SS nodes to function as full members of the WiMAX network. The virtual WiMAX frames follow the WiMAX frame structure as closely as possible. Communications between the BS and CHR nodes will consist of WiMAX packets encapsulated within the underlying protocol, as shown in Figure15.

Table 5 WIMAX RELAY HEADER FIELDS

Name	Length (bits)	Description
RMI	1	Relay mode indication
		1 = Use relay MAC header
		0 = Use GMH
CID	16	CID
		Tunnel CID or BCID of the RS
Priority	3	Priority of the tunnelled MPDU
HCS	8	Header check sequence
HT	1	Header type (use 0 for relay header)
		1 = Bandwidth request header
		0 = Relay header
LEN	11	Length
		Length of the MAC PDU in bytes
Rsv		Reserved for future use. Set to 0
Type	6	Flags to indicate the type of payload attached

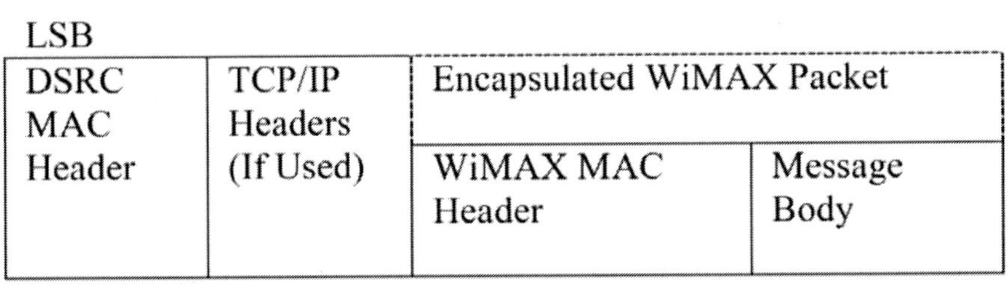

Figure 15 VIRTUAL WIMAX FRAME STRUCTURE.

In the proposed system, we assume that the clusters are successfully formed and that the cluster members know which node is the cluster head and can communicate with them. The nodes communicate using TCP/IP, where all nodes in the system have an IP address. The allocation of the addresses is done by the underlying DSRC protocol. SS nodes are aware of the IP address of the serving CHR node. The CHR node offers a WiMAX service at a known port. When an SS node wants to communicate with the WiMAX network, it will connect to the service on the CHR node. Once this TCP/IP connection is established, WiMAX messages in the packet structure form described in the next subsection are sent across the connection.

Connection differences with WiMAX

The WiMAX over DSRC system attempts to replicate the WiMAX system as close as possible for the sake of compatibility. However, due to differences in network topology, a few minor changes have to be made. During a network entry using a standard WiMAX channel, system configuration parameters (such as the Base Station ID) are obtained by passively listening to the channel and reading the Frame Control Header (FCH). Since the point-to-point nature of WiMAX over DSRC precludes this, a new WiMAX command called SYNC is created to provide system configuration information. Using a management message of type 255, a SYNC message provides the BSID of the CHR node and the WiMAX over DSRC protocol version. This message is sent from the CHR node to the SS automatically upon establishment of a connection. The message has the structure seen in Figure16, with the fields described in more detail in Table6.

LSB

Generic MAC Header (See Figure 10)	**0xFF** (Message Type for SYNC)	**System Version** (2-4 bytes – TBD)	**BSID** (4 bytes)

Figure 16 VIRTUAL WIMAX FRAME STRUCTURE.

Table 6 SYNC MANAGEMENT MESSAGE FIELDS

Field	Description
Version	The protocol version of the "Virtual WiMAX" system. This ensures that the client software is compatible with the protocol used by the RS node
BSID	The base station ID of the RS node. Used for the network entry messages, etc.

The WiMAX over DSRC system uses an abbreviated ranging process, as physical ranging is handled by the underlying DSRC layer. The ranging command is only used once and the reply is used to supply all the important information supplied in a WiMAX ranging session, including the basic and PMCIDs. Finally, no bandwidth allocation is used, and the WiMAX frames are transmitted as a series of messages across the underlying DSRC connection.

Architecture interoperation and behavior

In this section, we provide details regarding the interoperation and behavior of the components within SS, CHR, and BS nodes. The interconnections between components are presented in unified modeling language (UML) deployment diagram format and are intended to show a high level view that can be referred to when implementing our proposed design. It is important to note that although we show the SS node and CHR node as being separate entities, they are in fact simply two different states of the same system. This is bccausc thc SS nodc can changc to a CHR operating node and vice-versa. Also, there is potential for many active SS nodes and CHR nodes for each BS. The UML deployment diagram in Figure17 uses the notation by the use of 1.* to indicate a one-to-many relationship between SS, CHR, and BS nodes. In this diagram, the operation mode is assumed to be SS nodes accessing WiMAX over DSRC through CHR nodes for accessing a WiMAX facilitated services through BS nodes. Components are drawn within the nodes, and important subcomponents appear inside of other components. Dashed line arrows are used to represent the flow of control information between components. Interestingly, because we are modeling a communication system, control information generally flows in the same direction as data because the information is stored in headers attached to the data. Relationships between the nodes are shown with a solid line, and the communication physical interface between the nodes is shown with the stereotypes < <DSRC> > for OFDM-based DSRC connection, <<WiMAX> > for OFDMA-based WiMAX connection, or < <Backend Connection> > for reliable backend data connection.

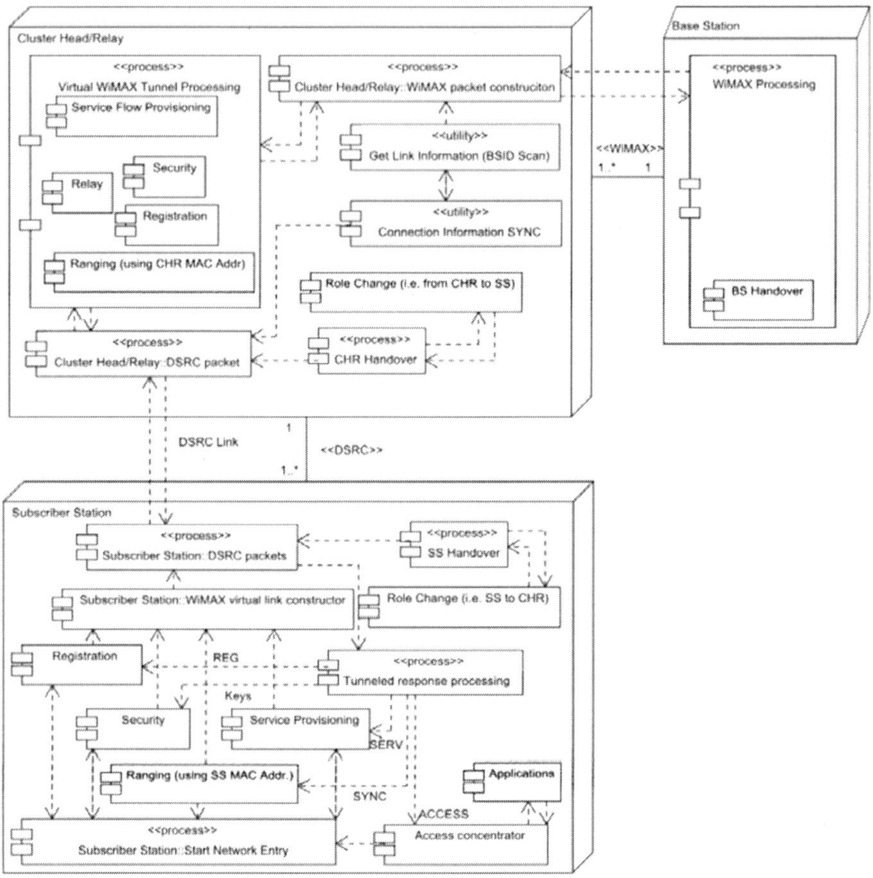

Figure 17 DEPLOYMENT DIAGRAM.

In Figure17, the locations of various important components are documented showing their relationship between each other. The WiMAX network entry processes in the CHR node include registration, security, service provisioning, and ranging, which are the same operations from standard WiMAX network entry[3]. The virtual WiMAX tunnel processing is similar to multi-hop WiMAX amendment[5] except that the client side of the tunnel is connected to a DSRC transmitter/receiver instead of a WiMAX transmitter/receiver. The network entries over DSRC processes in the SS node are modified versions of the WiMAX network entry processes, and are described in detail in the following section with accompanying UML diagrams. The interconnection between the two CHR and SS nodes operate over the DSRC interface, but the SS nodes look like multi-hop WiMAX nodes to the WiMAX BS node. Handover operations are very important to proper system operation, where the occurrence of handovers are influenced by the physical environment, relative positions of other nodes, and other factors as covered in Section "Handover process". The connectivity of each process/component is provided to illustrate interoperation of the components to aid in understanding of internal system structure. The role change processes in the SS and CHR nodes are used to transition the node's operation mode when a handover occurs, i.e., SS nodes changing its role to become a CHR node, or vice-versa. The next section begins the descriptions of the processes that are distinct in our system from Figure17, with additional functional descriptions presented in UML diagrams.

Network entry process

The network entry process is important because it informs the BS node of which WiMAX connected nodes it is responsible for serving and receiving communication content. As mentioned earlier, network entry is the process of establishing connection between a BS and a WiMAX node, which consists of Registration (REG), Ranging (RNG), Get Link Information, Provisioning of Service Flow, and Private Key Management (PKM). In respect to our proposed system, the network entry over DSRC must still perform these same operations to be able to communicate with a BS node, but must do so using the CHR node as a man-in-the middle for connection tunnelling. In Figure17, we saw how the processes interacted in the actual deployment, but it is also important to see the direct relationship between the processes that are similar to each other because the DSRC network entry processes are cousins to their WiMAX only version.

The use case diagram of Figure18 shows the relationship between the standard network entries that a node would use when communicating directly with a BS, and a network entry that would be performed by an SS over a DSRC connection with a CHR node. It is clear that the network entry over DSRC is directly a tunnelled version of the WiMAX network entry. The following section goes into detail of the network entry over DSRC processes with full description, and importantly highlights the differences between the tunnelled and non-tunnelled versions. Before getting into full detail, the general network entry operations for CHR to BS, and SS to CHR to BS network entries, over both WiMAX and DSRC are presented.

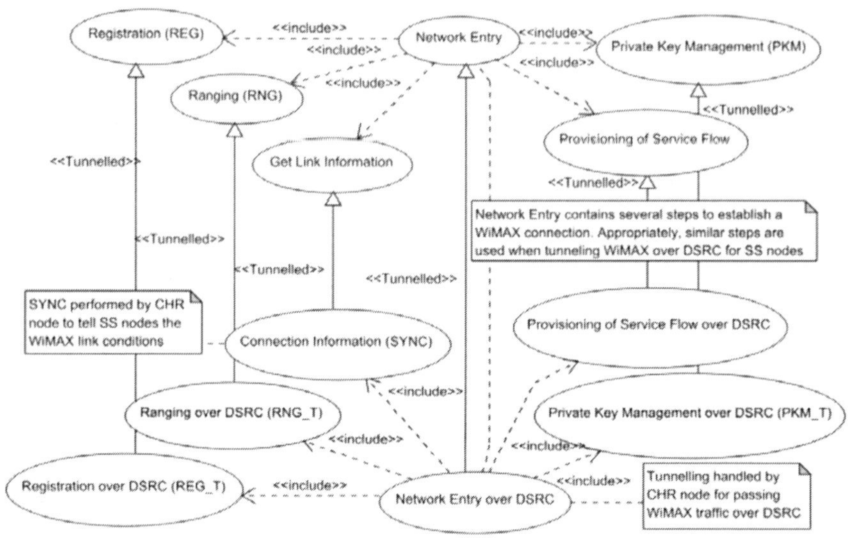

Figure 18 NETWORK ENTRY USE CASE DIAGRAM.

Establishing CHR-BS connection over WiMAX PHY

The connection handshake process follows the process laid out in IEEE 802.16j. This process authenticates CHR nodes with the WiMAX network and sets them up to accept connections with SS nodes, which will be tunnelled and processed by the network. The entry process is shown in Figure19 in the form of a sequence diagram, and the important operations of network entry are described in subsequent paragraphs. The BS node maintains a routing table of active connections within the WIMAX cell. This includes SS and CHR nodes connected directly through the WIMAX interface, as well as SS nodes connected indirectly through CHR nodes registered with the BS. Each entry stores the current status of the connection and encryption information for secure communication.

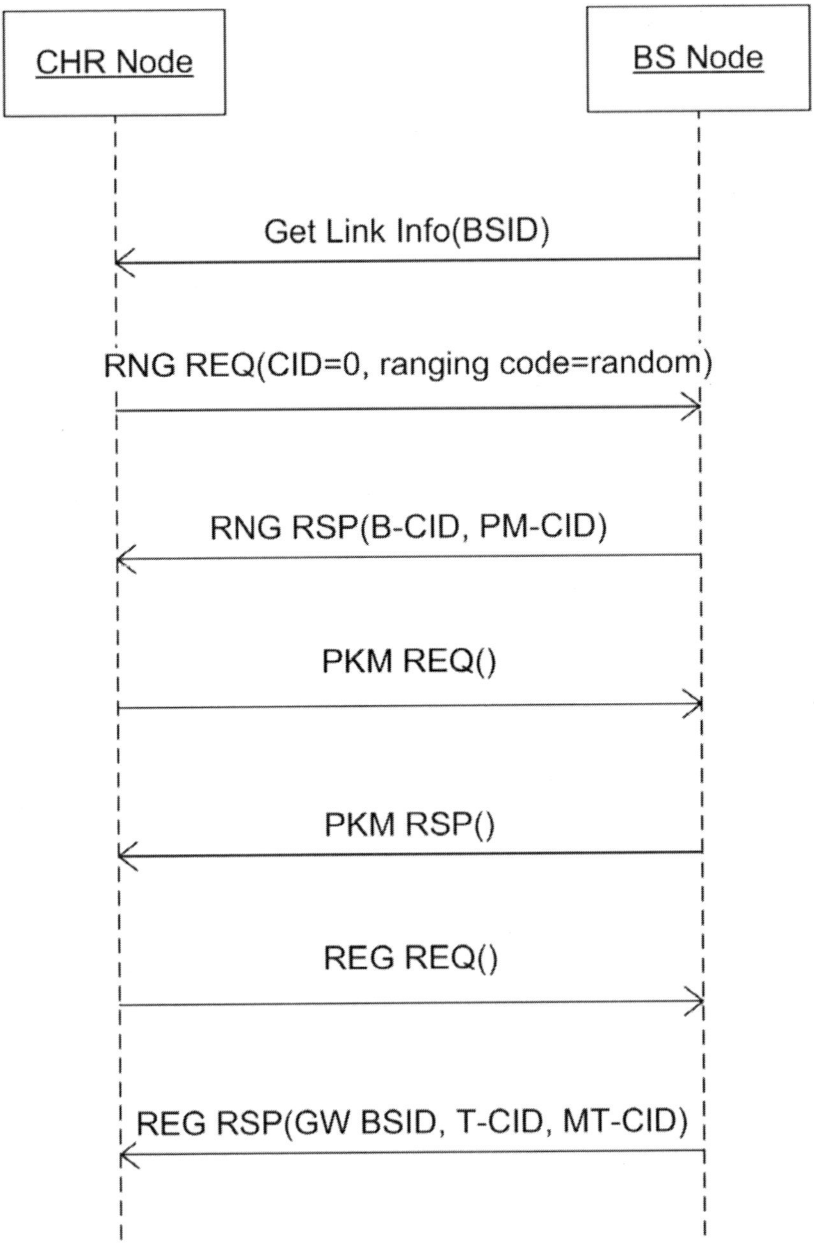

Figure 19 CHR NETWORK ENTRY SEQUENCE DIAGRAM.

Figure19 presents the network entry sequence diagram which contains the required procedures as shown in Figure17 for connecting a CHR node to a BS over WiMAX. The BS nodes use Get Link Information operation, which includes broadcast of its BSID. This operation is the traditional method of entry for WIMAX nodes over an S-OFDMA air interface. This process is used to passively retrieve channel parameters about the WiMAX cell. The CHR node scans to find the closest BS node. It then synchronizes with the WiMAX frames and listens for the channel MAPs and channel descriptors. The BSID of the BS is determined and it indicates the appropriate location within the frame to send the initial registration message. CHR nodes can issue a Range (RNG) request (REQ), for determining signal quality between the BSIDs in its region. The ranging process is the first communications between the CHR node and the BS node it is registering with. This process is conducted according to the normal access process specified in IEEE 802.16 and IEEE 802.16j. One of the parameters transmitted with the "RNG REQ" message is the CHR's MAC address. It is in this process that the concept of the CHR node having two MAC addresses becomes important. This is useful because the CHR node can tunnel its own traffic through the same connection as the SS traffic it is tunnelling through to the BS station. The response of the BS node to the RNG REQ is a range response (RNG RSP). After this response, security keys are exchanged using PKM. The "PKM" process performs the initial authentication of the BS node using public keys. It also establishes the session keys for encrypted communications between the CHR and BS nodes. This will be used by the CHR node to secure communications of tunnelled information. Afterwards, registration request (REG REQ) by the CHR node, and registration response (REQ RSP) by the BS node take place. This is the final step in the network entry for the CHR node. At some stage, it was determined that the node is a CHR. The BS node provisions three values required for the CHR role at this stage. The first is a valid BSID to be used by the gateway for accepting clients. This is the effective "BS" that each client will be using, even though all traffic is ultimately forwarded by the CHR to the BS node responsible for controlling a cell. This BSID is unique within the WiMAX system. The other two values are the management tunnel connection identifier (MT-CID) and tunnel CID (T-CID). These values are specified by IEEE 802.16j to provide the functionality for tunnelling of SS traffic data between the CHR and BS nodes. These CIDs will indicate to the BS that encapsulated traffic is being transmitted (as opposed to management messages between the BS and CHR nodes). At the completion of the handshaking process, the CHR node has been assigned a number of parameters. From the initial ranging, it has a BCID and PMCID number. After the network registration, it is also assigned a BSID value, as well as the tunnel and management tunnel CIDs.

Establishing SS-CHR-BS connection over DSRC and WiMAX PHY

This section details how an SS node enters the WiMAX network across a DSRC connection and using a CHR node as its network entry point. This process is shown in Figure20 in the form of a sequence diagram, and the important operations for WiMAX network entry over DSRC are described in subsequent paragraphs. This is based on the client network entry process described in IEEE 802.16j[5]. The CHR node maintains a forwarding table to track the connections across the WiMAX/DSRC gateway interface. This enables the CHR node to properly route traffic as it crosses through the gateway interface, as well as tracking the status of the connections. The BS node maintains a table of all SS connections in the system, including the CHR node currently serving each one. The virtual WiMAX connection uses a substitute message SYNC to obtain initial connection information from the CHR node. This message is sent upon initial connection between the SS node and the CHR service on the CHR node. A new entry in the CHR routing table is also created. Upon initial connection, this connection is empty (other than perhaps physically connection information, version number, physical address, etc.).

Next, ranging over DSRC operation is conducted. This process is similar to the ranging process described in IEEE 802.16j, except the information is not being transmitted across a physical WIMAX link, and message fields such as power control can be ignored. This difference effectively shortens the ranging process, which normally requires several iterations across a WiMAX physical layer. In the ranging request message, the SS prepares a WIMAX control packet with a CID of 0. Within the RNG request, a random ranging code is selected. Other information transmitted includes the WIMAX MAC address of the SS node. This is the MAC address of the WIMAX radio, which is not being used for the DSRC communications but will be used if the node later directly connects with the BS node. When this packet is received at the CHR, it is recognized as a ranging packet by the CID of 0 (as well as the command message packet header). The CHR updates the routing table entry with the ranging code, and then forwards it along in a tunnelled connection. The packet is received at the BS and is processed like any SS entering the system, as defined in the IEEE 802.16j specification. As the packet is forwarded, message fields such as power settings are once again ignored. The BS provisions a B-CID and PM-CID and creates a new entry for the SS in its internal routing table, as is the assigned CHR for that node. The reply message (RNG RSP) is sent back across the management tunnel to the CHR. When the reply is received at the CHR, the entry is looked up in the routing table by the ranging code value. The ranging code value is removed and replaced with the B-CID and PM-CID values in the ranging reply. The message is then forwarded along to the SS node. The message is finally received back at the SS node. The B-CID and PM-CIDs are used for further communications.

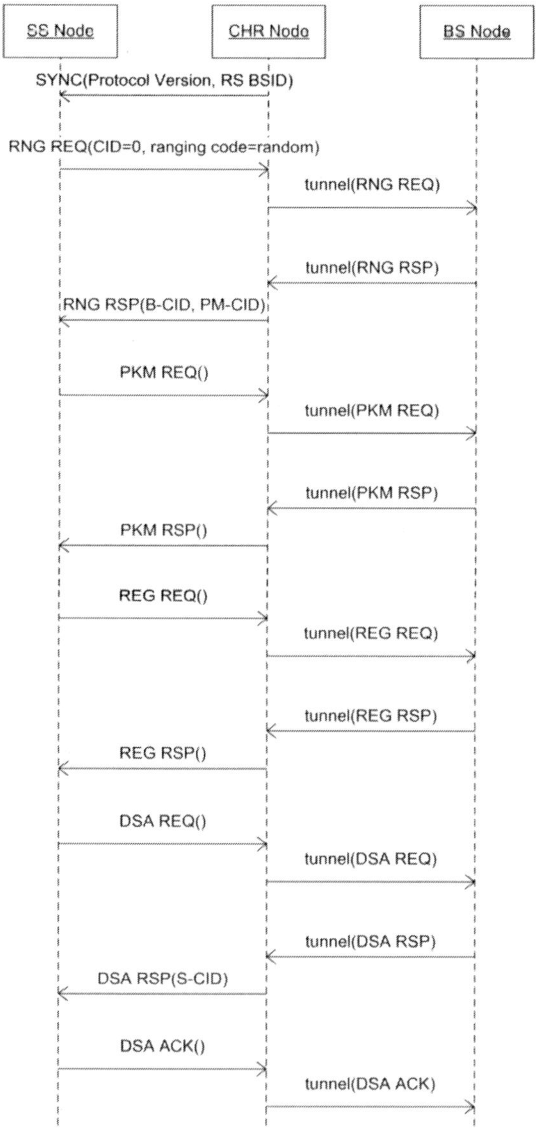

Figure 20 SS NETWORK ENTRY.

This stage ensures that the SS node is authorized to use the system. It involves a security key exchange between the SS and BS node. If the authentication is accepted, the BS node replies back with an authentication reply message. This includes the session key, sequence number, and timer information, encoded in the SS's public key. Encryption is important because data traffic from SS nodes is tunnelled through CHR nodes, which may contain sensitive information. After being authenticated, the node registers on the WiMAX network using the REG commands by tunnelling a REG REQ through its CHR node to the BS. The CHR node relays the REG RSP to the SS node over DSRC. This finishes the initial handshake process. At this point, the node is ready to send and receive traffic across the WiMAX network.

Once the system is registered, it needs to obtain a service flow for transporting the higher level network data. There are two methods that this can be accomplished with WiMAX. WiMAX systems can support "managed connections", where IP support is intrinsic to the connection. This IP configuration is not optimized for mobility, where users will get a new IP upon handing over, and existing IP connections are broken. The other option is to provide unmanaged connections. From an Internet access perspective, this unmanaged connection can be made to resemble a simple serial-like link layer. The service flow is provisioned using the dynamic service add (DSA) series of commands. These specify the QoS/service parameters describing the Internet access link service requested. After the DSA handshake is complete, the node will be assigned a service CID (S-CID), which is then ready for use for Internet access. Upon receipt of the DSA RSP message (with the assigned CID), the CHR node will update its routing table with the S-CID value. It also updates the status of the connection to indicate that the SS node is registered and ready to handle user traffic. At this point, communications between the individual SS nodes and the BS node through the CHR node has been established. Now that the methodology of establishing a connection has been described, we can now go into detail on how to maintain that connection in the vehicular ad hoc network scenario.

Handover process

The handover process is an important part of the WiMAX over DSRC system because it must be handled properly in order to achieve reliable and robust connections across the two mediums. There are four types of handovers that can occur in our proposed architecture as shown in Figure21, namely the standard WiMAX BS-BS handovers, CHR-BS handovers, CHR-CHR handovers, and BS-CHR handovers. The relationship between SS, CHR, and

BS nodes in relation to the handovers are shown in the use case diagram Figure21.

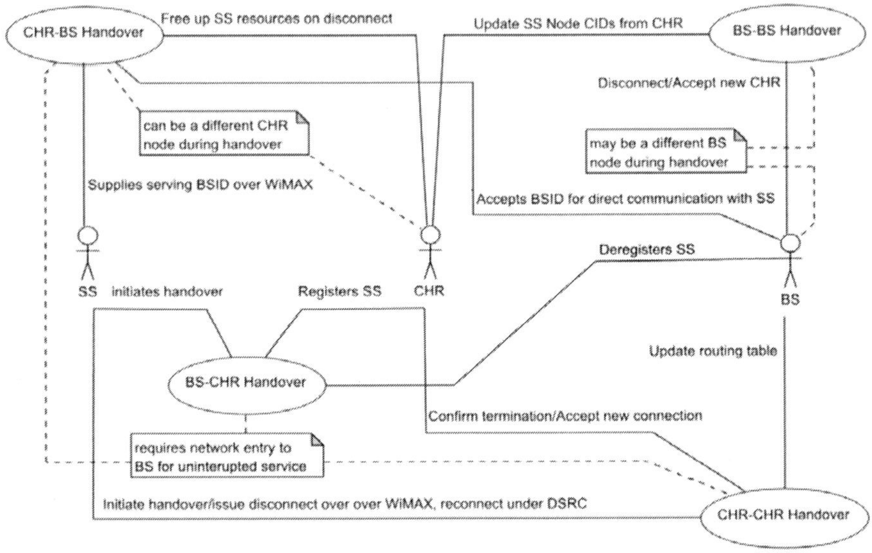

Figure 21 HANDOVER USE CASE DIAGRAM.

Due to the mobile nature of the proposed architecture, mobile nodes cannot remain connected to the same node for services. Figure21 shows the use case diagram for achieving the four types of handovers encountered in our proposed architecture with respect to maintaining WiMAX connectivity. Handovers occur when a node leaves one service area and enters into another. When going through a handover, the node has to connect and authenticate with the new service area and then transfer its session from the old one. In the context of the described system, there are several kinds of handovers that will occur. These are divided up into SS node handovers and CHR node handovers. CHR node handovers are somewhat more complicated, as it involves the indirect handover of several different connections at the same time. The SS is subject to three handover conditions. The first is moving from one DSRC cluster to another (CHR–CHR). The other two involve entering a cluster mode (BS-CHR) or exiting cluster mode (CHR-BS). Finally, the CHR and BS nodes will experience a handover condition when a mobile CHR node hands over from one BS node to another (BS–BS). More specific detail regarding the network entry processes involved in the handovers are shown in the expanded use case diagram in Figure22. This diagram is a more detailed form of Figure21, with the DSRC and WiMAX network entry included as required for each respective handover.

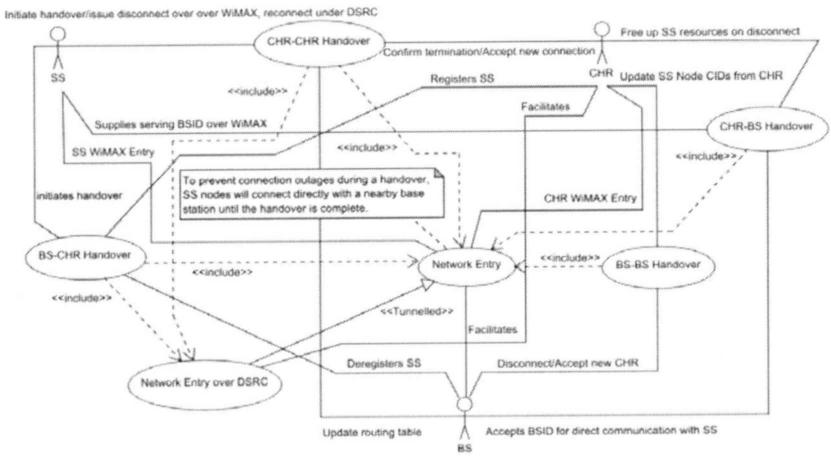

Figure 22 DETAILED HANDOVER USE CASE DIAGRAM.

As can be seen, CHR-BS handover and BS–BS handover only require a WiMAX connection to perform the handover operation, while CHR–CHR and BS-CHR handovers require both WiMAX and DSRC connections to transition between each other. In all cases, we are assuming that cross-layer information is not available (i.e., the system is not aware of an impending handover situation), which will result in an abrupt termination of connections. The nodes will perform the appropriate handover process depending on its situation and environment. Detailed descriptions follow in the next sections for the four handovers, namely the CHR–CHR, BS-CHR, CHR-BS, and BS–BS handovers.

SS performing a CHR–CHR handover

This handover occurs when a node leaves one DSRC cluster and enters another. This can be a vehicle going faster or slower than the cluster head, or a new node is elected as the cluster head. Section 6.3.21.2 of the IEEE 802.16 standard discusses the way standard Mobile WiMAX performs a handover. This begins with an SS node issuing a MOB_BSHO-REQ command to alert the current BS node that it is leaving. In the DSRC link, this is not possible due to the assumption of no cross-layer information, and instead the link termination will be abrupt. On the other hand, the termination of the underlying link between the SS and CHR node will provide confirmation that the connection is terminated. When CHR–CHR handover occurs, the underlying connection between the SS and the CHR will be broken. Both nodes will be aware of the connection being terminated. Both nodes will go into a handover state and will start a timer. If the handover does not occur in a timely manner, the connection will be terminated. All connections will be terminated and all system resources (such as CIDs) will be returned to the pool for reallocation. When the SS node connects to the new CHR, it will go through the network entry process from the previous section. The difference is that the node supplies a "serving BSID" value in its RNG REQ message. This alerts the new CHR node that the SS node is handing over from another RS. This information is forwarded along to the serving BS node. If the serving BS node is the same (one CHR to another within the same BS cell), the information is updated in the BS's routing tables. If the node is changing serving BS nodes, the new BS node will obtain session information from the old BS node, as the CHR node's BSID is unique and routable to a specific BS node.

SS performing BS-CHR handover

There are two special cases in the client handover—a client that is directly connected to the BS joining the DSRC cluster system, and a client leaving the clustered system for a direct WiMAX connection. In realistic terms, this could represent a client entering or leaving the freeway or a major road for a less populated side road. When a node is entering a DSRC system, it performs the handover as described in Section 6.3.21.4 of the IEEE 802.16j proposal. However, instead of performing a full scan, it begins the connection with the CHR node while maintaining a connection with the BS node using the WiMAX radio. Once the connection with the CHR node is complete, it finishes the handover and terminates the direct WiMAX connection with the BS node.

SS performing CHR-BS handover

This handover occurs when the SS node leaves a cluster situation and starts direct communication with the BS. Similar to a CHR–CHR handover, this will likely be abrupt, giving the node no time to prepare for the handover. Much like a CHR–CHR handover, the node starts by establishing a connection with the BS node across the WiMAX interface using the network entry process described in Section 6.3.9 of the IEEE 802.16 specification. Similar to the CHR–CHR handover, the SS node supplies a "serving BSID" value in its RNG

REQ. This allows the BS node to carry over the session and continue any established sessions uninterrupted.

CHR performing a BS–BS handover

As the CHR node itself is going to be mobile, it will also be subject to frequent handovers. This is probably the most frequent handover that will be seen in the proposed system. The mobile RS handover process is described in Section 6.3.21.4 of the IEEE 802.16j draft standard. The CIDs for the SS nodes will be updated according to the MS CID process which informs the new BS of its new CHR node(s) and accompanying SS nodes. Hence in our proposal, the CHR node swaps the CIDs of the SS nodes for new CIDs, and then forwards the MAC information over to the SS nodes over DSRC.

In summary, the goal of the system was to provide a similar experience to WiMAX (long range, managed latency, high bandwidth), while solving the efficiency issue when there are a high number of simultaneous connections. By utilizing DSRC over WiMAX, handovers are performed within the WiMAX system as simply a relay-to-relay or relay-to-BS handover, with CHR nodes functioning as IEEE 802.16j relay nodes. As the WiMAX connection does not have to be re-established, the complexity is going to be minimized and compatibility with the WiMAX system is maximized. DSRC-based ad hoc routing is notorious for wildly variable latency, as the connection paths are constantly changing and it's hard to fix the routing length. By using WiMAX as the backhaul technology, as compared to just DSRC and ad hoc routing, the number of hops (and latency) will be better controlled, while still benefiting from the increased capacity resulting from concatenated connections. The wide coverage area of WiMAX also ensures better reliability of service from a purely ad hoc system. This system has been demonstrated as a high-level prototype; hence, we presented the use cases listing the detailed steps/layout of the complete link layer infrastructure. The actual performance measurements will be confirmed once vehicle testing with real-world scenarios is started.

Simulation results

Since the proposed DSRC/WiMAX system has never been tested as one whole integrated system, the simulation section presented in this article is to demonstrate that it does in fact improve both efficiency and throughput relative to that of the current WiMAX-only system in use today. Another important use of these results will be as a benchmark for the building of the actual prototype. Using the mathematical model we presented in Section "Proposed combined WiMAX/DSRC system", we were able to test different configurations of the system, including the number of active connections, the ratio of uplink symbols to downlink symbols and the modulation and coding used for the data being sent. The simulation results of Figures23,24,25, and Table7 demonstrate the "System Efficiency by Cluster Size", "WiMAX Effeciency Improvement with DSRC Clustering Integration", "Proposed System's Percentage Throughput Improvement", and "System Throughput Per User", respectively. To evaluate the benefits of our proposed system, we use MATLAB to simulate our model with different cluster sizes and parameter values shown in Table8. These values are taken from the Mobile WiMAX standard, presented in Tables1 and2. The number of clusters range from 1 to 70 clusters. The frame duration, T_{frame}, is 5 ms, while the symbol duration, T_{symbol}, is 102.9 µs. The data carriers per channel, N_{SC}, is 48. The total symbols dedicated to downlink and uplink, $N_{DL\ Symbols}$ and $N_{UL\ Symbols}$, and data subcarriers in the downlink and uplink, $N_{DL\ SC}$ and $N_{UL\ SC}$, and the data carriers/symbol, N_{SD}, are obtained for WiMAX channel bandwidth of 10 MHz.

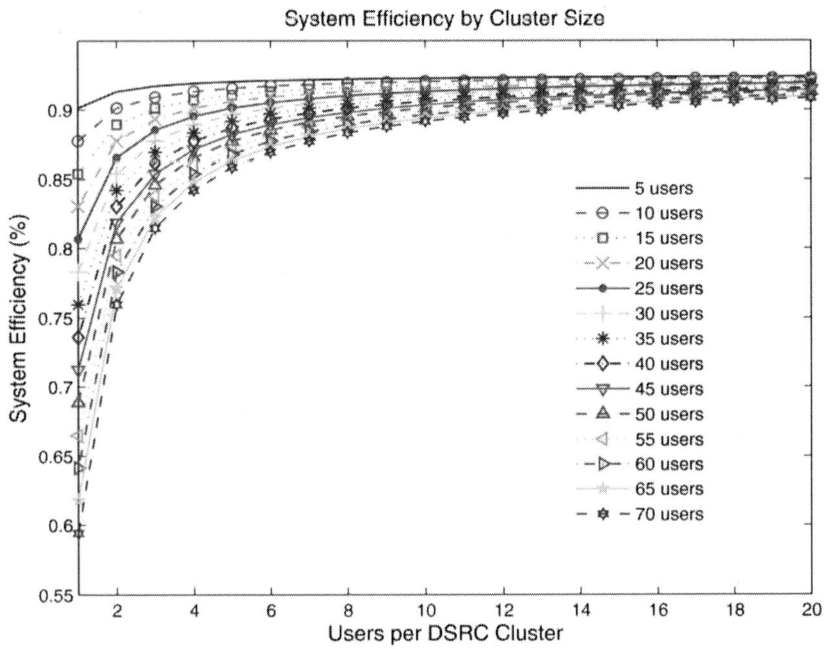

Figure 23 SYSTEM EFFICIENCY BY CLUSTER SIZE.

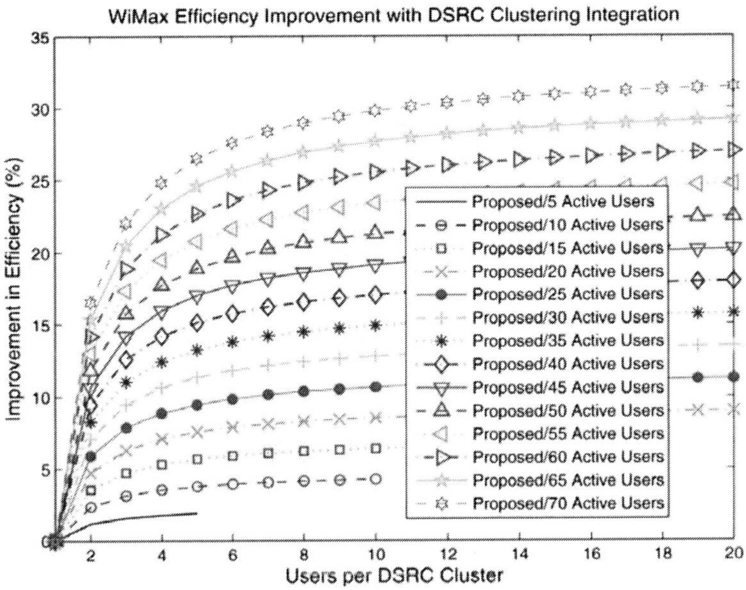

Figure 24 WIMAX EFFICIENCY IMPROVEMENT WITH CLUSTERING.

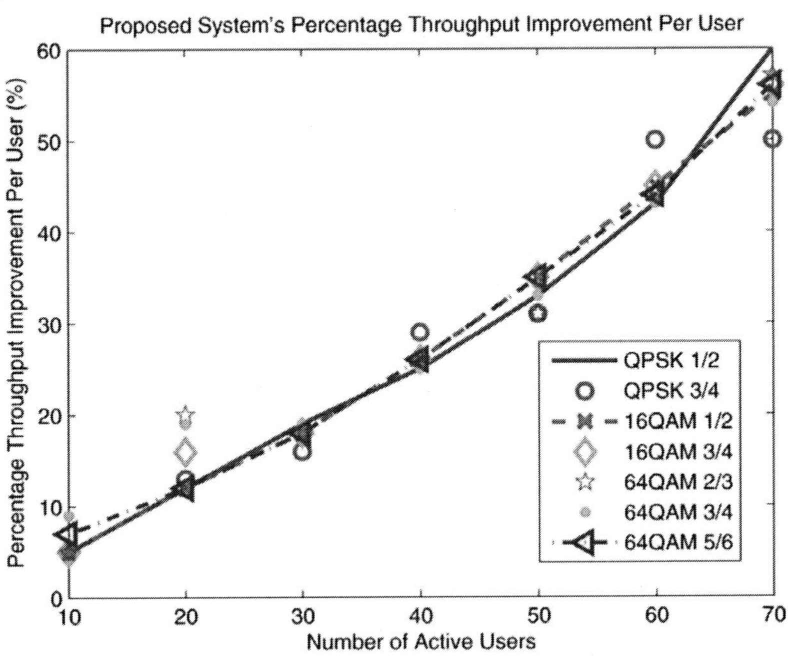

Figure 25 SYSTEM THROUGHPUT PER USER.

Table 7 SYSTEM THROUGHPUT PER USER

Modulation and coding rate	Number of bidirectional active connections n_c	WiMAX throughput (Mbps)	Proposed system throughput (Mbps)	Percentage throughput improvement (%)
QPSK 1/2	[10 20 30 40 50 60 70]	[0.55 0.26 0.16 0.120.09 0.07 0.05]	[0.58 0.29 0.19 0.15 0.12 0.10 0.08]	[5 12 19 25 33 43 60]
QPSK 3/4	[10 20 30 40 50 60 70]	[0.83 0.39 0.25 0.17 0.13 0.1 0.08]	[0.87 0.44 0.29 0.22 0.17 0.15 0.12]	[5 13 16 29 31 50 50]
16-QAM 1/2	[10 20 30 40 50 60 70]	[1.1 0.52 0.33 0.23 0.17 0.13 0.11]	[1.16 0.58 0.39 0.29 0.23 0.19 0.17]	[5 12 18 26 35 46 55]
16-QAM 3/4	[10 20 30 40 50 60 70]	[1.66 0.78 0.49 0.35 0.26 0.20 0.16]	[1.75 0.87 0.58 0.44 0.35 0.29 0.25]	[5 16 18 26 35 45 56]
64-QAM 2/3	[10 20 30 40 50 60 70]	[2.2 1.0 0.65 0.46 0.35 0.27 0.21]	[2.3 1.2 0.77 0.58 0.46 0.39 0.33]	[5 20 18 26 31 44 57]
64-QAM 3/4	[10 20 30 40 50 60 70]	[2.4 1.10 0.73 0.52 0.39 0.30 0.24]	[2.62 1.31 0.87 0.65 0.52 0.43 0.37]	[9 19 19 25 33 43 54]
64-QAM 5/6	[10 20 30 40 50 60 70]	[2.70 1.30 0.82 0.580.43 0.34 0.27]	[2.90 1.46 0.97 0.73 0.58 0.49 0.42]	[7 12 18 26 35 44 56]

Table 8 SIMULATION PARAMETERS

Coding rate Rc	Modulation schemes (bits) $\log_2 M$	Channel block size (bits) (BpS_m)	Number of bidirectional active connections n_c	Total symbols dedicated to downlink and uplink $N_{DL\ Symbols}$: $N_{UL\ Symbols}$	Data subcarriers in downlink and uplink $N_{DL\ SC}$: $N_{UL\ SC}$	Data carriers/symbol (down/up) N_{SD}
1/2	2	48	[10 20 30 40 50 60 70]	35:12	30:35	720/560
3/4	2	72	[10 20 30 40 50 60 70]	35:12	30:35	720/560
1/2	4	96	[10 20 30 40 50 60 70]	35:12	30:35	720/560
3/4	4	144	[10 20 30 40 50 60 70]	35:12	30:35	720/560
2/3	6	192	[10 20 30 40 50 60 70]	35:12	30:35	720/560
3/4	6	216	[10 20 30 40 50 60 70]	35:12	30:35	720/560
5/6	6	240	[10 20 30 40 50 60 70]	35:12	30:35	720/560

Figure23 shows the effect of clustering on system efficiency with various levels of loading. Regardless of the amount of active users, the system showed an improvement with an introduction of clustering. However, the greatest improvement with clustering is shown with the heavily loaded systems (ones with higher numbers of active nodes).

Figure24 shows the proposed system's percentage improvement in efficiency over the current WiMAX system for varying amount of active users. We have tested the efficiency using the parameters shown in Table8; however, the efficiency results did not present a difference with different coding rate and modulation scheme, so we show the results for R_c =1/2 and for QPSK modulation scheme. The greatest improvement is seen with introducing clusters of 2 to 4-users per cluster, and eventually converges to a constant gain. The main reason lies in the explanation discussed in Section "Proposed combined WiMAX/DSRC system" about the overhead problem found in the frame structure of WiMAX (shown in Figure1). With this overhead problem, as we increase the number of connections, the WiMAX-only system decreases in efficiency. This is also confirmed in Figure23 simulation result. It is observed from Figure24 that the proposed system yields a significant improvement in overall system efficiency.

More specifically, Figure24 shows what will be the projected improvement when DSRC link is used with clustering for out-of-band relaying compared to WiMAX-only links used by nodes. For example, on the horizontal comparison, where the number of users in a cluster is not fixed but with the total number of users being fixed, and if we look at 60 active users scenario and look at two example users per cluster scenarios; 5 users per cluster in this scenario (12 clusters of 5 users each) improves the efficiency by 23%, whereas 10 users per cluster (6 clusters of 10 users each) provides an efficiency increase of around 25% over WiMAX-only system. This concludes that for any given active users, having more users per cluster improves the system efficiency, again for the same overhead problem reason that exists in WiMAX-only system with an increased number of connections. A problem not found in the proposed system.

On the vertical comparison of the same figure, namely when we fix the number of users per cluster, the scenarios that have more active users in the system yield the better performance. For example, if there are 70 active users in the system with 10 users per cluster, our proposed system exhibits an improvement of around 30% in efficiency relative to the same amount of active users for WiMAX-only system; whereas for 30 active users, an improvement of around 14% is observed. This is due to the same reason, i.e., as the number of users is increased, the more the overhead problem is evident, which in turn the use of the proposed WiMAX/DSRC integration result in the most efficiency improvement.

Table7 contains the parameters used to obtain the throughput results for WiMAX-only and our proposed WiMAX/DSRC clustered network systems *per user*, as well as the percentage throughput improvement of our proposed system relative to that of WiMAX. The channel bandwidth is 10 MHz, hence the number of data subcarriers in the downlink and uplink, $N_{DL\ SC}$ and $N_{UL\ SC}$ will be 30 and 35, respectively. The total symbols dedicated to the downlink and uplink, $N_{DL\ Symbols}$ and $N_{UL\ Symbols}$, is 35 and 12, respectively. We have shown in Figure3 that this ratio outperforms the remaining downlink to uplink ratios. The frame duration, T_{frame}, is 5 ms. The data carriers per channel, N_{SC}, is 48.

It can be seen from Table7 that the proposed system's throughput outperforms that of the WiMAX system in all cases. It is important to note that the throughput in Mbps is obtained for only one user. Hence, the total throughput for all users in the proposed system is more significant, which is desired in vehicular ad hoc systems. Additionally, Table7 shows that as the number of active users increases, so does the throughput improvement per user for our proposed system relative to the WiMAX system. For a better visual, this percentage throughput improvement due to our proposed system is provided in Figure25. It is noted that the proposed system has shown an increase of around 60% for the 70 active users' case. This improvement will become more significant as we increase the number of active users.

CONCLUSION

This article proposed a novel link layer infrastructure for vehicular ad hoc networks for the purpose of enabling WiMAX network service to a large number of vehicles using both WiMAX and DSRC networking. Having a large number of active WiMAX connections created a bottleneck in the system because of the amount of overhead associated with OFDMA reservation scheme. In order to more effectively address this problem in vehicular ad hoc networks, our proposed system using DSRC link relay by employing CHR nodes as mobile RSs to SS nodes allowed WiMAX to maintain its full bandwidth without subdivision. This article provided the complete link layer infrastructure details that are important in ensuring standard compliance with the two existing technologies, while illustrating the functionality of the combined system. Simulation results provided the performance of the proposed system, and have shown to achieve its goal of reducing the number of simultaneous WiMAX connections with an increase of overall throughput performance. An increase in WiMAX efficiency by over 30% was observed. The proposed system performed the best under high load scenarios where it reduced the number of connections the most, where individual users' throughput can be improved by more than 60%. A detailed handover mechanism is provided for the DSRC and WiMAX systems' transitions between each other; great consideration was taken into providing this seamless operation. In order to aid in design and understanding, we separated the traditional WiMAX network entry into two parts, namely the WiMAX network entry, and WiMAX network entry over DSRC, of which the differences were explained.

Quantifying the pseudo-linear nature of congested vehicular traffic and development of a CHR selection protocol, based on real-world traffic data will allow us to model the overhead introduced by cluster handovers under a variety of traffic conditions and determine the situations where clustering will be more efficient than flat network architecture, the appropriate cluster size, and determine which nodes should be included in clustering. Combination of other road data (such as destination information from an onboard GPS unit) could allow for even more efficient clustering. In addition to simulation, physical test beds will allow evaluation of the system under real-world conditions, and allow optimization of the combined WiMAX/DSRC system.

REFERENCES

1. Doyle NC, Jaber N, Tepe KE: Improvement in vehicular networking efficiency using a new combined WiMAX and DSRC system design. IEEE Pacific Rim Conference on Communications, Computers and Signal Processing (PacRim) Victoria, BC, 23–26 Aug. 2011, p. 42-p. 47.
2. Doug G: Mobile WiMAX – Part I: a technical overview and performance evaluation. (WiMAX Forum. August 2006). Accessed July 2010 http://www.wimaxforum.org/technology/downloads/Mobile_WiMAX_Part1_Overview_and_Performance.pdf
3. IEEE Std. 802.16-2009 – IEEE Standard for Local and metropolitan area networks Part 16 WG802.16 - Broadband Wireless Access Working Group. Air Interface for Broadband Wireless Access Systems 2009, pp. C1-pp. C2004.
4. IEEE Std. 802.11p-2010 – IEEE Standard for Information technology – Telecommunications and information exchange between systems – Local and metropolitan area networks – Specific requirements Part 11: Wireless LAN Medium Access Control (MAC) and Physical Layer (PHY) Specifications – Amendment 6: Wireless Access in Vehicular Environments, WG802.11 - Wireless LAN Working Group. 2010, pp. 1-pp. 51.
5. IEEE 802.16j-2009 – IEEE Standard for Local and metropolitan area networks Part 16 WG802.16 - Broadband Wireless Access Working Group . Air Interface for Broadband Wireless Access Systems – Amendment 1: Multiple Relay Specification 2009, pp. C1-pp. C290.
6. Peters SW, Heath RW: The future of WiMAX: multihop relaying with IEEE 802.16j. IEEE Commun. Mag 2009, 47(1):104-111.
7. Tao Z, Teo K, Zhang J: Aggregation and concatenation in IEEE 802.16j mobile multihop relay (MMR) networks. IEEE Mobile WiMAX Symposium Orlando FL, 22–29 Mar. 2007, p. 85-p. 90.
8. Lin CR, Gerla M: Adaptive clustering for mobile wireless networks. IEEE J. Sel. Areas Commun. 1997, 15(7):1265-1275. 10.1109/49.622910
9. Shang Y, Cheng S: A stable clustering formation in mobile ad hoc network. International Conference on Wireless Communications, Networking and Mobile Computing Wuhan, 2005, 2: p. 714-p. 718.

10. Foroozan F, Tepe K: A high performance cluster-based broadcasting algorithm for wireless ad hoc networks based on a novel gateway selection approach. International Workshop on Performance Evaluation of Wireless Ad Hoc, Sensor, and Ubiquitous Networks (PE-WASUN '05) Montreal, 10–13 Oct. 2005, p. 65-p. 70.

11. Anitha VS, Sebastian MP: Scenario-based diameter-bounded algorithm for cluster creation and management in mobile ad hoc networks. IEEE/ACM International Symposium on Distributed Simulation and Real Time Applications (DS-RT '09) Singapore, 25–28 Oct. 2009, p. 97-p. 104.

12. Wu H, Qiao C, De S, Tonguz O: Integrated cellular and ad hoc relaying systems: iCAR. IEEE J. Sel. Areas Commun. 2001, 19(10):2105-2115. 10.1109/49.957326

13. Sakhaee E, Jamalipour A: Stable clustering and communications in pseudolinear highly mobile ad hoc networks. IEEE Trans. Veh. Technol. 2008, 57(6):3769-3777.

14. Amjad M, Zaman A, Sakib K: Efficient scalable clustering scheme for pseudolinear mobile ad hoc network. International Conference on Wireless Communications, Networking and Mobile Computing (WiCOM 2011) Wuhan, 23–25 Sept. 2011, p. 1-p. 4.

15. Jamalipour A, Sakhaee E: U.S. Patent US 2009 0092074 Al, Aeronautical ad-hoc networks International filing date: 20 Nov. 2006, International pub. date 31 May 2007, pub. number WO/2007/059560, U.S. filing date: 20 Nov. 2006. U.S. pub; date 9 Apr. 2009.

16. Roccetti M, Marfia G, Amoroso A, Gerla M, Pau G, Lim JH: Cognitive cars: constructing a cognitive playground for VANET research testbeds. ACM International Conference on Cognitive Radio and Advanced Spectrum Management (CogART 2011) Article 29 (Barcelona, 26–29, Oct. 2011), p. 1-p. 5. 10.1145/2093256.2093285

17. Sherman M, McNeill K, Conner K, Khuu P, McNevin T: A PMP-friendly MANET networking approach for WiMAX/IEEE 802.16. IEEE Military Communications Conference (MILCOM 2006) Washington, DC, 23–25 Oct. 2006, p. 1-p. 7.

18. Sichitiu M, Kihl M: Inter-vehicle communications systems: a survey. IEEE Commun. Surv. Tutor. 2008, 10(2):88-105.

19. Jiang D, Delgrossi L: IEEE 802.11p: Towards an international standard for wireless access in vehicular environments. IEEE Vehicular Technology Conference Singapore, 11–14 May 2008, p. 2036-p. 2040. 10.1109/VETECS.2008.458

20. IEEE Std. 802.11-2007 – IEEE Standard for Information Technology – Telecommunications and Information Exchange between Systems – Local and Metropolitan Area Networks – Specific Requirements Part 11: Wireless LAN Medium Access Control (MAC) and Physical Layer (PHY) Specification. WG802.11 - Wireless LAN Working Group 2007, pp. 1-pp. 1076. 10.1109/IEEESTD.2007.373646

21. Jiang D, Taliwal V, Meier A, Holfelder W, Herrtwich R: Design of 5.9 GHz DSRC-based vehicular safety communications. IEEE Wirel. Commun. Mag 2006, 13(5):36-43.

22. Abdulhamid H, Tepe KE, Abdel-Raheem E: Performance of DSRC systems using conventional channel estimation at high velocities. Int. J. Electron. Commun. (AEU) 2007, 61: 556-561. 10.1016/j.aeue.2006.10.005

23. Chang R, Gibby R: A theoretical study of performance of an orthogonal multiplexing data transmission scheme. IEEE Trans. Commun. Technol. 1968, 16(4):529-540. 10.1109/TCOM.1968.1089889

24. Leontiadis I, Marfia G, Mack D, Pau G, Mascolo C, Gerla M: On the effectiveness of an opportunistic traffic management system for vehicular networks. IEEE Trans. Intell. Transp. Syst. 2011, 12(4):1537-1548.

25. Redana S, Lott M, Capone A IEEE Vehicular Technology Conference, vol. 5 . In Performance evaluation of point-to-multi-point (PMP) and mesh air-interface in IEEE Standard 802.16a. Los Angeles, CA; 26–29 Sept. 2004:p. 3186-p. 3190.

26. Doyle NC, Jaber N, Tepe KE Ad Hoc Networks . Complete architecture and demonstration design for a new combined WiMAX/DSRC system with improved vehicular networking efficiency In press, available online 14 February 2012

8
CHAPTER

PERFORMANCE EVALUATION OF WIMAX BROADBAND FROM HIGH ALTITUDE PLATFORM CELLULAR SYSTEM AND TERRESTRIAL COEXISTENCE CAPABILITY

Z. Yang, A. Mohammed, and T. Hult

School of Engineering, Blekinge Institute of Technology, 372 25 Ronneby, Sweden

ABSTRACT

The performance obtained from providing worldwide interoperability for microwave access (WiMAX) from high altitude platforms (HAPs) with multiple antenna payloads is investigated, and the coexistence capability with multiple-operator terrestrial WiMAX deployments is examined. A scenario composed of a single HAP and coexisting multiple terrestrial WiMAX base stations deployed inside the HAP coverage area (with radius of 30 km) to provide services to fixed users with the antenna mounted on the roof with a directive antenna to receive signals from HAPs is proposed. A HAP cellular configuration with different possible reuse patterns is established. The coexistence performance is assessed in terms of HAP downlink and uplink performance, interfered by terrestrial WiMAX deployment. Simulation results show that it is effective to deliver WiMAX via HAPs and share the spectrum with terrestrial systems.

1. INTRODUCTION

Delivering worldwide interoperability for microwave access (WiMAX) services in the 3.5 GHz band from HAPs is an effective way to provide wireless broadband communications. HAPs, recently proposed novel aerial platforms operating at an altitude of 17–22 km, have been suggested by the International Telecommunication Union (ITU) for providing communications in mm-wave broadband wireless access (BWA) and the third-generation (3G) communication frequency bands [1–3]. Investigations on HAPs have therefore been mainly concentrated in mm-wave band and code division multiple access (CDMA) schemes delivered from HAPs. HAP systems have many characteristics including high receiver elevation angle, line of sight (LOS) transmission, large coverage area, and mobile deployment. These characteristics make HAPs to be competitive to conventional terrestrial and satellite systems, and furthermore contribute to a better overall system performance, greater system capacity and cost-effective deployment.

Many countries have made significant efforts in the research of HAPs and potential applications. Some well-known projects are: (1) the HeliNet and CAPANINA projects funded by the European Union (EU) [4], (2) the SkyNet project in Japan [1], (3) a HAP project started by ETRI and KARI in Korea [5], (4) a series of research and demonstrations of HAP practical applications carried out in the U.S. by Sanswire Technologies Inc. (Fort Lauderdale, USA), and Angel Technologies [6] (St. Louis, USA). These projects mainly focus on international mobile telecommunications 2000 (IMT-2000) services, IEEE802.1x services and fixed broadband wireless access (FBWA) in different frequency bands.

WiMAX is a standard-based wireless technology for providing high-speed, last-mile BWA up to 50 km for fixed stations and 5–15 km for mobile stations in frequency bands ranging from 2 to 66 GHz [7]. In contrast, the wireless fidelity (WiFi/802.11) wireless local area network (WLAN) standards are limited in most cases to only 100–300 feet (30–100 m). WiMAX has been regarded as one of the most promising standards for delivering broadband services in the next few years and a strong competitor to the 3G system. Its standards based on IEEE 802.16a offer the potential to deliver a significantly enhanced nonline of sight (NLOS) coverage area from HAPs in the frequency bands below 11 GHz, which leads to a more favorable propagation path due to its unique position compared with traditional base stations located on mountains or tall buildings. Providing WiMAX from HAPs is a novel approach, and some preliminary research has been done to show its effectiveness [3, 8, 9]. In this paper, we focus on the application scenario for delivering WiMAX IEEE802.16a from HAPs. In our scenarios, we assume fixed users with the directive antenna mounted on the roof to receive signals from HAPs. It is anticipated that providing WiMAX from HAPs is a competitive approach with a low deployment complexity of broadband services.

Terrestrial cellular architectures described by Lee in [10] are based on a division of the coverage area into a number of cells which are assigned to different channels with respect to adjacent cells, in order to manage cochannel interference and achieve frequency reuse. Conventionally, cells are grouped into clusters of three, four, seven, or nine cells, with all the available frequency bands allocated between them. A cluster with a larger number of cells has a greater reuse distance but fewer number of channels per cell. Previous works [8] have initially examined the fundamental performance achievable from a single-HAP and a single-terrestrial base station without considering cellular deployment for both systems. In a HAP cellular system with multiple antennas, interference is mainly caused by antennas serving cells on the same channel employing the terrestrial frequency reuse schedule [11].

This paper focuses on the system performance in different cellular reuse schemes and investigates coexistence performance with terrestrial WiMAX deployments. The paper is organized as follows. In Section 2, the description of the HAP WiMAX cellular system model, the signal path loss, and antenna models considered for HAPs and that of the terrestrial deployments is described. In Section 3, the downlink and uplink HAP WiMAX cellular system performance is evaluated in terms of criteria such as carrier to interference ratio (CIR) and carrier to interference plus noise ratio (CINR). In Section 4, a coexistence scenario is proposed to investigate the coexistence capability of HAP and terrestrial systems. In Section 5, an approach to improve HAP system performance by increasing the frequency reuse factor is shown. Conclusions are given in Section 6.

2. HAP WIMAX SYSTEM

Most of the research on HAPs considers the stratospheric platform equipped with a multiantenna payload projecting a number of spot beams within its coverage area. HAPs are employed as a group of base stations in terrestrial communication systems. A spot beam antenna architecture is able to rapidly provide a high system capacity to a number of users with a narrow beamwidth [1]. Consequently, we assume that the HAP is fitted with WiMAX base stations onboard.

2.1. Hap Cellular System

A scenario including a HAP cellular system is proposed in Figure 1. It consists of a single HAP with multiantenna payload at an altitude of 17 km serving multiple cells. The radius of the HAP coverage area and a HAP cell is typically 30 km and 8 km, respectively. We assume that cells are hexagonally arranged and clustered in different frequency reuse patterns to cover the HAP service area.

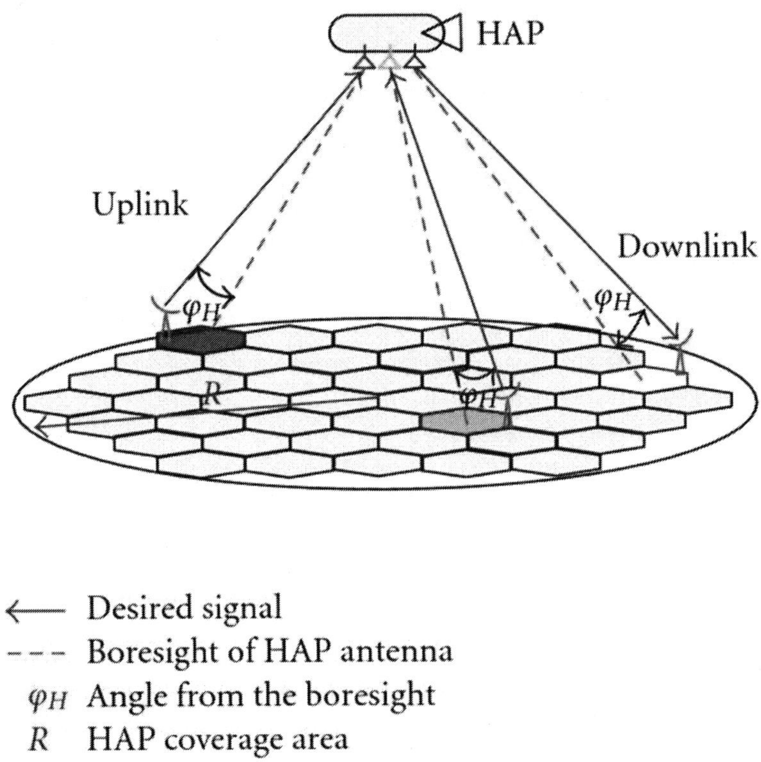

Figure 1 HAP CELLULAR SYSTEM WITH A MULTIANTENNA PAYLOAD SERVING MULTIPLE CELLS.

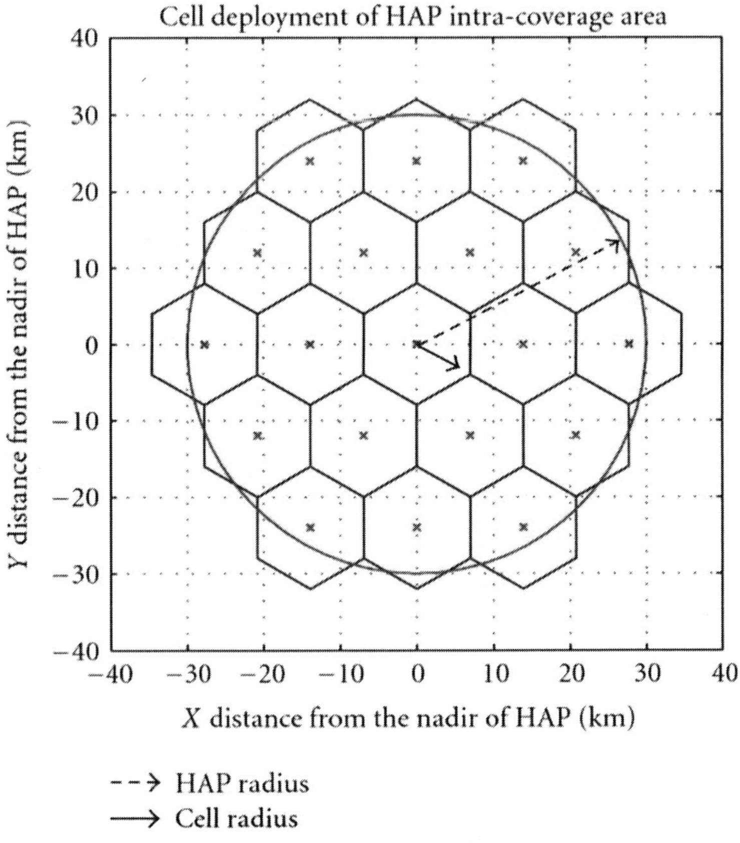

Figure 2 PLANE VIEW OF CELL DEPLOYMENT OF HAP COVERAGE AREA.

The receiver shown in Figure 1, which we refer to as a "user", is assumed to be located on the ground on a regular grid with one kilometer separation distance. This allows the performance of the coverage plot to be evaluated. After the performance is evaluated at one point, the user is moved to the next point and the same simulation test is carried out again. At anytime only one user is considered to be involved in the simulation, so interference between multiple users is not taken into account. The user is located on the grid points, spaced of one kilometer in the horizontal and vertical direction. The reason of choosing the one kilometer spacing distance is that CNR or CINR does not change significantly over the distances of less than one kilometer, and also to ensure that the computation burden is not heavy especially when the coverage area is extended further.

Figure 2 shows a plane view of a HAP ground cellular deployment. Hereby, we assume a single cell radius of 8 km, which is a typical value in WiMAX system. It results in 19 cells inside the HAP coverage area. The "*x*" markers in Figure 2 indicate the footprints of boresight of the HAP antennas.

2.2. Hap Antenna and User Antenna Patterns

An antenna radiation pattern is an important and critical design factor in determining the performance of radio communication systems. Ideally in cellular system, the antenna pattern would radiate uniform power across its serving cell and no power should fall outside. In practice, there is unavoidably power spilling outside the coverage area, which can cause interference to other cells.

In this paper, we employ a directive antenna pattern in [11], which can ensure more power radiated in the desired directions and decrease the power radiated towards undesired directions, on both the HAP and ground user. Antenna models are presented in (1) and (2), respectively. The gain $A_H(\varphi)$of the HAP antennas at an angle $\varphi$with respect to its boresight, and that of the ground receiver antenna $A_U(\theta)$at an angle away from its boresight are approximated by a cosine function raised to a power roll-off factor n and a notional flat sidelobe level s_f. G_H and G_U represent the boresight gain of the HAP antenna and user antenna, respectively:

$$A_H(\varphi) = G_H(\max[\cos(\varphi)^{n_H}, s_f]), \tag{1}$$

$$A_U(\theta) = G_U(\max[\cos(\theta)^{n_U}, s_f]). \tag{2}$$

The antenna peak directivity, which is usually achieved in the direction of the boresight, is assumed to be achieved at the centre of each HAP cell corresponding to its serving antenna. The boresight gain is calculated as

$$G_{boresight} = \frac{32\ln 2}{2\theta_{3\,dB}^2}. \tag{3}$$

In this paper, the HAP antenna payload is composed of multiple antennas with the same pattern. The beamwidth $\varphi_{10\,dB}$is initially set to be equal to the subtended angle ψ_{edge}edge at the subplatform point (SPP) with a circular beamwidth pattern as illustrated in Figure 3. This allows antenna directivities to be specified independently of the angle of the cell edge. Here, $\theta_{3\,dB}$ is the 3-dB antenna beamwidth at which the directivity curve, controlled by a roll-off factor n, is 3 dB lower than its maximum value.

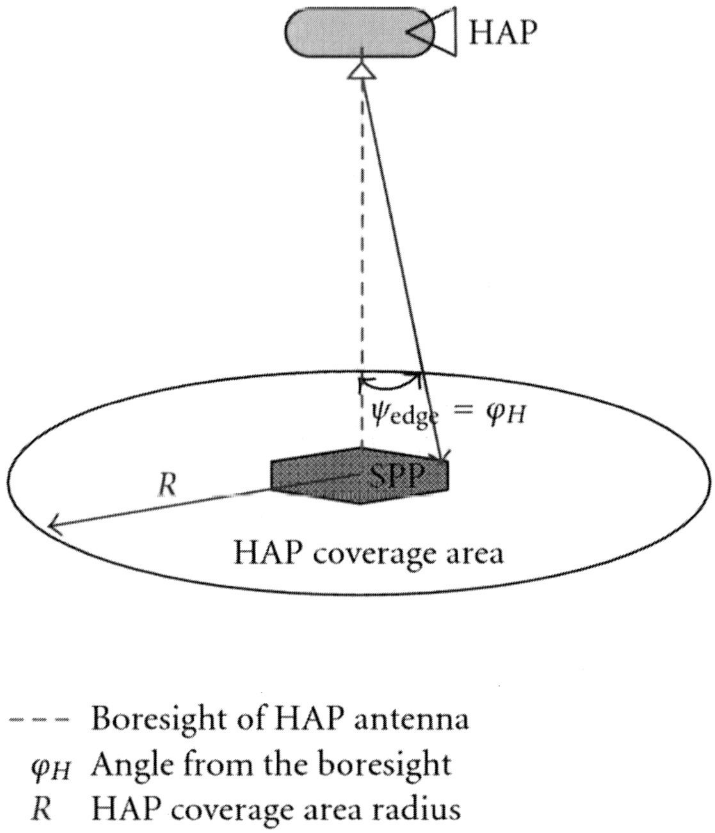

Figure 3 HAP ANTENNA BEAMWIDTH DEFINITION.

Figure 4 shows HAP, and user antenna radiation masks defined above. Figure 5 shows the performance of the HAP antenna payload on the ground. It illustrates that the best performance is achieved at the centre, where the boresight of antenna is pointing. Since all the antennas employ the common beamwidth of the antenna serving the central cell illustrated in Figure 3, cells further away from the centre have a better performance due to a smaller subtended angle at the cell edge from its boresight.

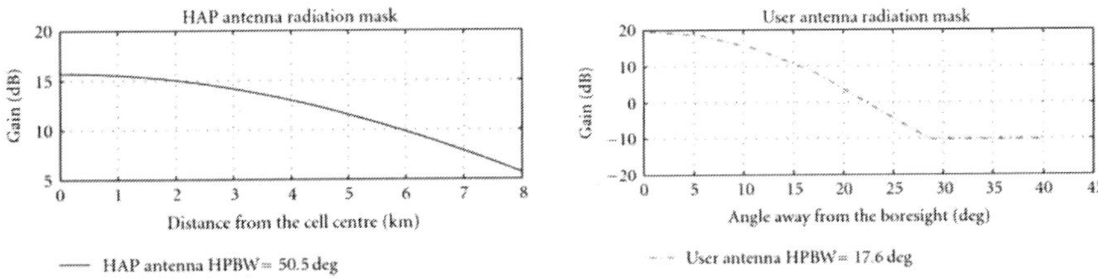

Figure 4 HAP AND USER ANTENNA RADIATION MASKS.

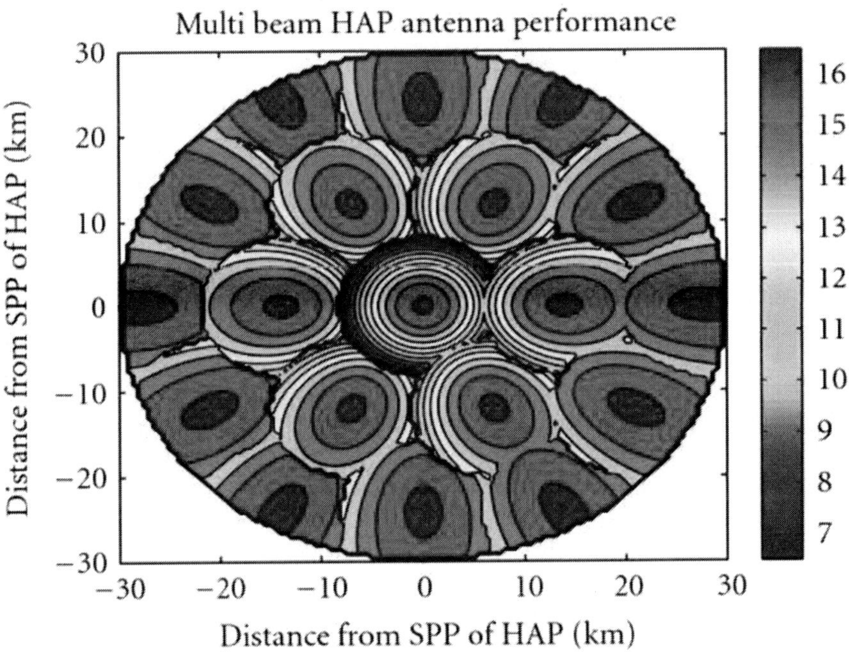

Figure 5 HAP CELLULAR ANTENNA PAYLOAD PERFORMANCE.

2.3. Hap Cellular Cochannel Interference Evaluation

In a HAP cellular system, the interference is due to HAP antennas serving cells in the same frequency band. A cochannel interference scenario is proposed in Figure 6. The user is assumed to be located in HAP cell *i*. When it communicates with the HAP antenna *i*, the user is interfered by the HAP antenna *j* and *k*, which are assumed to operate in the same frequency band as the antenna serving the cell *i*. Since each HAP antenna has its boresight footprint in the centre of the corresponding cell, angles from the boresight can be calculated separately in order to access the interference.

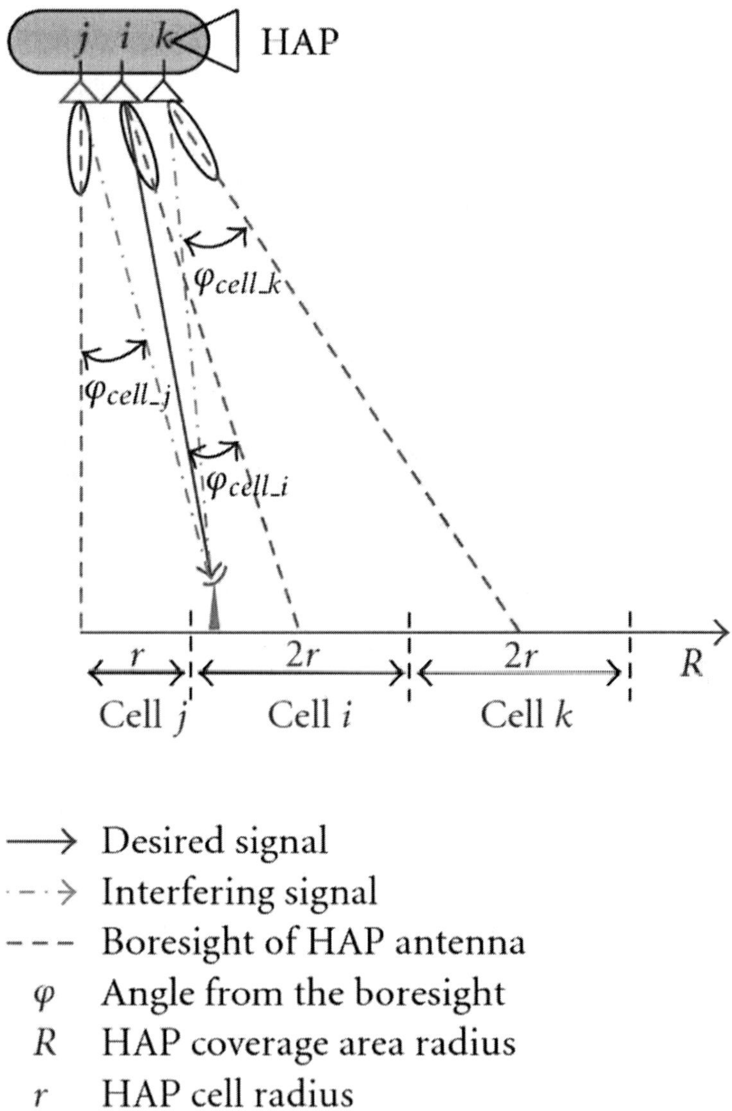

Figure 6 HAP CELLULAR COCHANNEL INTERFERENCE EVALUATION SCENARIO.

2.4. Hap Cellular Reuse Pattern

The WiMAX performance from a HAP cellular system is evaluated by assuming that the user inside the HAP coverage area communicates with the HAP and is interfered by the cochannel HAP antennas. In practice, a precise hexagonal pattern cannot be generated, due to topographical limitations, local signal propagation conditions, and practical limitations on sitting antennas [12].

In this paper, we use a circular shape to approximately cover the ideally proposed hexagonal pattern in HAP coverage area. The frequency reuse pattern in HAP cellular system consists of *N* cells assigned the same number of frequencies, which are defined as cochannel cells. *N* is termed as a reuse factor, which decides the number of cells in a repetitious pattern and is defined in [12]

$$N = I^2 + J^2 + (I \times J), \quad I, J = 0, 1, 2, 3, \ldots \tag{4}$$

Hence, we test 3, 4, and 7 as possible values of *N*. Accordingly, the minimum distance between the cochannel cells is given as [12]

$$D = r_{cell} \cdot \sqrt{3N} \,. \tag{5}$$

Figure 7 shows examples of frequency reuse patterns in HAP cellular system. Cochannel cells are depicted with the same colour. Frequency reuse allows the use of same frequency already employed in other cells nearby, thus allowing frequencies to be used for multiple simultaneous communications.

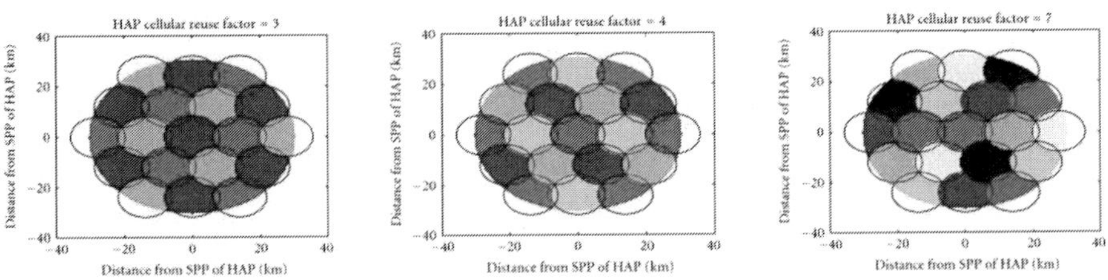

Figure 7 HAP CELL LAYOUTS WITH THE REUSE FACTOR **N** AT 3, 4, AND 7.

2.5. Path Loss Calculations and Simulation Parameters

Currently, most HAP research papers adopt the free space path loss (FSPL) presented in (6) as the propagation model used for HAP transmission, where d is distance from the transmitter and λ is the signal wavelength. Until now, no specific propagation model has been established for HAPs at 3.5 GHz, and therefore FSPL has been widely used in current research. Propagation models have been developed for HAPs in mm-wave band at 47/48 GHz, but they are not applicable in the 3.5 GHz frequency band. It should also be noted that directional user antennas are likely to be installed at a fixed location with this scenario. High elevation angles owing to the relatively small radius of HAP coverage also mean that the LOS propagation to the HAP is a reasonable assumption. Therefore, FSPL is used in this article, and diffraction and shadowing are not explicitly considered without loss of general validity:

$$PL_H = \left(\frac{4\pi d}{\lambda}\right)^2 . \tag{6}$$

The propagation pathloss model PL_T is shown in (7) for the terrestrial signal propagation model as presented in [13, 14]:

$$PL_T = PL_m + \Delta PL_f + \Delta PL_h . \tag{7}$$

PL_T is composed of a median path loss PL_m, the receiver antenna height correction term $\Delta PL_h$and the frequency correction term ΔPL_f. The two-correction terms $\Delta PL_h$and ΔPL_f[15] are added in order to make PL_T more accurate by accounting for the antenna heights and frequencies. In this paper, parameters in suburban environment (category C [13]) are used for simulations of a terrestrial deployment environment. We assume that the HAP carries the multiantenna payload with a radiation pattern described in (1) and the terrestrial base stations use omnidirectional antennas. Table 1 shows the most important system parameters for downlink (DL) and uplink (UL) simulations.

Table 1 Important system simulation parameters.

Parameters	HAP	Terrestrial
Coverage radius	30 km (R_H)	7 km (R_T)
Transmitter height	17 km (H_H)	30 m (H_T)
Transmitter power	40 dBm (P_H)	40 dBm (P_T)
Antenna efficiency	80%	
User roll-off rate	58 (n_U)	
User boresight gain	18 dB (G_U)	
Sidelobe level	−30 dB (s_f)	
Bandwidth	7 MHz/1.75 MHz (DL/UL)	
Frequency	3.5 GHz	

3. HAP CELLULAR SYSTEM PERFORMANCE

3.1. Hap Downlink System Performance Evaluation

A user in a location (x, y) is considered to be communicating with its serving HAP antenna and to be receiving interference signals from other antennas operating in the same frequency band. Performance can be evaluated as a function of CIR and CINR in (8) and (9), respectively:

$$\mathrm{CIR}_H(x, y) = \frac{P_H A_H A_U \mathrm{PL}_{HU}}{\Sigma_{i=1}^{N_H} P_{H_i} A_{H_i} A_{U_i} \mathrm{PL}_{H_i U}}, \tag{8}$$

$$\mathrm{CINR}_H(x, y) = \frac{P_H A_H A_U \mathrm{PL}_{HU}}{N_F + \Sigma_{i=1}^{N_H} P_{H_i} A_{H_i} A_{U_i} \mathrm{PL}_{H_i U}}, \tag{9}$$

where

i. P_H is the transmission power of the HAP transmitter;
ii. P_{Hi} is the transmission power of the interfering HAP antennas;
iii. A_H and A_U are the antenna gains of the HAP and the user, and they depend on the angular deviation from the boresight;
iv. PL_{HU} is the propagation pathloss from HAP to user;
v. N_H is the total number of cochannel cells in the HAP system;
vi. N_F is the noise power (-100.5 dBm).

Figure 8 illustrates the CIR performance in the HAP coverage area when considering the cochannel interference. Contours inside each cell have approximately the same shape as that in Figure 5, which demonstrates that cellular performance is not susceptible to cochannel interference.

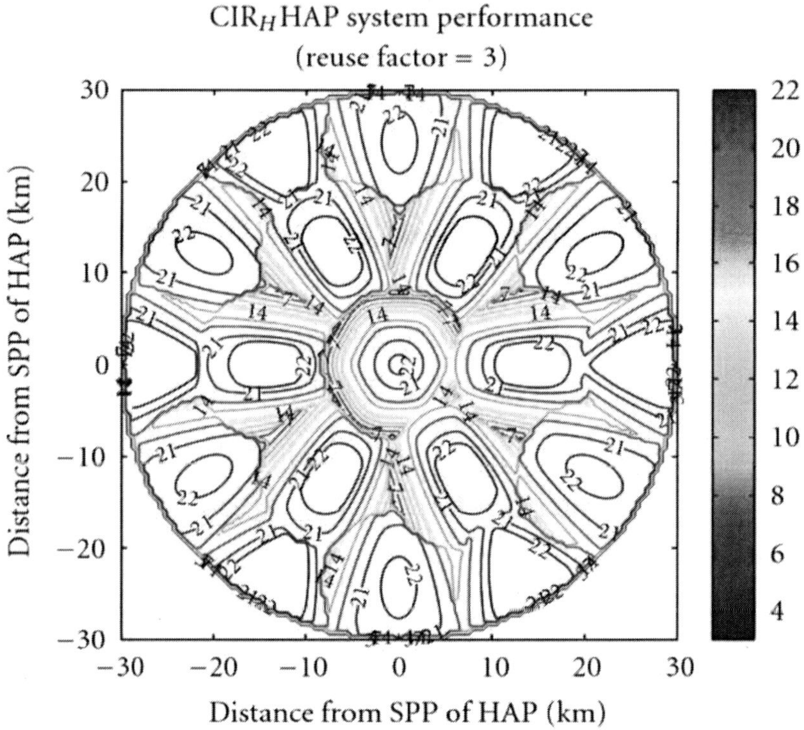

Figure 8 CIR PERFORMANCE OF A HAP WIMAX SYSTEM WITH REUSE FACTOR = 3.

Figure 9 shows the cumulative distribution function (CDF) of CIR$_H$ and CINR$_H$ of a HAP WiMAX system. It can be seen that WiMAX services can be provided on average 21 dB in both scenarios with values of CIR or CINR larger or equal to 21 dB. Interference from cochannel is dominant compared to noise in the system since the curves in Figure 9 are overlapping. Hence, strategies for system performance improvement should mainly focus on reducing excess power from cochannel interference.

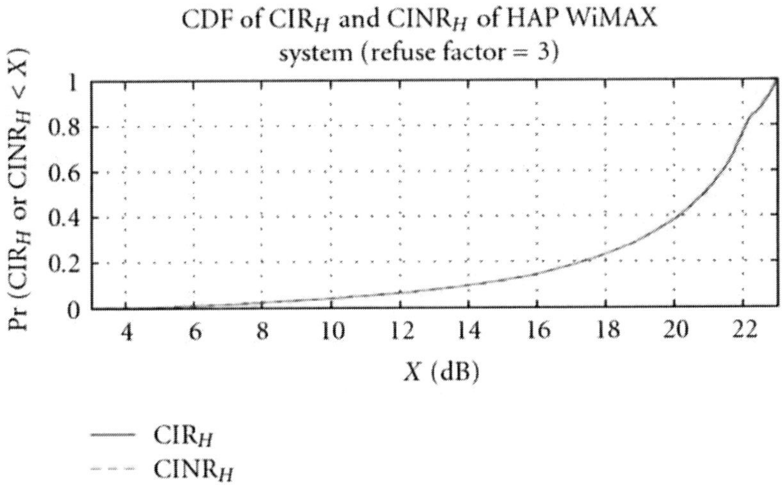

Figure 9 CDF OF CIR AND CINR PERFORMANCE OF A HAP WIMAX SYSTEM WITH REUSE FACTOR = 3.

3.2. Hap Uplink System Performance Evaluation

Uplink performance of a HAP WiMAX system can be evaluated by considering a user in a location (x, y)communicating with its serving HAP antenna and other antennas serving other cells in the same frequency band. CIR and CINR can be expressed as

$$\mathrm{CIR}_{UH}(x,y) = \frac{P_U A_U A_H \mathrm{PL}_{UH}}{\Sigma_{i=1}^{N_H} P_U A_{U_i} A_{H_i} \mathrm{PL}_{UH_i}},$$
$$\mathrm{CINR}_{UH}(x,y) = \frac{P_U A_U A_H \mathrm{PL}_{UH}}{N_F + \Sigma_{i=1}^{N_H} P_U A_{U_i} A_{H_i} \mathrm{PL}_{UH_i}}, \quad (10)$$

where

- P_U is the transmission power of user in the target cell (30?dBm);
- PL_{UH} is the propagation pathloss from user to HAP;
- N_F is the noise power (-106.5 dBm).

Figure 10 shows the CDF of uplink CIR and CINR of HAP WiMAX system. It can be seen that WiMAX uplink services can be provided averagely around

22 dB in both cases. Cochannel interference is also dominant compared to noise.

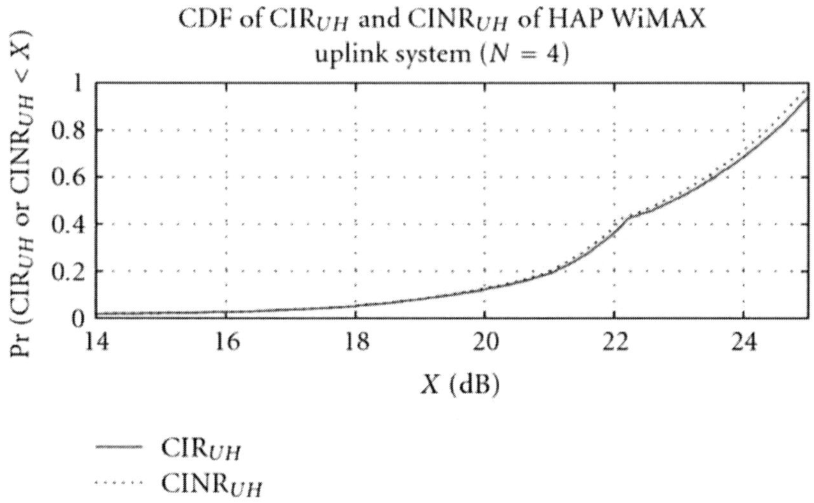

Figure 10 CDF OF UPLINK CIR AND CINR PERFORMANCE IN A HAP UPLINK SYSTEM WITH REUSE FACTOR = 4.

4. PERFORMANCE OF A HAP COEXISTENCE SCENARIO

Providing WiMAX from HAPs is a novel means to deliver broadband services. Thus, it is vital to consider its coexistence capability with current terrestrial WiMAX system in order to get an assessment of the performance. In this paper, we mainly focus on evaluating interference from terrestrial WiMAX to the HAP system.

4.1. Hap Coexistence Scenario

The considered coexistence model is depicted in Figure 11. We assume that the terrestrial WiMAX system employs the same cellular configuration as the HAP system. There are therefore 19 terrestrial base stations considered in the scenario, and all the base stations are located in the centre of HAP cells. A user communicating with the HAP in an arbitrary HAP cell is interfered by the HAP cochannel antennas and the terrestrial base station located in the centre of the same cell. This scenario, in which a user always receives interference from its nearest terrestrial base station, can be regarded as the worst case, since interfering terrestrial base station is further away from the user if any different reuse pattern was adopted.

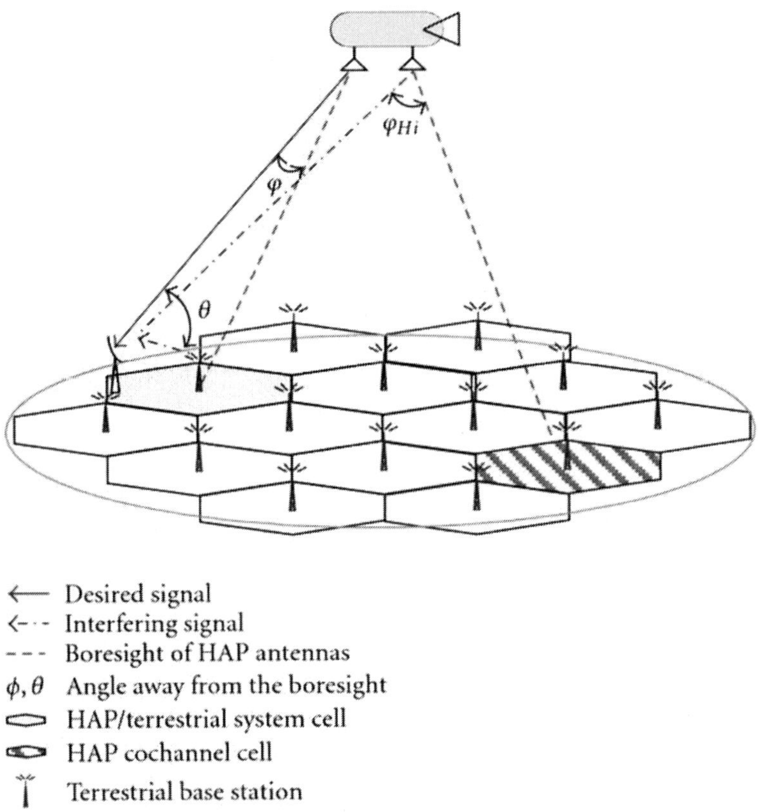

Figure 11 COEXISTENCE MODEL OF HAP AND TERRESTRIAL WIMAX SYSTEM.

4.2. Hap Downlink Coexistence Scenario

Downlink coexistence performance of HAP and terrestrial systems can be assessed by evaluating the CIR as

$$\mathrm{CIR}_{HT}(x,y) = \frac{P_H A_H A_U \mathrm{PL}_{HU}}{\sum_{i=1}^{N_H} P_{H_i} A_{H_i} A_{U_i} \mathrm{PL}_{H_i U} + P_T A_T A_{UT} \mathrm{PL}_{TU}}, \quad (11)$$

where

i. P_T is the transmission power of the interfering terrestrial base station;

ii. PL_{TU} is the pathloss from the terrestrial base station to user;

iii. A_{UT} is the user antenna gain at an angle away from its boresight.

In Figure 12, the uplink contour plot clearly shows that in most of the cell areas HAP system can provide stable services to the users, regardless of interference from the terrestrial system.

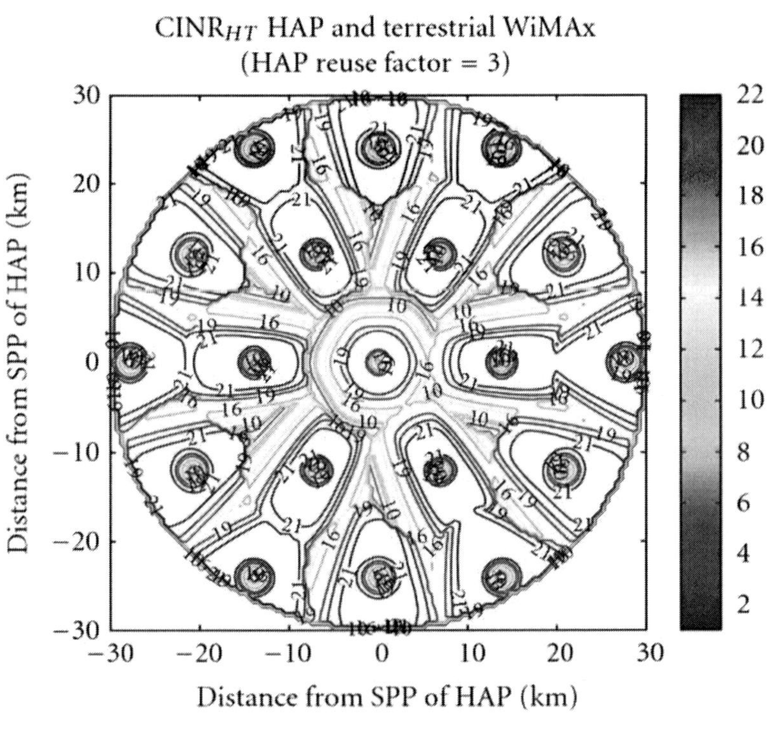

Figure 12 DL HAP SYSTEM PERFORMANCE OF USER INTERFERED BY TERRESTRIAL DEPLOYMENTS.

Figure 13 shows the CDF of the CIR value in the HAP and the coexistence scenario, respectively. On average, the HAP system can be operated with a CIR larger or equal to 21 dB, but a slight decrease in performance in the coexistence scenario can be observed because of the interference from the terrestrial system.

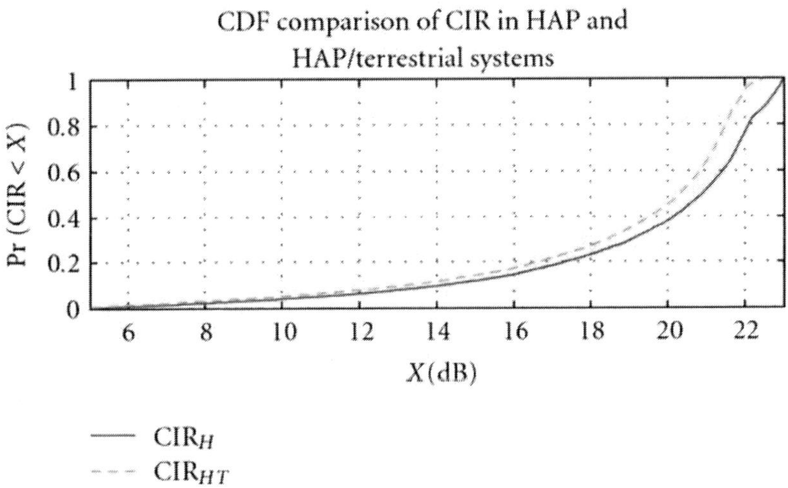

Figure 13 COMPARISON OF THE DL CIR VALUES IN HAP (CIR_H) AND HAP/TERRESTRIAL (CIR_{HT}) WIMAX COEXISTENCE SCENARIOS.

4.3. Hap Uplink Coexistence Evaluation

Uplink coexistence performance of HAP and terrestrial systems can be assessed by evaluating the CIR as

$$\mathrm{CIR}_{HT}(x,y) = \frac{P_U A_U A_H \mathrm{PL}_{UH}}{\sum_{i=1}^{N_H} P_U A_{U_i} A_{H_i} \mathrm{PL}_{UH_i} + P_U A_U A_T \mathrm{PL}_{UT}}, \tag{12}$$

where

i. PL_{UT} is the propagation path loss from the user to the terrestrial base station.

Figure 14 shows the contour plot of the uplink coexistence scenario of the HAP system. In most of cell areas, the HAP can provide stable uplink services to users, which are not susceptible to interference from the terrestrial system.

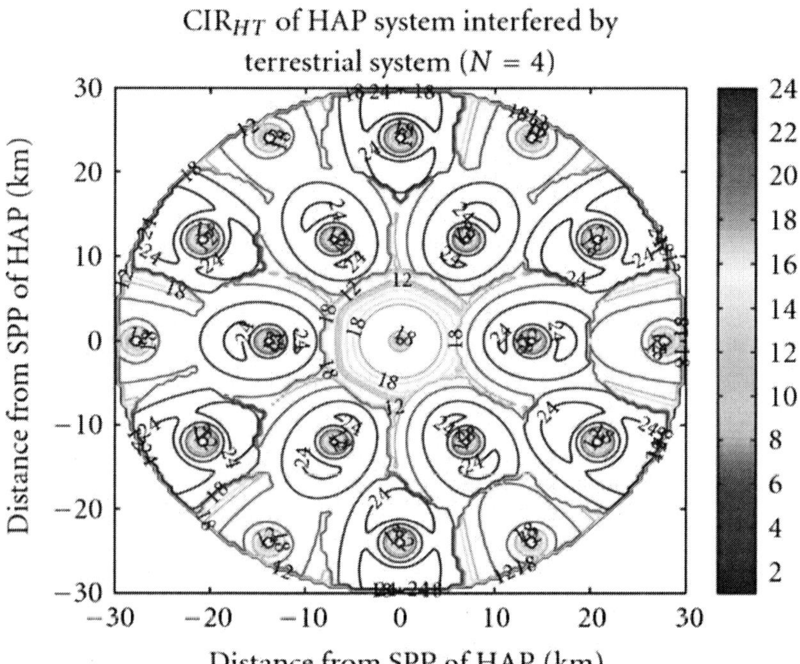

Figure 14 UL CIR VALUE OF HAP SYSTEM (CIR $_{HT}$) INTERFERED BY THE TERRESTRIAL SYSTEM.

5. HAP WIMAX SYSTEM IMPROVEMENT

A number of approaches have been used to improve the cellular system performance, for example, adding new channels, cell splitting, cell sectoring. Increasing *D*, the minimum distance between cochannel cells, is an effective approach, since it can keep the cochannel cells further away from each other and therefore decrease the interference without requiring more spectrum.

Figure 15 shows the downlink CIR_H performance of the HAP system for different values of *D*, obtained by increasing the reuse factor *N*. It shows that a CIR increase of approximately 2 dB can be achieved with each increment of *N* in the figure (from 3 to 4, or from 4 to 7). Usually the total number of frequencies allotted to the system is constant, so increasing *D*, on the other hand, decreases available channel resources in each cell of the system.

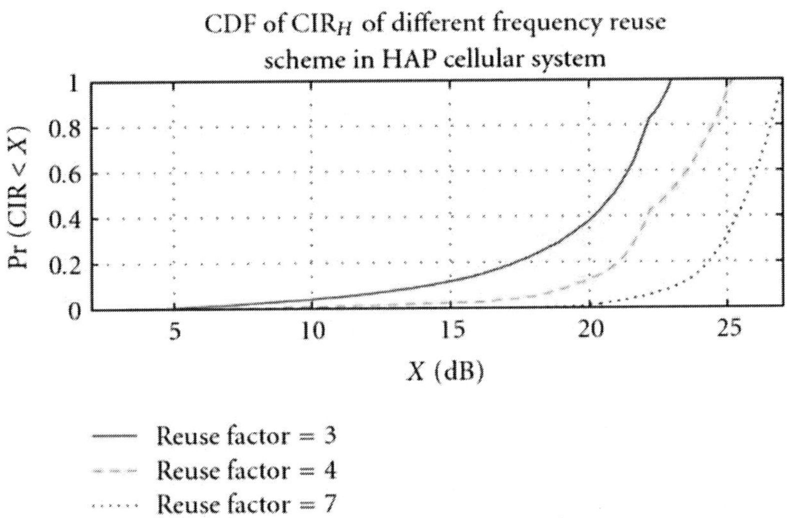

Figure 15 DL CIR $_H$ PERFORMANCE OF HAP SYSTEM WITH DIFFERENT VALUES OF THE REUSE FACTOR **N** (3, 4, AND 7).

6. CONCLUSIONS

In this paper, we have shown the performance of both downlink and uplink WiMAX broadband standard transmitted from a HAP cellular system in the 3.5 GHz band across a coverage area of 30 km radius, while operating in the same frequency band with terrestrial WiMAX deployments based on a proposed coexistence scenario. A cellular configuration has been proposed for the HAP WiMAX system based on the typical WiMAX terrestrial system. The HAP coverage area was divided into 19 individual cells served by multiantenna payload. WiMAX broadband system performance of individual HAP system was evaluated both separately and when taking into account the cochannel interference from the antennas operating in the same frequency band. Coexistence capability was investigated based on a proposed coexistence scenario and examined by considering interference from the nearest terrestrial base station to HAP system. Simulation results clearly demonstrate that the internal cochannel interference was dominant when delivering WiMAX via HAPs, and HAP system can effectively share the spectrum with terrestrial WiMAX systems.

REFERENCES

1. Huang J-J, Wang W-T, Ferng H-W: Uplink capacity enhancement for an integrated HAPS-terrestrial CDMA system. IEEE Communications Letters 2007, 11(1):10-12.
2. Elshaikh ZEO, Islam R, Ismail AP, Khalifa OO: High altitude platform for wireless communications and other services. Proceedings of the 4th International Conference on Electrical and Computer Engineering (ICECE '06), December 2006, Dhaka, Bangladesh 432-438.
3. Ahmed BT: WiMAX in high altitude platforms (HAPs) communications. Proceedings of the 9th European Conference on Wireless Technology (ECWT '06), September 2006, Manchester, UK 245-248.
4. Grace D, Mohorcic M, Oodo M, Capstick MH, Pallavicini MB, Lalovic M: CAPANINA—communications from aerial platform networks delivering broadband information for all. Proceedings of the 14th IST Mobile and Wireless and Communications Summit, June 2005, Dresden, Germany
5. Park J-M, Ku B-J, Kim Y-S, Ahn D-S: Technology development for wireless communications system using stratospheric platform in Korea. Proceedings of the 13th IEEE International Symposium on Personal, Indoor and Mobile Radio Communications (PIMRC '02), September 2002, Lisbon, Portugal 4: 1577-1581.
6. Hong TC, Ku BJ, Park JM, Ahn D-S, Jang Y-S: Capacity of the WCDMA system using high altitude platform stations. International Journal of Wireless Information Networks 2006, 13(1):5-17. 10.1007/s10776-005-0014-x
7. IEEE802.16 Broadband Wireless Access Working Group : IEEE802.16-2004 part 16: air interface for fixed broadband wireless access system. 2004.

8. Likitthanasate P, Grace D, Mitchell PD: Coexistence performance of high altitude platform and terrestrial systems sharing a common downlink WiMAX frequency band. Electronics Letters 2005, 41(15):858-860. 10.1049/el:20051930

9. Yang Z, Mohammed A, Hult T, Grace D: Optimizing downlink coexistence performance of WiMAX services in HAP and terrestrial deployments in shared frequency bands. Proceedings of the 3rd International Waveform Diversity and Design Conference (WDD '07), June 2007, Pisa, Italy 79-82.

10. Lee WCY: Spectrum efficiency in cellular. IEEE Transactions on Vehicular Technology 1989, 38(2):69-75. 10.1109/25.61338

11. Thornton J, Grace D, Capstick MH, Tozer TC: Optimizing an array of antennas for cellular coverage from a high altitude platform. IEEE Transactions on Wireless Communications 2003, 2(3):484-492. 10.1109/TWC.2003.811052

12. Stallings W: Wireless Communications and Networking. Prentice-Hall, Upper Saddle River, NJ, USA; 2002.

13. IEEE Standard 802.16a-2003 : Modifications and additional physical layer specifications for 2-11 GHz. 2003.

14. Erceg V, Greenstein LJ, Tjandra SY, et al.: An empirically based path loss model for wireless channels in suburban environments. IEEE Journal on Selected Areas in Communications 1999, 17(7):1205-1211. 10.1109/49.778178

15. IEEE802.16 Broadband Wireless Access Working Group : Channel models for fixed wireless applications. 2003.

9

CHAPTER

COEXISTENCE OF WIFI AND WIMAX SYSTEMS BASED ON PS-REQUEST PROTOCOLS

Jongwoo Kim [1], Suwon Park [1,*], Seung Hyong Rhee [2], Yong-Hoon Choi [3], Young-uk Chung [4] and Ho Young Hwang [5]

[1] *Department of Electronics and Communications Engineering, Kwangwoon University, Seoul 139-701, Korea*

[2] *Department of Electronics Convergence Engineering, Kwangwoon University, Seoul 139-701, Korea*

[3] *Department of Information Control Engineering, Kwangwoon University, Seoul 139-701, Korea*

[4] *Department of Electronic Engineering, Kwangwoon University, Seoul 139-701, Korea*

[5] *Department of Computer Engineering, Kwangwoon University, Seoul 139-701, Korea*

ABSTRACT

We introduce both the coexistence zone within the WiMAX frame structure and a PS-Request protocol for the coexistence of WiFi and WiMAX systems sharing a frequency band. Because we know that the PS-Request protocol has drawbacks, we propose a revised PS-Request protocol to improve the performance. Two PS-Request protocols are based on the time division operation (TDO) of WiFi system and WiMAX system to avoid the mutual interference, and use the vestigial power management (PwrMgt) bit within the Frame Control field of the frames transmitted by a WiFi AP. The performance of the revised PS-Request protocol is evaluated by computer simulation, and compared to those of the cases without a coexistence protocol and to the original PS-Request protocol.

Keywords: coexistence; mutual interference; power saving mode; PS-Request; duration field

1. INTRODUCTION

Many wireless communication systems have been developed and deployed to support various kinds of wireless data services. Many of the developed frequency bands are becoming more congested, especially in the densely populated urban areas due to the scarcity of commercially usable frequency resources and the explosive increasing desire for wireless data service. Consequently, it is essential that several heterogeneous wireless communication systems share a frequency band if possible, such as the 2.4 GHz ISM (Industrial, Scientific, and Medical) band. The spectrum sharing of several heterogeneous systems is termed coexistence. Coexistence can improve the utilization of the frequency band, because all the participating systems are not fully using their frequency band. However, the coexistence of heterogeneous wireless communication systems sharing a spectrum simultaneously and in the same geographic region causes their mutual interference, the coexistence problem. This degrades the performance of each wireless communication system. Sharing in the unlicensed bands, including 2.4 GHz ISM band, has been studied for a long time. Well known wireless communication systems in the ISM band are WiFi for wireless local area network (WLAN) systems, Bluetooth or ZigBee for wireless personal area networks (WPANs), and the like.

There has been growing demand for broadband connections, along with the proliferation of wireless data services. WiMAX (Worldwide Interoperability for Microwave Access) for wireless metropolitan area network (WMAN) systems is an attractive technology to address this need. WiMAX is being considered as a candidate system operated in the ISM band, because it can provide low price or free high speed wireless internet service. If the WiMAX system is newly deployed in the ISM band, like a kind of home base station, it inevitably generates interference with the other wireless communication systems that exist in the ISM band. WiFi is a popular wireless communication system operated in that band, and deployed in many indoor or outdoor places. Therefore, the coexistence of WiFi and WiMAX systems sharing a frequency band should be studied, when they operate in an adjacent or the same area, as shown in Figure 1.

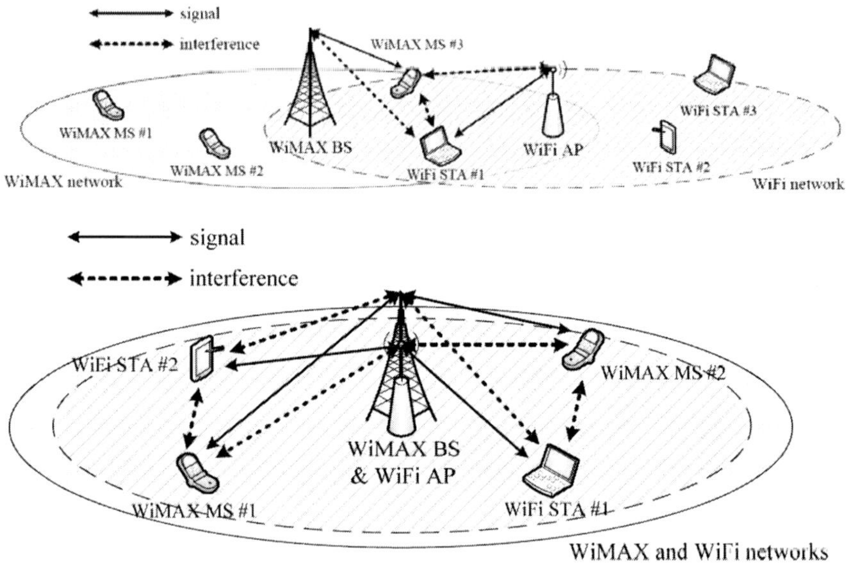

(b)

Figure 1. Coexistence of WiFi and WiMAX systems. (**a**) WiFi AP and WiMAX BS are not collocated; (**b**) WiFi AP and WiMAX BS are collocated.

Table 1 defines the four types of coexistence of heterogeneous wireless communication systems. In this paper, we focus on the case in which WiFi system based on IEEE 802.11g and WiMAX system based on IEEE 802.16e are collaborative. We want to make them coexist, based on their minimum collaboration and the minimum change of their specifications.

Table 1. Types of system for coexistence.

		Collaboration	
		Not collaborative	**Collaborative**
Collocation	**Not collocated**	Out of scope	Figure 1(a)
	Collocated		Figure 1(b)

In Section 2, we explain related work on coexistence. In Section 3, we introduce the time division operation of WiFi and WiMAX systems to avoid their mutual interference. In Sections 4 and 5, we introduce the previously proposed PS-Request protocol of the WiFi system and coexistence zone within the WiMAX frame structure. In Section 6, we propose a revised PS-Request protocol to resolve drawbacks of the original PS-Request protocol. In Section 7, we evaluate by simulation the performance of the revised PS-Request protocol. Finally, we will conclude in Section 8.

2. RELATED WORK

Many studies have been conducted on the coexistence of heterogeneous wireless communication systems. The coexistence of WiFi and Bluetooth systems was dealt in [1–3], that of WiFi and ZigBee systems was discussed in [4,5]. Chiasserini [1] proposed two solutions for the coexistence of WiFi and Bluetooth systems. One involves WiFi STAs estimating the interference pattern, and transmitting a frame of the minimum payload size, 500 bytes, comparable to the duration of a single slot Bluetooth packet. This may reduce the probability of collisions between the two systems. However, it is hard to estimate the interference pattern. The second deals with frequency hopping, a suggested packet length and restricts the length of packets. Thus, the allowable data rate decreased. Golmie [2] introduced a backoff strategy and adaptive frequency hopping (AFH). This also reduced the collision probability of the two systems. The backoff strategy does not require changes to the Bluetooth specification. However, changing the frequency hopping pattern requires some changes to the Bluetooth specification. The benefits of AFH may not be as obvious for delay jitter and packet loss constrained applications, such as voice or video services. Hsu [3] studied a dynamic WiFi system fragmentation technique. A WiFi station (STA) divides one frame into several fragments, and sends them. The WiFi STA should be acknowledged fragment by fragment. If the transmission of a specific fragment fails, the WiFi STA only retransmits the failed fragment. This decreases the allowable data rate, because it reduces the length of the collided fragment. Sikora [4] evaluated the effects of the mutual interference, and quantity coexistence issues. Wei [5] introduced several coexistence models between ZigBee and WiFi systems, and evaluated their interference effects.

References [6–11] deal with the coexistence of WiFi and WiMAX systems. Dongeun Kim [6] proposed a solution based on a circulator to reduce the mutual interference between them. It showed a width of the overlapped frequency band to tolerate the mutual interference. Even though both systems can endure a certain amount of mutual interference, it does not avoid but rather mitigates the mutual interference. Berlemann [7,8] and Siddique [9] proposed solutions based on time division multiplexing (TDM) of WiMAX systems based on IEEE802.16 and WiFi systems based on IEEE802.11a/e. They improved spectral efficiency. However, no protocol controls the transmission of WiFi STAs. Even if a certain period is allocated to the WiFi system, a WiFi STA may send frames beyond the period. This can cause the mutual interference. Jing [10] studied the coexistence between WiMAX systems based on IEEE802.16a and WiFi systems based on IEEE802.11b. A WiFi access pointer (AP) senses WiFi channels and selects a channel with the minimum received signal strength indicator (RSSI). The WiFi AP broadcasts the selected WiFi channel, and all of WiFi STAs move to it. However, if an interferer is not located within the range of a WiFi AP, the WiFi AP cannot recognize it. Jing [11] also introduced a cognitive radio (CR) method. Each node announces information, such as frequency, power, modulation, duration, interference margin, and service type, through the common spectrum coordination channel. The method can reduce the influence of the mutual interference, but it needs much overhead. Two mechanisms termed Listen-Before-Talk (LBT) and Extended Quiet Period (EQP) are adopted to perform white space detection efficiently based on IEEE 802.16h [12]. A WiMAX BS should listen to a channel all the time to know whether the channel is occupied, and adopt LBT to determine if an EQP is needed after the end of a quiet period. If the channel is unoccupied, the WiMAX BS and WiMAX MSs transmit frames during an active period. The WiMAX BS and WiMAX MSs do not transmit during a quiet period. This information about the quiet period is included in the DL-MAP. It is extended if the channel is still occupied after the quiet period. Thus, it is termed Extended Quiet Period (EQP) in which there is no transmission. John [13] evaluated the performance of LBT and EQP protocols.

3. TIME DIVISION OPERATION OF WIFI AND WIMAX SYSTEMS

We exploit the time division operation (TDO) to avoid or reduce the mutual interference between WiFi and WiMAX systems, as shown in Figure 2. The mutual interference can be avoided, since the TDO means that only one of two systems transmits, and the other system pauses its transmission [14,15]. Master devices of the systems, such as a WiFi AP and a WiMAX base station (BS), should control the signal transmission of its terminals, such as

WiFi STAs and WiMAX mobile stations (MSs), respectively, to achieve the TDO, as shown in Figure 2. A little information exchange between WiFi and WiMAX systems is needed for timing synchronization of the transmission periods of two systems.

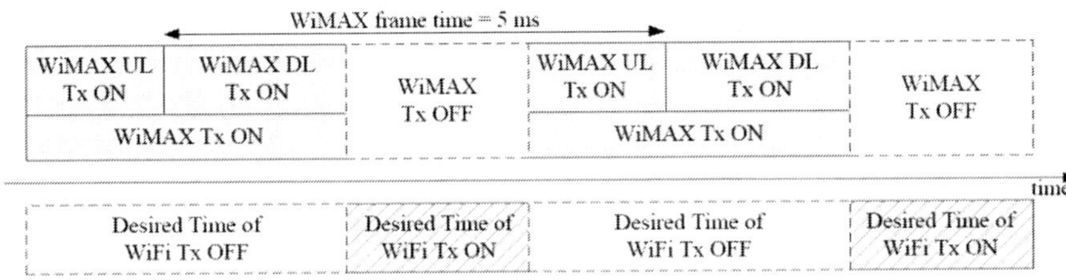

Figure 2. Desired TDO of WiFi and WiMAX systems.

In this paper, we only focus on the coexistence of collaborative WiFi and WiMAX systems. Thus, we assume the minimum collaboration between a WiMAX BS and a WiFi AP for timing synchronization of the transmission period of each system is acquired. Information exchange among the WiMAX BS and WiFi STAs, among WiMAX MSs and the WiFi AP, and among WiMAX MSs and WiFi STAs, is unnecessary. However, the WiMAX BS and the WiFi AP should independently provide information about the transmission period to their terminals. If the WiMAX BS and the WiFi AP are in one device, as shown in Figure 1(b), the collaboration can be easily implemented.

A WiMAX BS based on IEEE802.16e specification can control the signal transmission of all of its WiMAX MSs through the DL-MAP (Downlink map) and UL-MAP (Uplink map). However, a WiFi AP based on IEEE802.11g cannot control the signal transmission of all its WiFi STAs. Thus, the desired structure of the TDO, as shown in Figure 2, cannot be achieved at the WiFi STAs, even though the WiFi AP can be operated based on the TDO structure due to the minimum information exchanged with the WiMAX BS. We propose a PS-Request protocol, the so called original PS-Request protocol, which makes the WiFi AP control the signal transmission of all of its WiFi STAs, to solve the problem.

4. ORIGINAL PS-REQUEST PROTOCOL

Figure 3 shows the format of a general medium access control (MAC) frame of the WiFi system based on the IEEE802.11g specification. It comprises a set of fields that occur in a fixed order in all frames, as shown in Figure 3. The conventional WiFi system has three frame types: control, data, and management. Each of the frame types has several defined subtypes. The control frame consists of sixteen subtypes. Eight subtypes are being used, and eight subtypes are reserved. Detailed explanations of the types and subtypes are given in the IEEE802.11g specification [16].

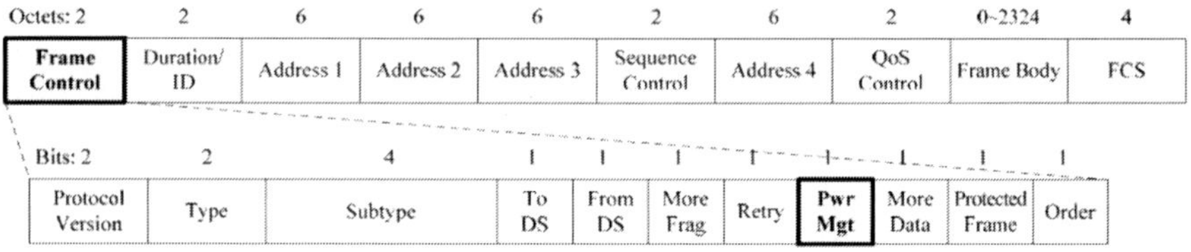

Figure 3. Format of general MAC frame of WiFi system.

The original PS-Request protocol is based on the use of the vestigial power management (PwrMgt) bit within the frames transmitted by a WiFi AP. In the case of an infrastructure configuration, the conventional WiFi system based on IEEE 802.11g specification can only support the power management mode for WiFi STAs. A WiFi STA can be in one of the two power management modes: an active mode (AM) or a power-saving mode (PSM). A WiFi STA in the AM is fully activated. It can send or receive frames at any time. A WiFi STA in the PSM can be in one of two states: a sleep state or an awake state. The WiFi STA in the PSM is usually in the sleep state, and periodically moves to the awake state to receive frames from a WiFi AP, such as beacon frames. The WiFi STA consumes much smaller energy than it does in the AM, because the WiFi STA in the PSM does not send any frames.

In the legacy WiFi system of an infrastructure configuration, a WiFi STA that wishes to move into the PSM shall inform its WiFi AP of an indication by setting the PwrMgt bit within the Frame Control field of the frame including the indication to '1'. The WiFi STA should be acknowledged by its WiFi AP, and move into the PSM. However, a WiFi AP cannot be in the PSM, because none of its WiFi STA can communicate for the PSM duration of the WiFi AP. Therefore, the PwrMgt bit within the Frame Control field of all frames transmitted by the WiFi AP is always set to '0'. This indicates that the WiFi AP is always in the AM.

If all WiFi STAs within the coverage of a WiFi AP are in the PSM, they do not transmit any signals. Thus, they do not generate interference to other systems during the PSM period. That is, if a WiFi system, including a WiFi AP and all of its WiFi STAs, do not transmit a signal during a specified period, the other wireless communication systems, such as a WiMAX system, can communicate without interference during the period. However, this can be achieved, only if the WiFi AP and all of its WiFi STAs move into the PSM. As mentioned before, the PSM of the conventional WiFi system can be controlled only by the WiFi STAs. That is, the WiFi AP cannot control the signal transmission of its WiFi STAs.

We previously proposed to use the vestigial PwrMgt bit within the Frame Control field of frames transmitted by a WiFi AP for the WiFi AP to control the signal transmission of all of its WiFi STAs. We named it the PS-Request protocol. The PS-Request frame transmitted by the WiFi AP commands all its WiFi STAs to move into the PSM to stop them transmitting a signal. Figure 4 shows a format of the proposed PS-Request frame. One of the eight reserved subtypes, such as 0000, can be allocated, and the PwrMgt bit is set to '1' in the Frame Control field. The Duration/ID field represents the value of a network allocation vector (NAV). This means that the WiFi AP will not transmit a signal during the PSM period of the NAV, and the WiFi AP stops its WiFi STAs from transmitting a signal during the period. Figure 5 shows an example of the original PS-Request protocol. There are three cases after the start time of WiFi Tx ON period equal to the end time of WiMAX Tx ON period, as shown in Figure 5.

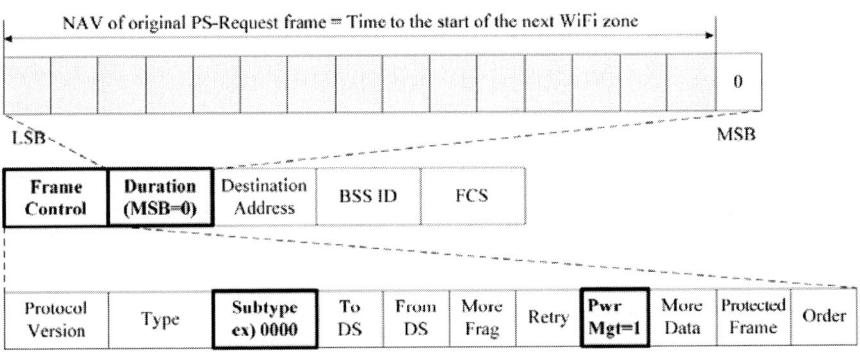

Figure 4. Format of original PS-Request frame.

Case (a):

A WiFi AP collaborating with a WiMAX BS senses a wireless medium during a distributed inter-frame space (DIFS) time. If the medium is idle, the WiFi AP begins the backoff procedure. When its backoff timer expires, the WiFi AP sends the proposed PS-Request frame. From then on, the PSM of all of WiFi STAs starts.

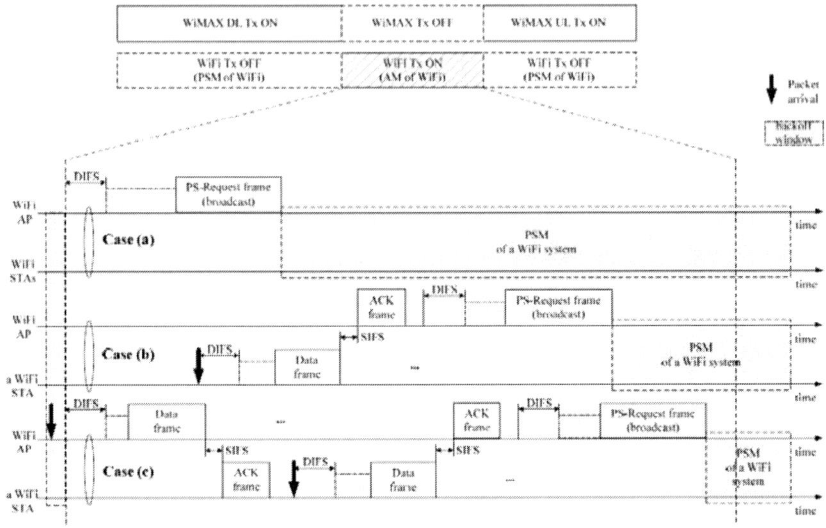

Figure 5. Example of the original PS-Request protocol.

Case (b):

Before the transmission of the original PS-Request frame, a WiFi STA receiving a packet sends its data frame after the CSMA (Carrier Sense Multiple Access)/CA (Collision Avoidance) procedure of the WiFi system, and receives an ACK frame from its WiFi AP. Then, the WiFi AP sends the proposed PS-Request frame, as in Case (a). Then, the PSM of all WiFi STAs starts.

Case (c):

Before the transmission of the original PS-Request frame, a WiFi AP receiving a packet sends its data frame after the CSMA/CA procedure of the WiFi system, and receives an ACK frame from a corresponding WiFi STA. Then, a WiFi STA receiving a packet sends its data frame after the CSMA/CA procedure of the WiFi system, and receives an ACK frame from its WiFi AP, as in Case (b). Then, the WiFi AP sends the proposed PS-Request frame, as in Case (a). Then, the PSM of all WiFi STAs starts. All WiFi STAs are in the PSM without sending an ACK frame to its WiFi AP, because the PS-Request frame is broadcast. The value of the Destination Address field within the frames transmitted by the WiFi AP determines if it is unicast or broadcast.

5. PROPOSED COEXISTENCE ZONE WITHIN WIMAX FRAME STRUCTURE

There are already several zones in the WiMAX system, as shown in Figure 6(a). The IEEE802.16e specification provides detailed explanations of these zones [17]. They may exist in the WiMAX frame structure to support some particular functions for a WiMAX system. The DL PUSC (Partial Usage of Sub-Channel) zone for FCH and DL-MAP is mandatory, and the UL PUSC is the default. We proposed a new zone, termed the coexistence zone, as shown in Figure 6(b).

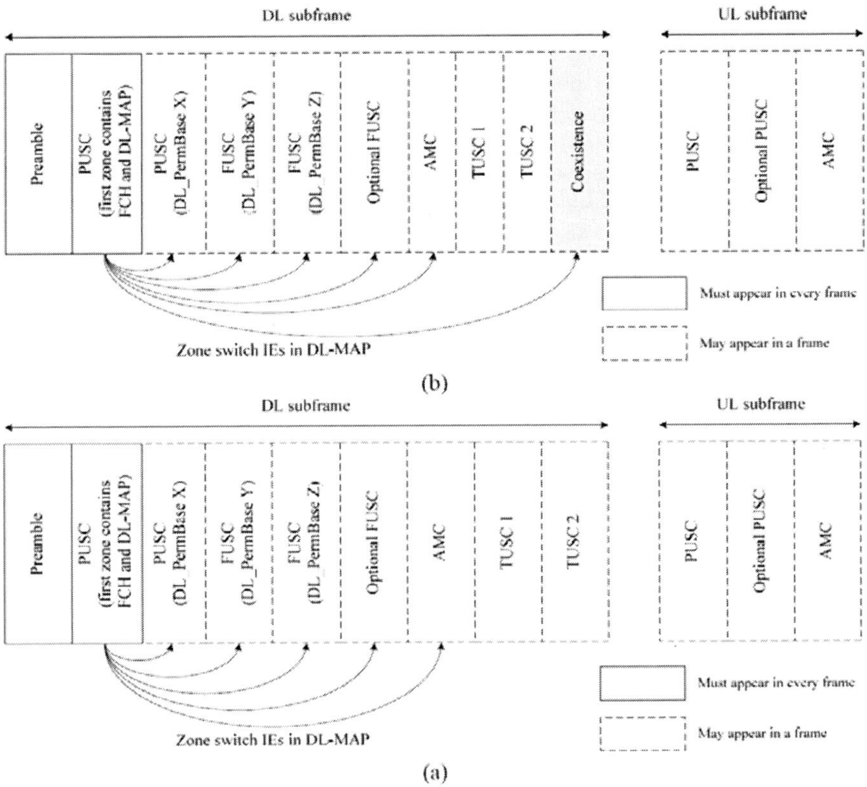

Figure 6. WiMAX zones. (**a**) Conventional WiMAX zones; (**b**) WiMAX zones, including the proposed coexistence zone.

Figure 7 shows the coexistence zone within the WiMAX frame structure. During the coexistence zone, the WiMAX system does not transmit any signal, because no OFDM symbols and subcarriers should be used for communication. We temporarily call the coexistence zone the WiFi zone, for the coexistence of WiFi and WiMAX systems, because the coexistence zone is used for WiFi systems in this paper. The PSM and the AM of WiFi STAs are alternated, based on the original PS-Request frame transmitted by a WiFi AP, as shown in Figure 7. The WiMAX system pauses its signal transmission during the WiFi zone. Thus, either the WiFi AP or its WiFi STAs can send frames without interference from the WiMAX system according to the CSMA/CA procedure of the WiFi system. The WiFi AP sends the original PS-Request frame to all its WiFi STAs to protect the WiMAX UL and DL transmission ON time from signals from WiFi STAs.

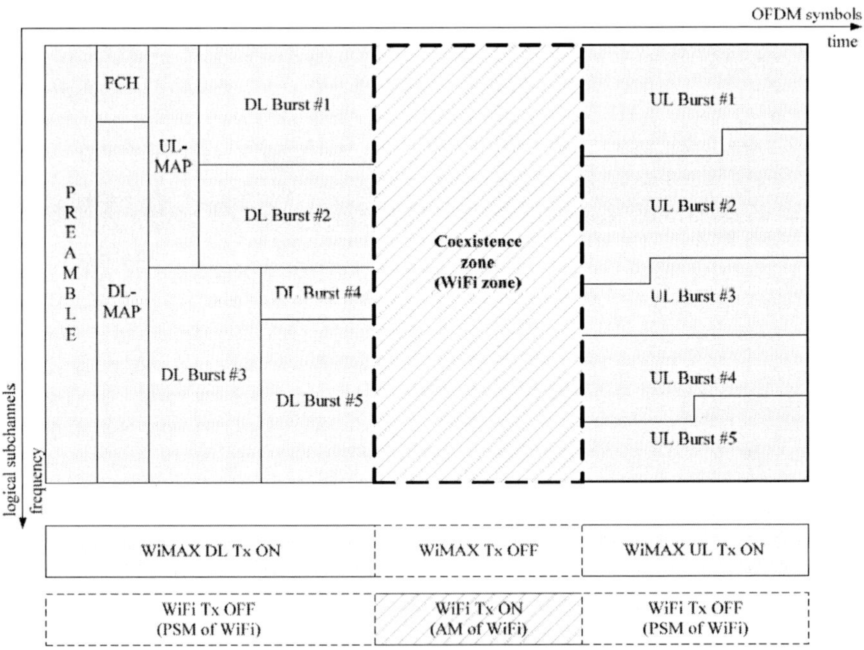

Figure 7. Proposed coexistence zone within WiMAX frame structure.

6. REVISED PS-REQUEST PROTOCOL

We proposed the PS-Request protocol, the so called original PS-Request protocol, to avoid the mutual interference between WiFi and WiMAX systems. However, we found that the original PS-Request protocol cannot completely solve the coexistence problem of WiFi and WiMAX systems. In the original PS-Request protocol, a WiFi AP only informs its WiFi STAs of the time to the start of the next WiFi zone, which is given in the Duration/ID field of the PS-Request frame, as shown in Figure 4. From this time, the WiFi AP and its WiFi STAs may compete to acquire the WiFi channel based on the CSMA/CA procedure of the WiFi system.

As shown in Figure 5, the WiFi STAs receiving the original PS-Request frame from its WiFi AP cannot know the time remaining in the current WiFi zone. They only know the time to the start of the next WiFi zone. Thus, the remaining time of the current WiFi zone can be wasted. A WiFi STA that acquires a wireless medium after competition within the next WiFi zone, may send a frame beyond the next WiFi zone, as shown in Figure 8, because it does not know the time to the end of the next WiFi zone. In that case, the mutual interference may be inevitable. This degrades the performance of both WiMAX and WiFi systems.

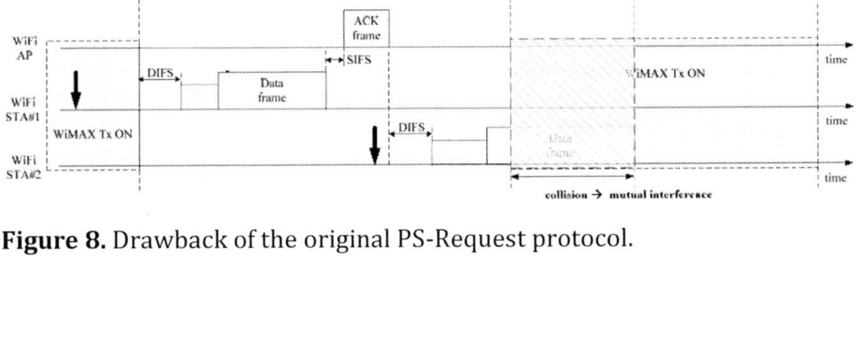

Figure 8. Drawback of the original PS-Request protocol.

If WiFi STAs can know the remaining time of the current WiFi zone after receiving a PS-Request frame, they can decide to send frames either at a time or at several times by segmentation. By doing so, the mutual interference, as shown in Figure 8, can be avoided. Thus, we propose a revised PS-Request protocol that delivers both the time to the end of the current WiFi zone and the PSM duration to the WiFi STAs, as shown in Figure 9. The Duration/ID field in the revised PS-Request frame represents two values of NAV. One is the time to the end of the current WiFi zone, and the other is the PSM duration. The sum is equal to the time to the start of the next WiFi zone of the original PS-Request frame. In this paper, we assume that seven bits of fifteen bits represent the time to the end of the current WiFi zone and the other eight bits represent the PSM duration. The number of bits for each part is determined according to the desired resolution and maximum value of the time.

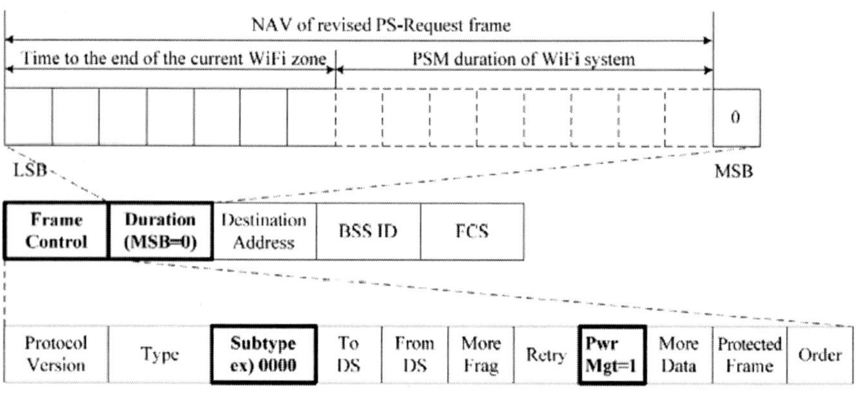

Figure 9. Format of revised PS-Request frame.

Figure 10 shows an example of the revised PS-Request protocol. A WiFi AP broadcasts a revised PS-Request frame through the transmission procedure of the WiFi system. WiFi STAs receiving the frame can send their frames until the time to the end of the current WiFi zone after a competition. After that time, all WiFi STAs are in the PSM, until the start of the next WiFi zone.

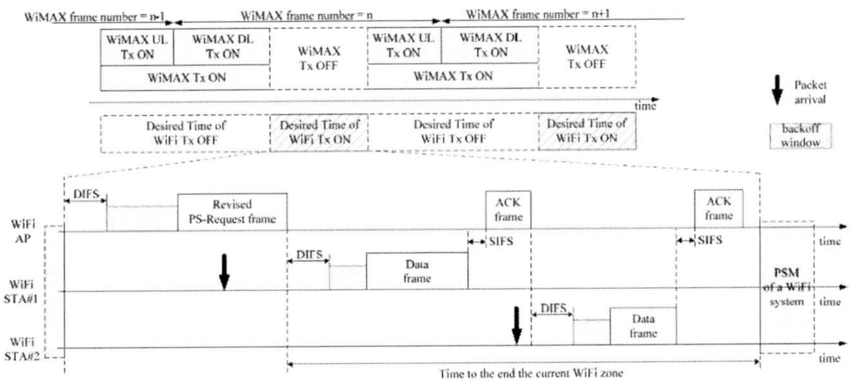

Figure 10. Example of the revised PS-Request protocol.

7. SIMULATION RESULTS

7.1. System Assumption

As mentioned before, the proposed two PS-Request protocols operate based on an information exchange only between a WiFi AP and a WiMAX BS. They need the minimum collaboration between the WiFi AP and the WiMAX BS, such as the timing synchronization, to know the start and end times of the coexistence zone within the WiMAX frame structure. However, they do not need an information exchange among WiFi AP and WiMAX MSs, among WiFi STAs and WiMAX BS, and among WiFi STAs and WiMAX MSs.

7.2. Performance Measures

Throughput of each system is considered as one of the performance measures to be evaluated by the simulation [18].

N_{WiFi}^{bit} and N_{WiMAX}^{bit} are the total number of successfully transmitted data bits of WiFi and WiMAX systems, respectively. $N_{WiMAX,DL}^{bit}$, and $N_{WiMAX,UL}^{bit}$ are the total number of successfully transmitted WiMAX data bits through DL (Downlink) and UL (Uplink), respectively:

$$N_{WiFi}^{bit} = \sum_{i=1}^{N_{STA}} N_{WiFi}^{ofdm} \times 48 \times M_{WiFi}^{i} \times C_{WiFi}^{i} \quad (1)$$

where i is the index of a WiFi STA and N_{STA} is the number of WiFi STAs. N_{WiFi}^{ofdm} is the number of transmitted OFDM symbols. Forty eight is the number of data subcarriers of 64 subcarriers for the WiFi system based on IEEE802.11g. M_{WiFi}^{i} and C_{WiFi}^{i} denote the modulation order and the code rate of WiFi STA of index i, respectively:

$$N_{WiMAX,DL}^{bit} = \sum_{j=1}^{N_{MS}} N_{WiMAX}^{slot,j} \times 48 \times M_{WiMAX}^{j} \times C_{WiMAX}^{j} \quad (2)$$

$$N_{WiMAX,UL}^{bit} = \sum_{j=1}^{N_{MS}} N_{WiMAX}^{tile,j} \times 12 \times M_{WiMAX}^{j} \times C_{WiMAX}^{j} \quad (3)$$

where j is the index of WiMAX MS, and N_{MS} is the number of WiMAX MS's. M_{WiMAX}^{j} and C_{WiMAX}^{j} denote the modulation order and the code rate for each data burst of WiMAX MS of index j, respectively. $N_{WiMAX}^{slot,j}$ and $N_{WiMAX}^{tile,j}$ denote the number of allocable resource units to the MS of index j for DL and UL, respectively. Forth eight and 12 denote the number of subcarriers in one slot and in one tile, respectively.

N_{WiMAX}^{frame} is the number of WiMAX frames transmitted in a simulation. T_{WiMAX}^{frame} is the frame time of the WiMAX system, and is typically 5 ms:

$$T_{WiMAX}^{frame} = \left(N_{DL}^{ofdm} + N_{CZ}^{ofdm} + N_{UL}^{ofdm} \right) \times T_{symbol}^{ofdm} + T_{TTG} + T_{RTG} \quad (4)$$

NofdmDL, *NofdmCZ* and *NofdmUL* denote the number of WiMAX OFDM symbols allocated to the DL subframe, the coexistence zone and the UL subframe within a WiMAX frame structure, respectively. *NofdmDL* includes one DL OFDM symbol for preamble, two DL OFDM symbols for FCH, DL-MAP and UL-MAP, and the other DL OFDM symbols for DL-Bursts of the WiMAX system. DL-Bursts of the WiMAX system are composed of DL OFDM symbols in multiples of two, and UL-Bursts of the WiMAX system are composed of UL OFDM symbols in multiples of three. *Tofdmsymbol* is one

OFDM symbol duration of the WiMAX system. T_{TTG} and T_{RTG} denote the transmit/receive transition gap (TTG) and the receive/transmit transition gap (RTG) within a WiMAX frame structure, respectively.

For the performance evaluation of each system, the average throughputs of WiFi and WiMAX systems are defined, as follows:

$$R_{WiFi} \triangleq \frac{N_{WiFi}^{bit}}{T_{WiFi}^{TxON} \times N_{WiMAX}^{frame}} = \frac{N_{WiFi}^{bit}}{N_{CZ}^{ofdm} \times T_{symbol}^{ofdm} \times N_{WiMAX}^{frame}} \quad (5)$$

$$R_{WiMAX} \triangleq \frac{N_{WiMAX}^{bit}}{T_{WiMAX}^{TxON} \times N_{WiMAX}^{frame}} = \frac{N_{WiMAX,DL}^{bit} + N_{WiMAX,UL}^{bit}}{(N_{DL}^{ofdm} + N_{UL}^{ofdm}) \times T_{symbol}^{ofdm} \times N_{WiMAX}^{frame}} \quad (6)$$

7.3. Simulation Environment

In this paper, WiMAX and WiFi systems are based on IEEE802.16e and IEEE802.11g specifications, respectively. For the simplicity of simulation and the performance evaluation of an extreme case, we assume that the WiMAX system consists of one WiMAX BS and only one WiMAX MS, and all of the radio resources are allocated to the WiMAX MS. The WiFi system also comprises one WiFi AP and one WiFi STA. Table 2 specifies the simulation parameters of the WiMAX system. Table 3 gives simulation conditions. The transmission period of each system is allocated based on the number of OFDM symbols of the WiMAX system.

Table 2. WiMAX system parameters.

	Conventional protocol	PS-Request protocols
N_{CZ}^{ofdm}	0	16
N_{DL}^{ofdm}	27(=1 + 2 + 2 × 12)	17(=1 + 2 + 2 × 7)
N_{UL}^{ofdm}	15(=3 × 5)	9(=3 × 3)
T_{TTG}	87.2 μs	87.2 μs
T_{RTG}	74.4 μs	74.4 μs
$N_{WiMAX}^{slot,1}$	360	180
$N_{WiMAX}^{tile,1}$	1050	840

Table 3. Simulation conditions.

	WiFi System (IEEE802.11g)	WiMAX System (IEEE802.16e)
Bandwidth (MHz)	20	8.75
Sampling frequency (MHz)	20	10
Oversampling	1	2
Sample rate (Msps)	20	20
FFT size	64	1,024
Subcarrier allocation	-	PUSC
Modulation	QPSK	QPSK
Pulse shaping filter	Square root raised cosine filter (Roll-off factor = 1.0)	Square root raised cosine filter (Roll-off factor = 1.0)
Channel coding	Convolutional code (R = 1/2, K = 7)	Convolutional code (R = 1/2, K = 7)
Matched filter	Square root raised cosine filter (Roll-off factor = 1.0)	Square root raised cosine filter (Roll-off factor = 1.0)
Channel	AWGN	AWGN
Allocated time interval, not including TTG and RTG	1.8432 ms	2.9952 ms
Allocated time interval, including TTG and RTG	1.8432 ms	3.1568 ms
SIFS	16 μs	-
DIFS	34 μs	-
Frame time	Variable	5 ms

For the case of $N_{DL}^{ofdm} + N_{CZ}^{ofdm} + N_{UL}^{ofdm} = 42$ and $T_{symbol}^{ofdm} = 115.2\ \mu s$, $T_{WiFi}^{TxON} = 1.8432\ ms = 16 \times 115.2\ \mu s$ and $T_{WiMAX}^{TxON} = 2.9952\ ms = 26 \times 115.2\ \mu s$.

7.4. Simulation Results

Figures 11 and 12 show the simulation results for the following three cases: without the proposed protocols, with the original PS-Request protocol, and with the revised PS-Request protocol. The Signal-to-Interference Ratio (SIR), denoted as $P_{interferee}/P_{interferer}$. $P_{interfree}$ is the average received signal power of the interferee system, and $P_{interferee}$ is the average received interference power from the interferer system. The case without the proposed protocols is evaluated for the following six conditions: without interference ($P_{interferee}$ = 0 [W], which is equivalent to $P_{interferee}/P_{interferee}$ = ∞ [dB]), $P_{interferee}/P_{interferee}$ = 12 [dB], $P_{interferee}/P_{interferee}$ = 9 [dB], $P_{interferee}/P_{interferee}$ = 6 [dB], $P_{interferee}/P_{interferee}$ = 3 [dB], $_{interferee}/P_{interferee}$ = 0 [dB]. If a WiFi AP can transmit a revised PS-Request frame within a WiFi zone, there will be no mutual interference in the following WiMAX Tx ON period equal to the PSM duration of the WiFi system. That is, $P_{interferee}/P_{interferee} = \infty$ [dB].

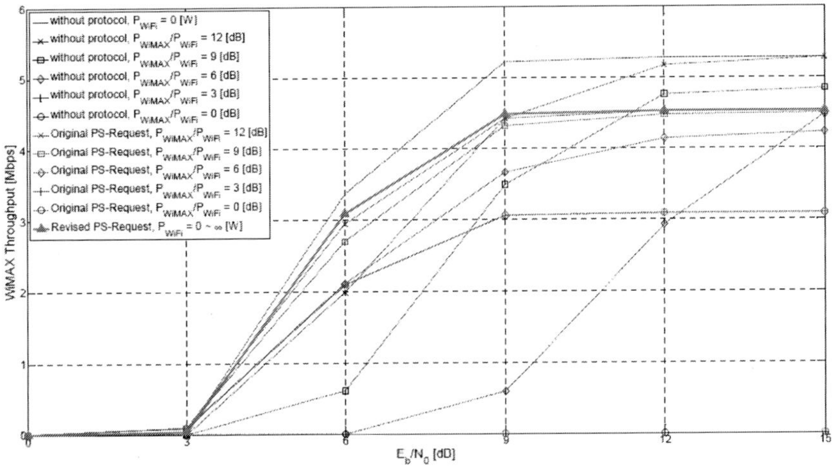

Figure 11. WiMAX throughput vs. received E_b/N_0.

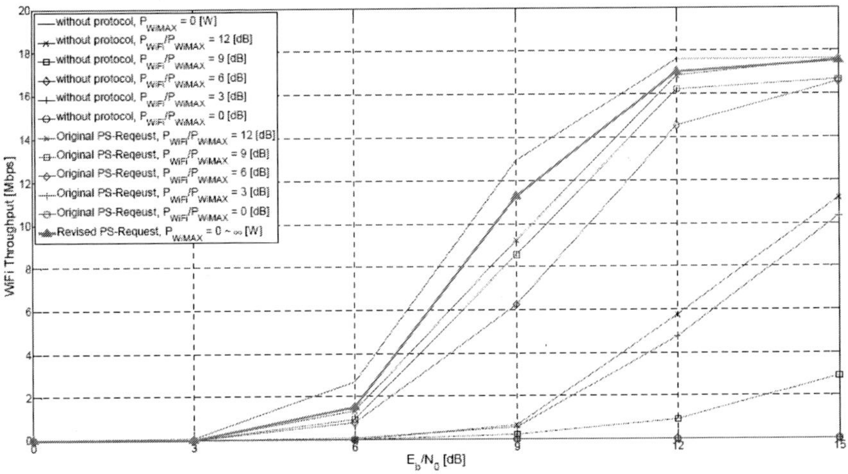

Figure 12. WiFi throughput vs. received E_b/N_0.

Figure 11 shows the throughput of a WiMAX system vs. the received E_b/N_0 under interference from a WiFi system sharing the same frequency band. Cases with the proposed PS-Request protocols perform better than the case without the proposed protocols. The performance of the revised PS-Request protocol does not depend on the interference power received from the WiFi system but the received E_b/N_0 of the WiMAX system, differing from the original PS-Request protocol. The revised PS-Request protocol has about 50% more throughput than the original PS-Request protocol for P_{WiMAX}/P_{WiFi} = 0 or 3 [dB] and E_b/N_0 = 12 [dB] at which the throughput for the case without the proposed protocols is zero. Protecting the transmission of WiFi STAs beyond the end of the WiFi zone, as shown in Figure 10, improves performance. For smaller or zero interference environments corresponding to large P_{WiMAX}/P_{WiFi}, the proposed PS-Request protocols may have worse performance, such as smaller throughput, because the allowed transmission interval for the WiFi system cannot be used to transmit data by the WiMAX system. However, for larger interference environments, corresponding to small P_{WiMAX}/P_{WiFi}, the proposed PS-Request protocols can have better performance, such as relatively higher throughput. For the worst case of $P_{WiMAX} = P_{WiFi}$ corresponding to P_{WiMAX}/P_{WiFi} = 0 [dB] at which the WiMAX throughput, the WiFi throughput and the sum of two throughputs for the case without the proposed protocols are zero, the revised PS-Request protocol has about 50% more WiMAX throughput than the original PS-Request protocol does.

Figure 12 shows the throughput of a WiFi system vs. the received E_b/N_0 under the interference from a WiMAX system sharing the same frequency band. The two proposed PS-Request protocols also perform better than the case without the proposed protocols. The performance of the revised PS-Request protocol outperforms the original PS-Request protocol. This also results from avoiding the overlap of the transmission time of the WiFi system and WiMAX system. The performance of the revised PS-Request protocol depends on the received E_b/N_0 of the WiFi system, rather than on the interference power received from the WiMAX system. For smaller or zero interference environments, corresponding to large P_{WiFi}/P_{WiMAX}, the proposed PS-Request protocols have worse performance, such as smaller throughput, because the allowed transmission interval for the WiMAX system cannot be used to transmit data by the WiFi system. However, for larger interference environments, corresponding to small P_{WiFi}/P_{WiMAX}, the proposed PS-Request protocols can have better performance, such as relatively higher throughput. For the worst case of $P_{WiFi} = P_{WiMAX}$ corresponding to P_{WiFix}/P_{WiMAX} = 0 [dB] at which the WiMAX throughput, the WiFi throughput and the sum of two throughputs for the case without the proposed protocols are zero, the WiFi throughput of the revised PS-Request protocol is similar to the WiFi throughput at the interference-free condition for the case without the proposed protocols, even if that of the original PS-Request protocol is still zero.

8. CONCLUSIONS

In this paper, we introduced the TDO for the coexistence of WiFi and WiMAX systems sharing a frequency band. We proposed a coexistence zone within WiMAX frame structure, and two PS-Request protocols to achieve the TDO. The two proposed PS-Request protocols are based on the vestigial power management bit within the Frame Control field transmitted by a WiFi AP. We showed using computer simulation that they outperform the case without the proposed protocols. The revised PS-Request protocol can resolve a drawback of the original PS-Request protocol. We evaluated and compared the performance for three cases: without the proposed protocols, with the original PS-Request protocol, and with the revised PS-Request protocol.

For smaller or zero interference environments, the proposed PS-Request protocols have worse performance, such as smaller throughput, because the transmission interval allowed to the other system cannot be used to transmit data by one system. However, for larger interference environments, the revised PS-Request protocols have the best performance of the three cases.

As a topic for further study, we plan to upgrade the simulator used to obtain more realistic and accurate performance measures for more combinations of simulation conditions. We also have a plan to solve the coexistence issue of WiFi and 3GPP LTE (Long Term Evolution) systems.

ACKNOWLEDGMENTS

This work was partly supported by Mid-career Researcher Program through the National Research Foundation of Korea (NRF) grant funded by the Ministry of Education, Science and Technology (MEST) (No. 2011-0027572). And the present research was partly conducted by a Research Grant of Kwangwoon University in 2011.

REFERENCES

1. Chiasserini, CF; Rao, RR. Coexistence mechanisms for interference mitigation between IEEE 802.11 WLANs and bluetooth. Proceedings of INFOCOM 2002, New York, NY, USA, June 2002; pp. 590–598.
2. Golmie, N; Chevrollier, N; Rebala, O. Bluetooth and WLAN coexistence: Challenges and solutions. IEEE Wirel. Commun **2003**, 10, 22–29.
3. Hsu, AC-C; Wei, DSL; Kuo, C-CJ. Coexistence mechanism using dynamic fragmentation for interference mitigation between WiFi and Bluetooth. Proceedings of MILCOM 2006, Washington, DC, USA, October 2006; pp. 1–7.
4. Sikora, A; Groza, VF. Coexistence of IEEE 802.15.4 with other systems in the 2.4 GHz-ISM-Band. Proceedings of IMTC 2005, Ottawa, ON, Canada, May 2005; pp. 1786–1791.
5. Wei, Y; Xiangyu, W; Linnartz, J-PMG. A coexistence model of IEEE 802.15.4 and IEEE 802.11b/g. Proceedings of the 14th IEEE Symposium, Delft, The Netherland, November 2007; pp. 1–5.
6. Kim, D; Kim, J; Park, S; Rhee, SH; Kang, C; Han, K; Kang, H. Circulator-based collocated system for coexistence of WiFi and WiMAX systems in shared bands. Proceedings of the 1st International Conference on Ubiquitous and Future Networks, Hong Kong, June 2009; pp. 214–219.
7. Berlemann, L; Hoymann, C; Hiertz, GR; Mangold, S. Coexistence and interworking of IEEE 802.16 and IEEE 802.11(e). Proceedings of Vehicular Technology Conference, Melbourne, Australia, May 2006; pp. 27–31.
8. Berlemann, L; Hoymann, C; Hiertz, GR; Walke, B. Unlicensed operation of IEEE 802.16: Coexistence with 802.11(a) in shared frequency bands. Proceedings of PIMRC 2006, Helsinki, Finland, September 2006; pp. 1–5.
9. Siddique, MM; Wenning, BL; Muehleisen, M. Spectrum sharing between IEEE 802.16 and IEEE 802.11 based wireless networks. Proceedings of WoWMoM 2010, Montreal, QC, Canada, June 2010; pp. 1–6.
10. Jing, X; Raychaudhuri, D. Spectrum co-existence of IEEE 802.11b and IEEE 802.16a networks using the CSCC etiquette protocol.

Proceedings of DySPAN 2005, Baltimore, MD, USA, November 2005; pp. 243–250.

11. Jing, X; Mau, S-H; Raychaudhuri, D; Matyas, R. Reactive cognitive radio algorithms for co-existence between IEEE 802.11b and 802.16a networks. Proceedings of GLOBECOM'05, St Louis, MO, USA, November 2005; pp. 2465–2469.

12. IEEE Std 80216h™-2010. IEEE Standard for Local and Metropolitan Area Networks—Part 16: Air interface for Broadband Wireless Access Systems; IEEE Computer Society: New York, NY, USA, 2010.

13. John, S; Amir, G. C80216h-07/017, Simulation of IEEE 802.16h and IEEE 802.11y Coexistence (Preliminary Report), Available online: http://ieee802.org/16/le/contrib/C80216h-07_017.pdf (accessed on 28 August 2011).

14. Kim, J; Park, S; Rhee, SH; Choi, Y-H; Hwang, H. Energy Efficient Coexistence of WiFi and WiMAX Systems Sharing Frequency Band; Springer: Berlin, Germany; December; 2010; Volume 6485, pp. 164–170.

15. Kim, J; Park, S; Choi, H-Y; Rhee, SH. Performance of WiFi system due to interference from WiMAX system in a shared frequency band. Proceedings of China-Korea Joint Conference on Information and Communications, Xi'an, China, August 2008; pp. 103–104.

16. IEEE Std 80211™-2007. IEEE Standard for Local and Metropolitan Area Networks-Specific Requirements, Part 11: Wireless LAN Medium Access Control (MAC) and Physical Layer (PHY) Specifications; IEEE Computer Society: New York, NY, USA, 2007.

17. IEEE Std 80216™-2004. IEEE Standard for Local and Metropolitan Area Networks-Part 16: Air Interface for Fixed Broadband Wireless Access Systems; IEEE Computer Society: New York, NY, USA, 2004.

18. Kim, J; Park, S; Rhee, SH; Choi, Y-H; Hwang, H; Chung, Y. Coexistence of WiFi and WiMAX systems based on coexistence zone within WiMAX frame structure and modified power saving mode of WiFi system. IEICE Trans Commun **2011**, E94-B, 1781–1784.

10
CHAPTER

A SMART CITY APPLICATION: A FULLY CONTROLLED STREET LIGHTING ISLE BASED ON RASPBERRY-PI CARD, A ZIGBEE SENSOR NETWORK AND WIMAX

Fabio Leccese *, Marco Cagnetti and Daniele Trinca

Dipartimento di Scienze, Università degli Studi "Roma Tre", Via della Vasca Navale 84, Rome 00146, Italy

ABSTRACT

A smart city application has been realized and tested. It is a fully remote controlled isle of lamp posts based on new technologies. It has been designed and organized in different hierarchical layers, which perform local activities to physically control the lamp posts and transmit information with another for remote control. Locally, each lamp post uses an electronic card for management and a ZigBee tlc network transmits data to a central control unit, which manages the whole isle. The central unit is realized with a Raspberry-Pi control card due to its good computing performance at very low price. Finally, a WiMAX connection was tested and used to remotely control the smart grid, thus overcoming the distance limitations of commercial Wi-Fi networks. The isle has been realized and tested for some months in the field.

Keywords: control systems for smart grids; lighting system; wireless networks for smart grids; ZigBee; sensors; Raspberry-Pi; ARM processor

1. INTRODUCTION

The Smart City (SC) paradigm helps renovate the traditional city concept. In fact, it is possible to realize and develop efficient demand-side strategies integrating the monitoring and automation features ensured by intelligent devices and their communication apparatuses typically used in many applications. Within this concept, public lighting, being a great electrical energy consumer, has recently been attracting the interest of the research community. Scientists, combining the SC paradigm with alternative energies and new lighting technologies, are conceiving systems previously unimaginable, which can increase the efficiency obtaining considerable energy consumption savings and consequently money savings [1–4].

This article shows how an isle of lamp posts, used to light a crossroad and placed in an area not reached by ADSL lines and 3G signals, can be much more efficiently managed by applying the smart city principles. The realized application changes the way of conceiving the isle: sensing the transit of a car approaching the crossroad and controlling it by an efficient management strategy, a typical power grid, like the isle of the lamp posts, is transformed in a smart city application.

The goal is obtained combining an efficient management strategy with new technologies. The strategy divides the complex structure of the isle into layers which are linked by communication layers. Each layer uses innovative technologies to make possible the application of the smart city principles. For example the lamp post exploits some innovative technologies both for the supply (photovoltaic panel) as well as for the control (ZigBee network and RaspBerry-Pi). Moreover, to provide a robust system also able to face and overcome the absence of Asymmetric Digital Subscriber Line (ADSL) or 3rd Generation (3G) communication, that in some parts of the city might not available, we used WiMAX technologies that allowed us to fully control the isle from a remote station.

The choice to test our system in this application is strongly joined to the road safety problem. In fact, the lighting of crossroads is fundamental to ensure the good night vision of typically hazardous points of the road where fatal crashes often happen. Based on this application, an economical analysis has also been realized to show how after the breakeven point is achieved, the system is more convenient with respect to older technology.

2. STATE OF THE ART

In this area, the efforts are focused on the use of alternative energies for the power supply of new lighting technologies [5–7], which allow obtaining considerable energy savings. Within lighting technologies, Light Emitting Diodes (LEDs) assure the possibility of switching on the lamp without the preheating typical of halogen ones; a very high lighting efficiency; low power consumption; a superior life time and quick switching times not comparable to those of older technologies (only incandescent lights have a lower lighting time, but with very big power consumption and the shortest time life); less sensitivity to transient phenomena, which have a big impact on other technologies, allowing thousands of lightings without the risk of lamp failure. These innovative characteristics allow development of a new remote-control system based on intelligent lamp posts that also send information to a central control system in order to simplify management and maintenance issues [1–4] also using holistic and bottom-up design strategies [1]. Within these researches, a relevant topic is the use of new sensor networks (wireless and not) and communication technologies, both to locally manage the lamp posts, but also sensor networks, and to send data towards a remote center [1–5,8–17].

Based on the best technologies and on the SG paradigm, our proposal describes a solution for an extremely diffused hostile scenarios, often neglected by the municipalities, but which have an enormous social relevance, and where both the mains and the internet signal provided by the most diffused communication technologies (ADSL and 3G) are not available.

3. THE LAYERS ARCHITECTURE

The architecture of the system has been conceived as a hierarchical layers structure for the full control of "unities" made up of sensors, actuators and local control cards. Each layer performs some control activities and transmits information to the other layers. The architecture foresees three layers called apparata, local control and remote control.

In the “apparata layer” we consider all the hardware and software designed for the autonomous and automatic control of the single apparatus or unit. This layer communicates with the higher one and vice versa using a ZigBee mesh wireless tlc network particularly suitable for closed devices and low bit rates [3]. The local control unit collects the data received from the single units to check their activity and functionality. It is realized with a RaspBerry-Pi control card, which collects and locally elaborates data from the single apparatus verifying that there are no problems. If a problem(s) occurs, the card sends an error message on the internet exploiting the upper layer. The latter is realized with a WiMAX trans-receiver, which is connected, via radio link, to the World Wide Web. Figure 1 shows the architecture of the system.

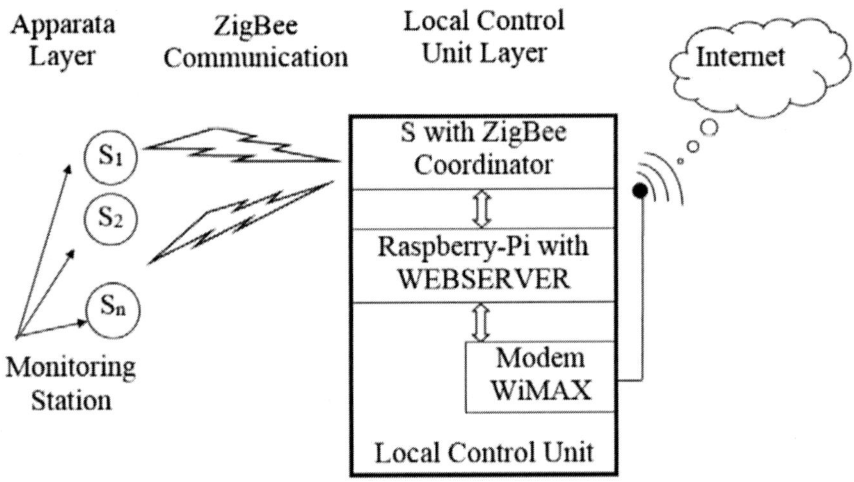

Figure 1. Block scheme of the system architecture.

4. TECHNOLOGIES AND DEVICES

Before analyzing the realized application, the three technologies and devices used in this work are briefly described.

4.1. ZigBee

ZigBee is a standard for wireless communication based on the IEEE 802.15.4 protocol. It is used to connect several devices together creating a mesh network with no central powerful device, but with distributed nodes able to transfer data throughout intermediate motes. Table 1 shows a comparison between the most famous technologies used as wireless communication. As the table shows, both Wi-Fi and the Bluetooth have higher bit rates than ZigBee, but the higher power consumption and the lower number of devices making up the network suggest that the ZigBee technology is the best candidate to create sensor networks for applications where the power consumption must be as low as possible. Furthermore, studies proved that the ZigBee is more affordable in terms of power consumption and costs than competitive technologies [3,18–20].

Table 1. ZigBee vs. some other wireless network.

	ZIGBEE	WI-FI	BLUETOOTH
IEEE standard	802.15.04	802.11b/g	802.15.01
Main application	Control	Broadband	Mobile devices
Number of network devices	Up to 65,000	32	7
Bit rate	20–250 kb/s	11/54 Mb/s	720 kb/s
Range	100 m	100 m	10 m
Battery life	100–1000 days	1–5 days	1–7 days

The communication between the lamp posts in the isle has been achieved using the ZigBee standard. This uses a mesh topology with a coordinator, routers (also called primary) and end device nodes (also called secondary); the first controls the formation and security of networks, the second extend the range of networks while the last perform specific sensing or control functions. Each lamp post has a Tx/Rx to get or send information and control commands. Furthermore, the lamp posts closer to the roads, used to approach the intersection, are also connected to a presence sensor card. The secondary nodes will send their status to the coordinator lamp post throughout the mesh network and receive back commands. Thus, the power consumption is limited to short time frames, since there is no continuous communication between nodes.

The networked communication is realized using the XBee radio modules by MaxStream which are already equipped with on-chip antenna [21–23]. These modules have been chosen because they have a good operational range for the application reaching tens of meters indoors and hundreds of meters outdoors. Thanks to its high radio sensitivity the XBee module has a low probability to receive corrupted packets (less than 1%). Furthermore the reported tests demonstrated that these performances are achieved without sacrificing power consumption. In fact, the current necessary to operate the two-way radio transmissio, is approximately 50 mA with a 3 V DC source. To ensure further power reduction, the sleep mode ensures a current draw of less than 10 μA.

4.2. RaspBerry-Pi

The role of the Coordinator Lamp Post is to integrate the computing part of the system. For this aim, the Raspberry-Pi card (or Pi following) has been adopted which ensures high computing power and interconnectivity with other devices such as the WiMAX modem [24–27].

Raspberry-Pi is a low-cost (≈25 €) basic computer contained on a credit-card size circuit board and featuring ports for HDMI, USB 2.0, composite video, analog audio, power, Internet and SD card. The computer runs entirely on open-source software and is able to run several applications such as: spreadsheets, word-processing, games, and high-definition video playback.

Other devices, as ARDUINO [28] or Beaglebone [29] could be used; Table 2 shows some principal characteristics of these devices. As usual, choosing which board you want depends on the type of intended project, and one's programming experience. If one wants to create a hardware project then the Arduino is by far the best choice. The analog inputs and PWM outputs add a whole spectrum of compatibility the Pi cannot provide natively. In addition, the large about of I/O pins let one connect multiple sensors and feedback components. The Arduino however is not as powerful as the Pi, so it has no proper audio, video or internet out-of-the-box (you can however add basic functionality for this). The Arduino can send data to a PC or Pi, over serial connections, and one can then create a program to read this data and do something useful.

If one wants to create a software project then the Pi and the Beaglebone are the way to go. The audio, video and internet capabilities make them the winners in this aspect. There is no need to attach external components, so there is no real need to learn electronics and they have the possibility to easily drive both 3G modems and WiMAX ones. Among them, the lower power consumption and lowest cost favor the RaspBerry-Pi, making it the best candidate for SC applications.

Table 2. Raspberry-Pi vs. Arduino and Beaglebone.

	RaspBerry	Arduino	Beaglebone
Base Price in US $	25	35	50
Power Draw	~150 mA @ 5V	~50 mA @ 5V	~250 mA @ 5V
Operating System	Linux	Custom	Linux
Suited for	Software	Hardware	Software
Number of I/O pins	8 Digital	14 Digital (6 PWM), 6 analog	65 Digital
Peripherals	2 USB Hosts, 1 Micro-USB Power, 1 10/100 Mbps Ethernet	None	1 USB Host, 1 Mini-USB Client, 1 10/100 Mbps Ethernet
Internet	Yes	*Via* Shield	Yes

The Raspberry is the hub of the system since it allows the visualization of the status of the entire lighting system. It is connected to the local monitoring station via the ZigBee network to manage the different lamp posts. Furthermore, it is able to receive information from the secondary lamp posts and from the presence sensor cards allowing monitoring the isle. Together, the Raspberry and the local monitoring station realize the Coordinator Lamp Post base control station, which receives information about the status of the lamp posts.

4.3. WIMAX

Worldwide Interoperability for Microwave Access (WiMAX) is the radio communication technology used to connect the Local Control Unit with the World Wide Web. It is based on the IEEE 802.16 standard, with the intent of deliver Internet connectivity to areas where a normal DSL service is still difficult and expensive to install or in parts of cities not reached by 3G signals [5,30].

The reason for choosing WiMAX is that in some parts of the city, e.g., very far from the downtown, sometimes, there is very scarce penetration of telecommunication lines or of the radio communication coverage. In the best cases with clean line of sight (LOS) the radio link can cover up to 70 km [30]. Other technologies could be used to realize the remote communication such as 3G, satellite communication, WiFi or ADSL, but, as Table 3 shows, none offer the specific abilities of WiMAX. Excluding ADSL that needs a cable whose placement could be extremely expensive in these areas, and WiFi whose coverage is extremely short (less than 100 m), satellite communications offer worldwide coverage, but its subscription is usually much more expensive than a WiMAX or a 3G. The latter, always because is not profitable for the provider to create a 3G network, is not always active in these areas. For these reasons, WiMAX is a good candidate for this application. This would also allow diffusion of the SC paradigm in areas not properly considered as "city".

Table 3. WiMAX vs. other communication technologies.

	WiMAX	ADSL	WiFi	3G	Satellite Communication
Wireless?	Yes	No	Yes	Yes	Yes
Distance Covered	Up to 70 km from the CPE	Everywhere with cable	Up to 100 m	Up to 5 km from the BS	Everywhere the signal satellite is on
Subscription Costs	About 20 €	About 20 €	About 20 €	About 20 €	About 100 €
Availability in rural area	Medium/High	No	No	Medium	Yes

In our system, we installed a base station (BS), spreading WiMAX single carrier TDD service at 3.4845 GHz center frequency and a bandwidth of 10 MHz. The BS transmitting antenna is an Argus tilt panel antenna model SPPX310M (65 deg Horizontal cut, 6.5 deg Vertical cut) [31].

The total transmission power is 55 dBm EIRP given by the output signal of 37 dBm and an antenna gain of 18 dBi. The antenna was placed on the roof of a building at 30 m from ground level. To communicate between the main lamp post and the BS, we used a Huawei branded HES-319M outdoor WiMAX Customer Premise Equipment (CPE) [32], equipped with a 3.5 GHz 45° Cross-Polarization built-in directional antenna, with a maximum power at antenna port of 26 dBm and 14 dBi antenna gain. This allowed us to send reliably data at high speed. Furthermore, it avoids any type of interference with the ZigBee standard due to its different operating frequency.

The WiMAX module uses 64QAM5/6 modulation for the downlink when the CINR is higher than 30 dB and 16QAM3/4 for the uplink [33,34]. Instead, when the CINR is lower than 10 dB, the link was still possible, but using a QPSK1/2 modulation because it is more robust at higher noise levels.

The CPE is Power over Ethernet (PoE) 802.3 a/f compliant and it runs at 48 VDC, so a DC-DC step-up boost has been required to make the CPE working in our system. We used a LTC3863 mounted on its demonstration circuit 1286 A in order to boost up the 12 V from solar panel to the 48 VDC needed at the CPE PoE connector.

5. THE APPLICATION

To widely test the idea, an isle of lamp posts, as shown in Figure 2, has been designed and realized.

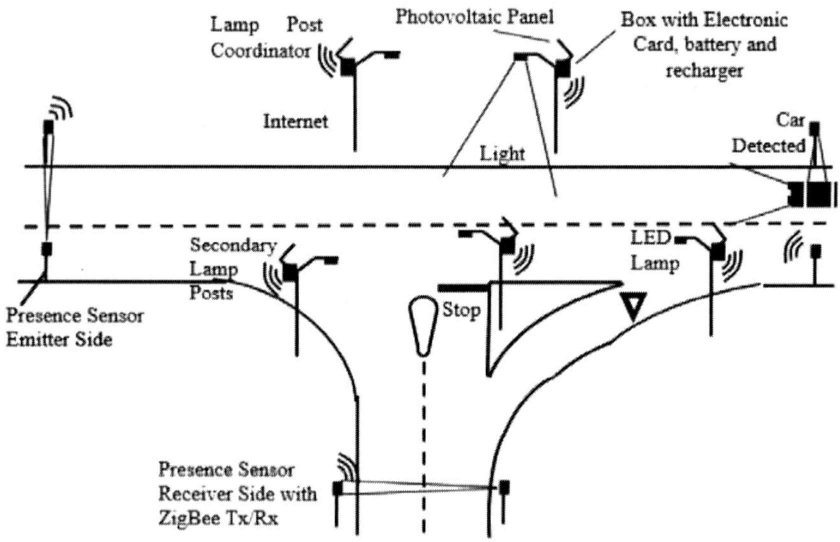

Figure 2. Schematic image of the on street system.

Through a flowchart, Figure 3 shows the management strategy of the isle.

The electronic card checks if the sunlight is lower than a fixed limit [3]; in this case, it waits until the sensors signal the presence of a car or a pedestrian. When this happens, the control unit switches on the light for a fixed time, the current sensor starts the measurements and, in case of a fault detection, an alarm is sent to the master lamp. If no fault is detected, the microcontroller stores the current values. During the night the Coordinator Lamp Post requests data from the secondary lamp posts and makes them available on the Internet. In the case of people in difficulty, which need light for their activities, they will be able to switch on the lights by pressing one of the emergency buttons located in specific locations at the crossroad. The electronic card, also called monitoring station, is used in two versions: the first detects the presence of a vehicle and is called Presence Sensor Electronic Card, while the second manages the lamp post and is called Monitoring Station for Lamp Posts.

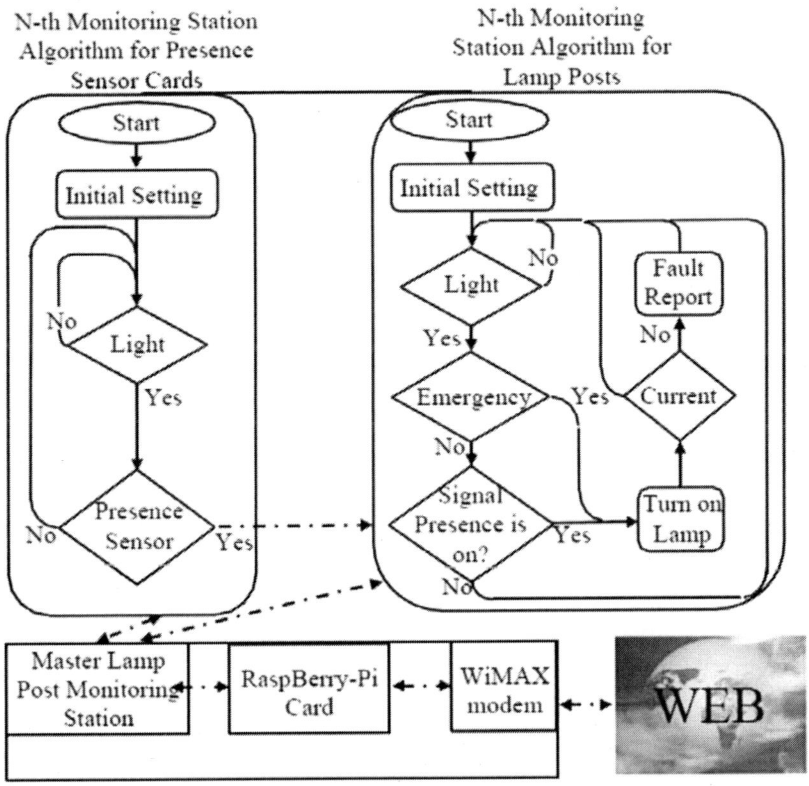

Figure 3. Management strategy of the Smart Grid.

The first are placed near the road at about 45 m from the crossroad. They are enabled when the sunlight is lower than a fixed limit. They read a photoelectric sensor and if its light beam is interrupted by the passage of a car, the control card sends a wireless signal to the isle of lamp posts by the ZigBee network. When this signal is received by the Monitoring Stations for Lamp Posts and it are equipped with a ZigBee Tx/Rx, they switch on the light.

The monitoring station is realized with the use of a PIC 16 f 688 microcontroller that also manages the wireless transmissions. The lamp post management is thus realized, but for intelligent management, the maintenance needs require a wider and deeper control of the lamp posts. To meet these requests, a current sensor, directly mounted on the lamp posts and read by the Monitoring Station, provides information about lamp post activity and power consumption. This information is stored in the PIC memory and, if requested by the next layer, is sent to it.

The second layer communicates with the first layer by the ZigBee network and collects data coming from each lamp post. The collected data are then available for the analysis of power consumption and for monitoring the lamp activity. Obviously, if the n-lamp post does not communicate or sends wrong data, this layer will send an intervention request toward a remote operator to solve the problem.

The second layer uses a RaspBerry-Pi card [24–27]. This is placed only in a lamp post called Coordinator, (another lamp post equipped with the RaspBerry-Pi may cause redundancy) and it is connected to the lamp post Monitoring Station. The latter makes possible the communication between the RaspBerry-Pi and the other lamp posts realizing the communication between the second and the first layer. The RaspBerry-Pi plays a role of webserver collecting data from each lamp post and displaying them. The third layer uses the RaspBerry-Pi to send these data to the World Wide Web. A graphical interface (Figure 4) shows the information on the lamp posts' activities.

Figure 4. Lamp control system GUI and measurement of power consumption.

All this makes possible an easy control of the lamp posts from a remote station and can allow an easy scheduling of any maintenance actions by the service engineer. Figure 5 shows the system placed in a shelter for laboratory tests.

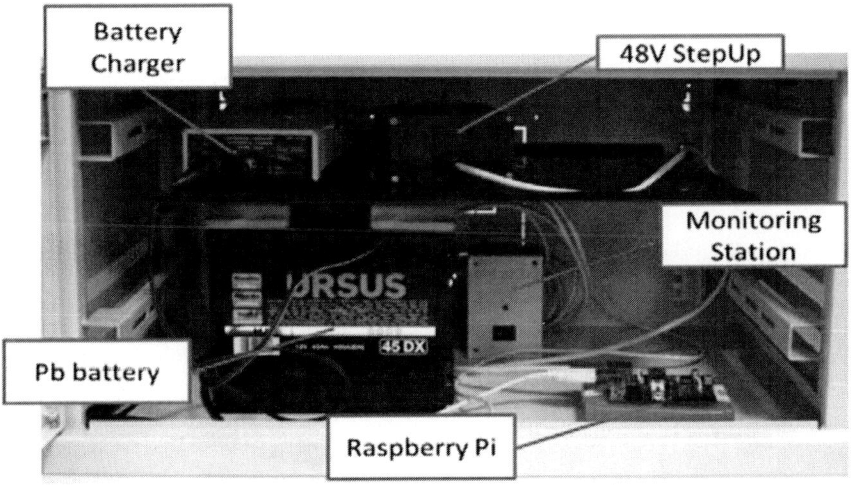

Figure 5. The system placed in a shelter for laboratory tests.

Figure 6 shows the operational test system working in real conditions.

Figure 6. Test system in the field.

6. THE LAMP POST

The lamp posts use, as new technologies, Light Emitting Diodes (LEDs) for the illuminating lamps and photovoltaic energy (PV) to supply the power. The use of these technologies is known in the literature, but a preventive traffic study allowed precisely defining the load allowing the correct dimensioning of the PV elements size.

The choice to supply the lamp posts of an alternative energy source is imposed by the absence of mains in the area where the isle is placed, and strongly suggests the use of this kind of energies in areas where the mains is far away. In fact, it would be very expensive (requiring copper wires and civil engineering works) to connect the area. Among the alternative energies, the absence of a constant wind, suggested to us that the best alternative energy to use is the photovoltaic one. PV systems are composed of a PV panel, a battery and a battery recharger [35].

The choice of the PV elements' size has been studied to match the operative conditions of the lamp posts. To evaluate the load, a preventive check of the car transit along the street has been made for four months during winter and spring when the nighttime has a longer duration. Every night we registered an average passage of about one hundred cars, often passing in group of two-four cars each time. For each transit of cars, we fixed 30 s for the lighting of the lamps so, the worst case provides that the total time the lamps need to be lit every night is about 25 min. Considering also anticipated some emergency situations, our system provides energy for one hour and a half. We also considered natural and weather conditions; in fact, we chose a larger battery to compensate for low sunlight for several days. Considering the loads, we have three different combinations:

(1)for the presence sensor card (0.2 A current consumption) the battery has a capacity of 6 Ah and the PV panel has a maximum power of 9 W–12 V; this assures a functioning of 10 h for three consecutively nights;

(2)secondary lamp posts use a 19 W–12 V PV panel and a battery of 10 Ah capacity able to ensure two hours per night of use for three consecutively nights;

(3)the coordinator lamp post has higher a consumption than the secondary ones because it has also the Raspberry-Pi (current consumption 0.5 A), which must be always connected to the Internet during the night and the WiMAX modem/router, which is activated only for two times during the night 15 min every time. To limit the consumption, the lamp section is managed like the secondary lamp posts, so the PV panel has a peak of 95 W–12 V and the battery a capacity of 48 Ah which assures an activity of 2 hours/night and the full activity for the hub for three consecutive nights. The modem average power consumption is about 20 W.

Obviously, the consumption, and consequently the PV panels, the batteries' sizes and the costs, would be at least three times higher without the intelligent management system; moreover, the lower weight produces less stress on the mounting poles.

7. TESTS AND RESULTS

The system has been realized and tested in the field for some months to verify the overall functionality under variable real-life conditions. The measurements allowed us to calculate energy and cost savings compared to other older technologies.

7.1. Range Tests

The article [3] has shown the reliability of the communication between two or more ZigBee modules under different environmental conditions is satisfactory. Nevertheless, following the procedure shown in [3], we performed some tests directly in the field where the lamp posts and the presence sensors are in line of sight (maximum distance about 45 m). Standard Xbee modules with patch antenna from Digi-MaxStream have been used because from datasheet [21] and from [3] they seemed to satisfy the range needs.

According to [3], we perfomed several tests, each one foreseeing 10,000 transmissions both during clear weather and during rain. Using an appropriate adapter to simulate the retransmission and using the X-CTU software provided by Digi-MaxStream, we checked the transmissions verifying that the test packetd, sent through the network by an Xbee module, arrived at the coordinator lamp and were correctly returned back. Using the minimum transmission power available, we obtained the results reported in Table 4, with an average reliability of 100%.

7.2. Local Transmission Tests

Other tests have been designed and realized to verify the overall communication ability between the coordinator, the secondary lampposts, and Internet. The first test verified how the system reacts when there is a fault in a lamp post or a presence sensor. Simulating the absence of one of them the hub software alerts of the malfunction on its graphical interface.

Table 4. ZigBee reliability tests.

XBEE STANDARD—Patch Antenna		
	Sunny	Rainy
	up to 45 m	up to 45 m
No obstacles	100%	100%
Vegetation	100%	100%

The second test verifies the case of break-downs of the coordinator lamp post. If the fault is either on the electrical section or in the local communication section, the hub software again shows the problem. Instead, if the fault is either on the RaspBerry-Pi card or on the WiMAX section, the impossibility to connect to the website means either breakage or theft. A secondary coordinator, activated on another lamp, could solve this problem.

7.3. WiMAX Test

The previously described WiMAX system was used to perform several reliability tests of the WiMAX technology. Due to the decay of the signal intensity with the increase in the working distance or with the presence of obstacles leading to multi-path fading, the IEEE 802.16.1 standard [36] does not provide any data on the distance achievable with WiMAX devices.

Because our goal is to also serve areas placed some kilometers away from the BS, we decided to test essentially Line of Sight (LOS) scenarios to verify our system by mounting the receiving antenna on top of an extendable tripod and placing the system on the roof of a building, whilst the BS was on the roof of a building itself. For the non-line of sight (NLOS) cases, three typical obstacle situations can be considered: when there is a hill between the BS and the client, when there are trees and when there is vegetation. As demonstrated by [37], vegetation and foliage in particular, cause unstable radio links and affect a lot the global throughput of the system obliging the CPE to work at lower modulations. However our aim is to verify the reliability of the radio link rather than the maximum bandwidth achievable.

The considered radio parameters are the received signal strength indicator (RSSI) and the carrier to interference plus noise ratio (CINR). The first is a measurement of the total received power of the frame preamble and is used by the subscriber station to determine the received signal level from the BS [38]. The latter represents a measure of the quality of the WiMAX signal. The higher the value of the CINR, the more throughput a link can maintain. The standard 802.16 supports the link adaptive modulation and the channel coding (AMC) and, in order to maintain a constant bit error rate (BER) of 1 part per million, it adapts to signal degradation by dropping to a lower modulation [39].

The AMC is controlled BS side and it applies the Modulation and Coding Schemes (MCS) which defines the most suitable modulation rate for current radio channel, and leads to the highest data rate possible. WiMAX mainly uses quadrature amplitude modulation (QAM) with its more powerful 64-QAM5/6 in downstream; it is optional in upstream [26]. Maximum modulation in upstream is usually 16-QAM3/4. In Table 5 the test shows no packet loss until 12.9 km.

Table 5. WiMAX uplink modulation in LOS.

UL Modulation	Packet Loss	CINR (dB)	AvgDistance from BS (km)
16-QAM3/4	0%	23	4.5
16-QAM1/2	0%	15	7.3
QPSK3/4	0%	13	10.1
QPSK1/2	0%	9	12.9

7.4. Measurements

As explained, the control system manages the isle providing the switch-on times (in minutes) and the absorbed currents of the lamps (in Amperes). Every control card of the lamp posts collects these data sending them to the coordinator when it requests them. After that, the coordinator control card provides to post the data on the website. To avoid data collisions, the coordinator calls the lamp posts one by one exploiting a specific identification number (ID) associated to each of them. Data collected during some autumn and winter months of 2013/14, showed the lamps absorb about 1.5 A during the activity for a period between 22.5 and 24 h per month.

7.5. Comparison with a Classic Lamp Posts

The absence of mains near the isle prevented us to making a direct comparison with a classical technology lamp post as for i.e., in [3], so we designed this comparison following two ways: in the first our lamp post (called L1) is directly compared with another (called L2), similar to the first, but without the presence sensor control, so it was always switched on during the night; in the second, we realize a simulation considering the L2 supplied by mains (this lamp post is called L3). The results are reported in Table 6.

Table 6. Current consumption and operating time of new and old lampposts.

Lamp ID	February				Plant Costs					
	Time (h)	I (A)	kWh	Cost h × kWh (0.200 €)	Battery Type and Cost	Solar Panel Type and Cost	Battery Recharger	Lamp €	Power Supplier	Electronic Devices €
New Lamp (L1)	21.36	1.48	0.367	0	5 A–16 €	19 W–50 €	10 €	18	0	40
New Lamp without light sensor (L2)	280	1.49	5.006	0	20 A–25 €	80 W–220 €	10 €	18	0	0
New Lamp without presence sensor and supplied by mains (L3)	280	0.09	5.796	1.159	None	None	0	18	30	0

We performed this test in February 2014 and both the lamp posts used a 18 W LED technology, 1550 lumen, 84 lm/W luminous flux, powered by a solar panel and a battery. For L3 we assumed the lamp post was supplied by mains through a 12 V AC/DC converter.

Considering fixed the cost of the kWh of power (in Italy ≈0.2 € [40]) and that the first and second lamp posts have no power consumption, to find the break-even we used the following formula [3]:

$$PC_Z - PC_Y + \sum_{i=1}^{x} \left(kWh_{Zi} - kWh_{Yi}\right) \cdot C_{kWh_i} = 0 \qquad (1)$$

where PC is the plant cost difference between different lamp posts and it is a fixed cost; kWh is the number of kilowatt per hour used by the specific lamp post day by day; C_{kWhi} is the cost of kWh; it can change during a long period. The term x is unknown and represents the activity days necessary to reach the breakeven between the two different choices of lamp posts. As imaginable the first solution is always more convenient than the second, while the first becomes more convenient than the third after 56 months.For the coordinator lamppost we need to add the cost of the WiMAX apparatus (80 €), while the Internet subscription cost is not considered because it is included in a WiMAX country public service. Moreover, there is a higher cost of the PV panel and of the battery equal to 80 €. Distributing this last cost between the five lampposts of the isle the breakeven is reached after 83 months. Obviously if the number of the lampposts of the isle is higher, the break-even will be reached earlier.

7.6. Management of Lamp Post Faults

In case of lamp post faults, to accelerate restore operations, the system has to inform the users as quickly as possible. To do it, the system has both a ZigBee communication control and a voltage battery one. The first checks both the breaking of the ZigBee element and the electronic card. In both cases the program shows, on its graphical interface, the problem. The system also checks the supply coming from the battery. If it goes down, a 4.5 V backup battery, integrated in the box, provides the necessary energy to work. The electronic card is moreover designed to check the backup battery voltage so, if this is too low, a battery fault signal will be sent to the coordinator program.

8. COMPARATIVE STUDIOS

To shows the advantages of our SC application, three other analysis have been conducted. The first shows a comparison between our isle and another realized with classical technology with lamp posts without presence sensors and supplied by mains. Considering the month of Table 6, the consumption of the classical isle composed of five lamp posts would be equal to 145.75 kWh with an energy cost of 29.15 € and a carbon production equal to 77.25 kg [41]. Obviously, with our isle the cost and the production of carbon dioxide is equal to zero.

The second shows an extension of the proposed system to other two isles of lamp posts realized in the meantime. Both these isles are composed by five lamp posts and are always placed in rural areas. Again, for both, WiMAX is the only technology usable for the Internet connection. The months considered are only November and December 2014 in which the three isles were contemporaneously active. If the isles had been realized with conventional technology, the overall power consumption was about 220 kWh with a cost of 44 € and a production of 116.6 kg of CO_2. Always in this case using our system the cost and the production of carbon dioxide is equal to zero.

Considering an annual use of the three isles equals about 67,725 kWh, the total estimated savings is equal to about 13,545 € and a production of about

35,894 kg of CO_2. It easy imaginable that a bigger number of isles allows one not only to increase the savings, but also to reduce the overall maintenance costs thus increasing the benefits.

Another comparison has been realized changing the scheme of the isle connecting in series eight lamp posts located 20 m away from each other. In this configuration, the presence sensor cards are placed at the start and at the end of the line. The overall functioning of the system is similar to the previous configuration, but, in this case, a transmission data problem surely happens. In fact, the ZigBee transmission range of XBee module is at maximum 100 m in LOS, so this obliges us to transform the mesh ZigBee architecture into a serial architecture where data sent by the n-lamp post, must be acquired by the n−1-lamp post, and so on up to the Coordinator. To face the problem it is necessary to modify the Monitoring Station Software creating a recursive program able to pack the data coming from the n-lamp post and send them together with data of the n−1-lamp post towards the n−2-lamp post and so on up to the Coordinator. On the other side, the requests from the Coordinator toward the secondary lamp posts have to be similarly packed and serially delivered to the lamp posts. Making these changes, the system correctly works. From an economical point of view, the general considerations made for a mesh configuration are equally valid.

9. CONCLUSIONS

The application of new technologies to a system, historically not the subject of much innovation, can transform it into an extremely efficient system allowing energy and money savings if compared with classical systems, as the Smart City paradigm teaches.

The example shown in this article is a clear demonstration of this thesis. An isle of lamp posts used to light a crossroads and placed far from the city where the Internet signal does not arrive, has been re-designed combining all new technologies available on the market: LEDs for the lamps, PV panels for the power supply and an intelligent management.

This last foresees an architecture that uses local sensors for intelligent lighting of the lamp, the storage of the functioning data, and their sharing by a local communication wireless mesh realized by ZigBee devices that send information to the coordinator lamp equipped with a RaspBerry-Pi card. The RaspBerry-Pi has been chosen for its low costs and for the possibility to drive also a WiMAX modem/router which allows to make the data system visible by a web site accessible by Internet also for areas very far from the city and not reached neither by the ADSL line nor by 3G signals.

Some analysis and comparisons testify the validity of the adopted choices. For its reliability, simplicity and low cost, the proposed system can also be used to update existing conventional lamp posts making it a serious candidate to efficiently manage a set of sensors applicable in different fields including monitoring of energy consumption, other Smart City application and smart grids which needs to diffuse sensors and actuators to realize an efficient management of the system under control.

AUTHOR CONTRIBUTIONS

Fabio Leccese conceived the idea of the system, all the authors realized the hardware and the software of the system. The Zigbee tests have been designed and performed by Fabio Leccese and Marco Cagnetti, while the test for the WiMAX have been conceived by Daniele Trinca and performed by all the three authors. Fabio Leccese wrote the paper.

CONFLICTS OF INTEREST

The authors declare no conflict of interest.

REFERENCES

1. Elejoste, P.; Angulo, I.; Perallos, A.; Chertudi, A.; García Zuazola, I.J.; Moreno, A.; Azpilicueta, L.; Astrain, J.J.; Falcone, F.; Villadangos, J. An Easy to Deploy Street Light Control System Based on Wireless Communication and LED Technology. Sensors **2013**, 13, 6492–6523.
2. Tan, Y.K.; Huynh, T.P.; Wang, Z. Smart Personal Sensor Network Control for Energy Saving in DC Grid Powered LED Lighting System. IEEE Trans. Smart Grid **2013**, 4, 669–676.
3. Leccese, F. Remote-Control System of High Efficiency and Intelligent Street Lighting Using a ZigBee Network of Devices and Sensors. IEEE Trans. Power Deliver. **2013**, 28, 21–28.

4. Pizzuti, S.; Annunziato, M.; Moretti, F. Smart street lighting management. Energy Effic. **2013**, 6, 607–616.

5. Lee, C.K.; Li, S.N.; Hui, S.Y. A Design Methodology for Smart LED Lighting Systems Powered by Weakly Regulated Renewable Power Grids. IEEE Trans. Smart Grid **2011**, 2, 548–554.

6. Costa, M.A.D.; Costa, G.H.; dos Santos, A.S.; Schuch, L.; Pinheiro, J.R. A High Efficiency Autonomous Street Lighting System Based on Solar Energy and LEDs. Proceedings of 2009 Power Electronics Conference, Bonito-Mato Grosso do Sul, Brazil, 27 September–1 October 2009; pp. 265–273.

7. Wang, Y.Q.; Hao, C.C.; Zhang, S.L.; Huang, Y.L.; Wang, H. Design of Solar LED Street Lamp Automatic Control Circuit. Proceedings of International Conference on Energy and Environment Technology, Guilin, China, 16–18 October 2009; pp. 90–93.

8. Gungor, V.C.; Lu, B.; Hancke, G.P. Opportunities and Challenges of Wireless Sensor Networks in Smart Grid. IEEE Trans. Ind. Electron. **2010**, 57, 3557–3564.

9. Nasirudin, M.A.; Za'bah, U.N.; Sidek, O. Fresh Water Real-Time Monitoring System Based on Wireless Sensor Network and GSM. Proceedings of 2011 IEEE Conference on the Open Systems (ICOS), Langkawi, Malaysia, 25–28 September 2011; pp. 354–357.

10. Peizhong, Y.; Iwayemi, A.; Zhou, C. Developing ZigBee Deployment Guideline under WiFi Interference for Smart Grid Applications. IEEE Trans Smart Grid **2011**, 2, 110–120.

11. Al-Ali, A.R.; Zualkernan, I.; Aloul, F. A Mobile GPRS-Sensors Array for Air Pollution Monitoring. IEEE Sens. J. **2010**, 10, 1666–1671. [Google Scholar]

12. Yang, Y.; Divan, D.; Harley, R.G.; Habetler, T.G. Power Line Sensornet—A New Concept for Power Grid Monitoring. Proceedings of 2006 Power Engineering Society General Meeting, Montreal, QC, Canada, 18–22 June 2006; pp. 1–8.

13. Rasin, Z.; Hamzah, H.; Aras, M.S.M. Application and Evaluation of High Power Zigbee based Wireless Sensor network in Water Irrigation Control Monitoring System. Proceedings of IEEE Symposium on Industrial Electronics & Applications (ISIEA 2009), Bandung, Indonesia, 4–6 October 2009; pp. 548–551.

14. Meng, X.; Xiao, S.; Xiong, Y.; Huang, H. ZigBee Based Wireless Networked Smart Transducer and its Application in Supervision

and Control System for Natural Gas Gate station. Proceedings of 4th International Conference on Computer Science & Education (ICCSE'09), Nanning, China, 25–28 July 2009; pp. 301–306.

15. Zhang, Y.; Wang, L.; Sun, W.; Green, R.C.; Alam, M. Distributed Intrusion Detection System in a Multi-Layer Network Architecture of Smart Grids. IEEE Trans. Smart Grid **2011**, 2, 796–808. [Google Scholar]

16. Leccese, F.; Cagnetti, M.; Calogero, A.; Trinca, D.; Di Pasquale, S.; Giarnetti, S.; Cozzella, L. A New Acquisition and Imaging System for Environmental Measurements: An Experience on the Italian Cultural Heritage. Sensors **2014**, 14, 9290–9312.

17. Chen, D.; Wang, M. A Home Security Zigbee Network for Remote Monitoring Application. Proceedings of 2006 IET International Conference on Wireless, Mobile and Multimedia Networks, Hangzhou, China, 6–9 November 2006; pp. 1–4.

18. Max Stream Inc. XBee™/XBee-PRO™ OEM RF Modules Product Manual v1.xAx - 802.15.4 Protocol. Available online: https://www.sparkfun.com/datasheets/Wireless/Zigbee/XBee-Manual.pdf (accessed on 21 October 2014).

19. Labiot, H.; Afifi, H.; de Santis, C. Wi-Fi™, Bluetooth™, ZigBee™ and WiMax™; Springer: Berlin, Germany, 2007. [Google Scholar]

20. Gislason, D. ZigBee Wireless Networking; Newnes: Burlington, MA, USA, 2008.

21. Digi. ZigBee RF Modules. Available online: http://www.digi.com (accessed on 21 May 2014).

22. EETIMES. ZigBee Application Note. Available online: http://www.eetimes.com (accessed on 21 May 2014).

23. ZigBee Alliance. ZigBee Application Note. Available online: http://www.zigbee.org (accessed on 21 October 2014).

24. Raspberry Pi®. Applications with Raspberry-Pi Card. Available online: http://www.raspberrypi.org/ (accessed on 21 October 2014).

25. Raspberry Pi®. Raspberry Pi Information Note. Available online: http://it.rs-online.com/web/generalDisplay.html?id=raspberrypi (accessed on 21 October 2014).

26. Raspberry Pi®. Raspberry Pi Tutorials. Available online: http://www.youtube.com/user/RaspberryPiTutorials/videos (accessed on 21 October 2014).

27. Wikipedia, Raspberry Pi. Available online: http://en.wikipedia.org /wiki/Raspberry_Pi (accessed on 21 October 2014).

28. How to Choose the Right Platform: Raspberry Pi or BeagleBone Black? Available online: http://makezine.com/magazine/how-to-choose-the-right-platform-raspberry-pi-or-beaglebone-black/ (accessed on 21 October 2014).

29. Arduino vs. Raspberry Pi Comparison. Available online: http://codeduino.com/information-and-news/hardware/arduino-vs-raspberry-pi/ (accessed on 21 October 2014).

30. Sivaneasan, B.; So, P.L.; Gooi, H.B.; Siow, L.K. Performance Measurement and Analysis of WiMAX-LAN Communication Operating at 5.8 GHz. IEEE Trans. Ind. Inf. **2013**, 9, 1497–1506. [Google Scholar]

31. COMMSCOPE. SSPX310M, Argus® Quad Antenna 3300–3800MHz, 65° Horizontal Beamwidth, Manual Electrical Tilt (Not Upgradeable). Available online: http://www.commscope .com/catalog/andrew/product_details.aspx?id=5280 (accessed on 21 October 2014).

32. ZyXEL. HES-319M Outdoor WiMAX CPE Manual. Available online: http://www.manualowl.com/m/ZyXEL/MAX318M/Manual/348786 (accessed on 21 October 2014).

33. UMTS World. WCDMA Link Budget. Available online: http://www.umtsworld.com/technology/linkbudget.htm (accessed on 21 October 2014).

34. WiMAX FORUM. Technical Specifications. Available online: http://www.wimaxforum.org/resources/technical-specifications (accessed on 21 October 2014).

35. Jutasolar. CMP12 JUTA Datasheet. Available online: http://www.jutasolar.com/ (accessed on 21 October 2014).

36. IEEE. IEEE Std 802.16.1-2012. Available online: http:// ieeexplore.ieee.org/xpl/mostRecentIssue.jsp?punumber=6297411 (accessed on 21 October 2014).

37. Ota, K.; Matsue, H.; Miyazawa, S.; Nanamatsu, S.; Hirata, Y.; Hasui, A.; Yamazaki, M.; Namera, T.; Fukui, H.; Harvey, M.; et al. Link

Budget Analysis and Field Experiments on Mobile WiMAX System in Azumino City. Proceedings of 2011 IEEE Global Telecommunications Conference, Houston, TX, USA, 5–9 December 2011; pp. 1–5.

38. Keysight Technologies. RSSI (8021.16 OFDMA). Available online: http://rfmw.em.keysight.com/wireless/helpfiles/89600B/WebHelp/subsystems/80216ofdma/Content/ofdma_symtblerrdata_rssi.htm (accessed on 21 October 2014).

39. Pidutti, M. Design How-To 802.16 Tackles Broadband Wireless QoS Issues. Available online: http://www.eetimes.com/document.asp?doc_id=1272053 (accessed on 21 October 2014).

40. Quale Tariffa. Quanto costa un kWh con Enel Energia? Available online: http://www.qualetariffa.it/quanto-costa-un-kwh-con-enel-energia/ (accessed on 21 October 2014).

41. Green. Green Jazz. Available online: http://www.rockhaus.it/berchidda%20rapporto%20co2.htm (accessed on 21 October 2014).

11

CHAPTER

DEFICIT ROUND ROBIN WITH FRAGMENTATION SCHEDULING TO ACHIEVE GENERALIZED WEIGHTED FAIRNESS FOR RESOURCE ALLOCATION IN IEEE 802.16E MOBILE WIMAX NETWORKS

Chakchai So-In [1], Raj Jain [2,*] and Abdel-Karim Al Tamimi [3]

[1] *Department of Computer Science, Faculty of Science, Khon Kaen University, Thailand*

[2] *Department of Computer Science and Engineering, Washington University in St. Louis, St. Louis, MO 63130, USA*

[3] *Department of Computer Engineering, Yarmouk University, Irbid, 21163, Jordan*

ABSTRACT

Deficit Round Robin (DRR) is a fair packet-based scheduling discipline commonly used in wired networks where link capacities do not change with time. However, in wireless networks, especially wireless broadband networks, i.e., IEEE 802.16e Mobile WiMAX, there are two main considerations violate the packet-based service concept for DRR. First, the resources are allocated per Mobile WiMAX frame. To achieve full frame utilization, Mobile WiMAX allows packets to be fragmented. Second, due to a high variation in wireless channel conditions, the link/channel capacity can change over time and location. Therefore, we introduce a Deficit Round Robin with Fragmentation (DRRF) to allocate resources per Mobile WiMAX frame in a fair manner by allowing for varying link capacity and for transmitting fragmented packets. Similar to DRR and Generalized Processor Sharing (GPS), DRRF achieves perfect fairness. DRRF results in a higher throughput than DRR (80% improvement) while causing less overhead than GPS (8 times less than GPS). In addition, in Mobile WiMAX, the quality of service (QoS) offered by service providers is associated with the price paid. This is similar to a cellular phone system; the users may be required to pay air-time charges. Hence, we have also formalized a Generalized Weighted Fairness (GWF) criterion which equalizes a weighted sum of service time units or slots, called temporal fairness, and transmitted bytes, called throughput fairness, for customers who are located in a poor channel condition or at a further distance versus for those who are near the base stations, or have a good channel condition. We use DRRF to demonstrate the application of GWF. These fairness criteria are used to satisfy basic requirements for resource allocation, especially for non real-time traffic. Therefore, we also extend DRRF to **support** other QoS requirements, such as minimum reserved traffic rate, maximum sustained traffic rate, and traffic priority. For real-time traffic, i.e., video traffic, we compare the performance of DRRF with deadline enforcement to that of Earliest Deadline First (EDF). The results show that DRRF outperforms EDF (higher achievable throughput under the promised delay latency) and maintains fairness under an overload scenario.

Keywords: deficit round robin; DRR; fragmentation; DRRF; temporal fairness, throughput fairness; generalized weighted fairness; GWF; WiMAX; IEEE 802.16; Mobile WiMAX; IEEE 802.16e; scheduling; resource allocation; quality of service; QoS; fairness; earliest deadline first; EDF; EDF-DRRF

1. INTRODUCTION

The IEEE 802.16e Mobile WiMAX standard [1] uses Orthogonal Frequency Division Multiple Access (OFDMA) in order to achieve high data rate, long distance coverage, and mobility supports. In Mobile WiMAX OFDMA systems, the entire channel (wireless spectrum) is divided into multiple subcarriers. The number of subcarriers is proportional to the channel spectral width. These subcarriers are also grouped into a number of subchannels. Each mobile station (MS) is assigned a group of subchannels for a certain amount of times, as shown by the two dimensional diagram in Figure 1.

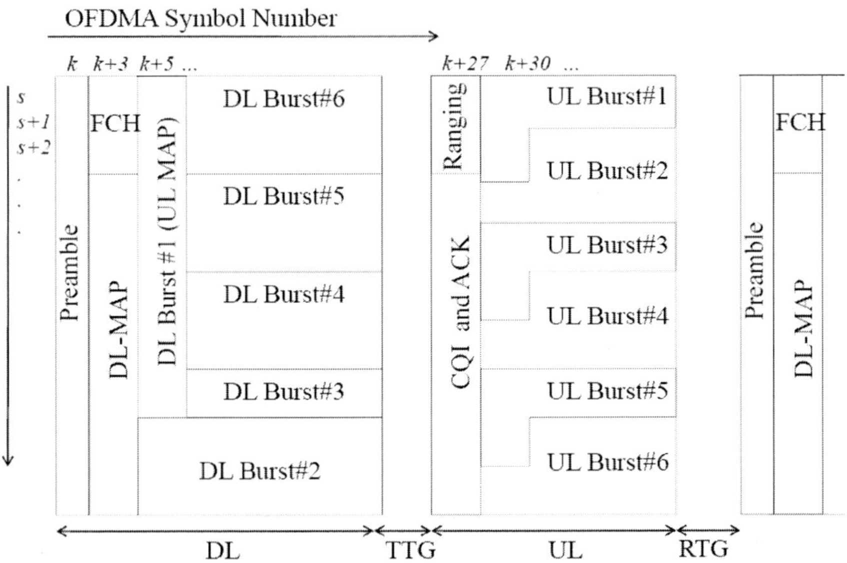

Figure 1. A Sample OFDMA Frame Structure: PUSC [3].

In Figure 1, the vertical axis is frequency (subcarriers or logical subchannels). The horizontal axis is time. The time is measured in units of OFDM (Orthogonal Frequency Division Multiplexing) symbols. Mobile WiMAX uses a fixed frame-based allocation. Basically, each frame is of 5 ms duration [2]. The frame starts with a downlink preamble and Frame Control Header (FCH) followed by downlink (DL) and uplink (UL) maps (DL-MAP and UL-MAP). These maps contain the information elements (IEs) that specify the profile for each burst which consists of a number of OFDM symbols and subchannels. The profile consists of burst-start time, burst-end time, modulation type, and Forward Error Control (FEC) used or to be used in each burst.

Figure 1 also shows a ranging region in the uplink subframe. This region is used to determine the distance between a base station (BS) and mobile stations (MSs) so that the transmission start times at various stations can be properly synchronized. This region helps to set the right transmit power level for each mobile station. A CQI&ACK region is used to send Channel

Quality Indication (CQI) feedback and acknowledgements (ACKs). TTG (Transmit to Transmit Gap) and RTG (Receive to Transmit Gap) separate the downlink and uplink subframes.

The resource allocation problem in IEEE 802.16e Mobile WiMAX Networks [3,4,5,6] is about how to schedule resources, i.e., number of slots, for each mobile user in each Mobile WiMAX frame. Note that each slot consists of one subchannel allocated for the duration of some numbers of OFDM symbols. The number of subcarriers in the subchannel and the number of OFDM symbols in the slot depend upon the link direction (uplink or downlink) and subchannelization modes.

The subchannelization mode or the mapping from logical subchannel to multiple physical subcarriers is called a permutation. Basically, there are two types of permutations: distributed and adjacent. The distributed subcarrier permutation is suitable for mobile users while adjacent permutation is for fixed (stationary) users. For example, in the Partially Used Sub-Channelization (PUSC) scheme, one of the distributed permutation modes, one slot consists of one subchannel over two OFDM symbol periods for downlink and one subchannel over three OFDM symbol periods for uplink. Others include Fully Used Subchannelization (FUSC) and Adaptive Modulation and Coding (band-AMC) [1,2]. In this paper, we focus on the PUSC mode, commonly used in a mobile wireless environment [2,4].

The IEEE 802.16e standard supports bi-directional communications as a frequency division duplexing (FDD) and a time division duplexing (TDD). For FDD, the uplink and the downlink use different frequency bands. For TDD, the uplink traffic follows the downlink traffic in the time domain. All scheduling schemes discussed in this paper can be used for both FDD and TDD systems. However, to keep the discussion focused, we use TDD throughout this paper.

Although the IEEE 802.16e standard allows several configurations, such as mesh networks and relay networks, our focus is only on point to multipoint networks. Thus, the resource allocation problem is basically that the base station is the single resource controller for both uplink and downlink directions for each mobile station. The mobile station has an agreed quality of service (QoS) requirement that is negotiated between the base station and the mobile station at the time of the connection setup. The base station grants transmission opportunities to various mobile stations based on their bandwidth requests and QoSs.

In IEEE 802.16e Mobile WiMAX Networks, several modulation and coding schemes (MCSs), such as Binary Phase Shift Keying (BPSK) and Quadrature Amplitude Modulation (QAM), are supported. BPSK results in 1 bit per symbol, and is used for channels in poor conditions. QAM results in more bits per symbol, and is used for more reliable channel conditions. Since MCS used for each user depends upon the user location, and varies with time, the slot capacity (number of bits in the slot) is also varied with time and location. In this paper, we use different MCSs to represent a variation of channel interferences.

The IEEE 802.16e standard defines five classes of service: Unsolicited Grant Service (UGS), extended real-time Polling Service (ertPS), real-time Polling Service (rtPS), non real-time Polling Service (nrtPS), and Best Effort (BE). Each of these has its own set of QoS parameters, such as minimum throughput, maximum allowable delay, and delay variation (delay jitter). In this paper, our focus is on non real-time traffic, i.e., BE and nrtPS classes. The QoS parameters are the throughput and the fairness criteria. However, we also investigate video traffic, an example of a real-time traffic class. The delay constraint (maximum allowable delay) is an additional QoS parameter for this class. For other classes of services, i.e., UGS, and ertPS, normally a fixed size allocation mechanism applied in every Mobile WiMAX frame and/or a simple earliest deadline first (EDF) algorithm can be used as a scheduling discipline [3,5,7].

To summarize, in this paper, our contributions are as follows: first we investigate two fair scheduling algorithms; namely, Generalized Processor Sharing (GPS) [8] and Deficit Round Robin (DRR) [9] in the context of IEEE 802.16e Mobile WiMAX Networks. Each algorithm has its own pros and cons; for example, both GPS and DRR result in perfect fairness, i.e., equal throughput for contending users. Since Mobile WiMAX allows a fragmentation, GPS can utilize full frame resource allocation; however, it creates high overhead because the allocation is usually distributed equally for all MSs in each Mobile WiMAX frame.

In contrast, because of the packet-based allocation framework, DRR results in the least overhead, but can leave some unused-space within the Mobile WiMAX frame. Thus, to achieve both full frame utilization and overhead reduction, we introduce a Deficit Round Robin with Fragmentation (DRRF) [10] by taking advantage of both approaches to best suit the scheduling of the best effort traffic, in which the goal is to maintain the fairness among competing flows.

Second, in wireless networks, the link capacity keeps changing over time and distance. Especially in wireless broadband networks where a longer distance means lower channel quality. The issue of defining fairness depends upon both service provider's and customers' points of views. As a result, in this paper, we also propose a Generalized Weighted Fairness (GWF) [11] which equalizes the weighted sum of service time units and transmitted bytes. In Mobile WiMAX, again the service time unit is represented as a number of slots allocated to each mobile station. Mobile WiMAX equipment manufacturers can implement the generalized fairness concept. The service providers can then set a weight parameter to any desired values, and achieve either slot fairness or throughput fairness or some combination of the two. We use DRRF to illustrate the effect of GWF.

In addition, we extend DRRF to support other QoS parameters for non real-time traffic (nrtPS class) in IEEE 802.16e Mobile WiMAX Networks, i.e., minimum reserved traffic rate, maximum sustained traffic rate, and traffic priority [1]. Third, for real-time traffic, i.e., video traffic, we demonstrate the effect of DRRF with deadline (maximum allowable delay) enforcement versus Earliest Deadline First (EDF) [12]. The results show that DRRF outperforms EDF, and maintains fairness under overload scenario [13].

This paper is organized as follows. In Section 2, we describe a simple scheduling technique imitating Generalized Processor Sharing, or GPS, along with Deficit Round Robin (DRR). In Section 3, we discuss a modification of DRR by allowing packet fragmentation, called DRRF. Section 4 presents a concept of Generalized Weighted Fairness (GWF) versus temporal fairness, or slot fairness, and throughput fairness, or byte fairness. Section 5 shows how to derive the quantum size for each mobile user with minimum reserved traffic rate constraint; and how to apply maximum sustained traffic rate and traffic priority parameters to DRRF. We describe scheduling techniques for real-time traffic in Section 6. Section 7 shows simulation results based on DRRF in both real-time and non real-time traffic scenarios. Finally, conclusions are drawn in Section 8.

2. FAIR SCHEDULING ALGORITHMS IN IEEE 802.16E MOBILE WIMAX NETWORKS

In this section, we describe two well-known scheduling algorithms, commonly used in wired networks. These are Generalized Processor Sharing (GPS) and Deficit Round Robin (DRR).

2.1. Generalized Processor Sharing (GPS)

GPS [8] is a simple throughput fair scheduling algorithm. The GPS concept is derived from the idea of CPU time sharing where each process receives equal amount of shared CPU time. Therefore, the resource allocation is unaware of packet boundary. In general, GPS can achieve ideal perfect fairness. In particular, full frame utilization can be accomplished in the context of a Mobile WiMAX frame-based allocation. Nevertheless, packet fragmentation may occur, resulting in reduced goodput—throughput after overhead—due to MAC (Media Access Control) header and fragmentation subheader overheads.

For example, suppose there is a 300-byte Mobile WiMAX frame capacity, and there are four active mobile stations (MSs). To simplify the derivation, assume that all packets are 125 bytes in size. Then, only one 300/4 = 75-byte fragmented packet will be transmitted in every frame for each MS. With a large number of MSs, the ideal GPS is infeasible due to the effect of fragmentation overheads.

2.2. Deficit Round Robin (DRR)

DRR [4] is one of the simple throughput fair allocations used in wired networks. DRR avoids packet fragmentation overheads by scheduling only a full packet. Basically, the packet is not scheduled, and the deficit (amount that would have been allocated) is stored in case a packet will result in exceeding the fair share.

Table 1 shows an example of DRR. In this example, there are four MSs, and each MS has only one flow. Each flow is mapped to a single queue. Suppose all packets are 125 bytes in size. Again, assuming the Mobile WiMAX frame capacity is 300 bytes. Then, to achieve a fair allocation, the fair share is set to 1/4th of 300 or 75 bytes; this is called the quantum size.

In this example, the packets are scheduled sequentially from each queue. Four frames are illustrated as an example. There are 2 rounds in the first and third frame. There are no rounds in the second and fourth frame. Note that notation N/M is used where N is the size of allowed transmission, and M is the cumulative deficit counter value at the end of that round.

Table 1. Updated Deficit Counter for DRR (Transmitted Packet Size/Deficit Counter) [10].

Frame No.	1st			2nd	3rd			4th
Round	1	2			3	4		
MS#1	0/75	0/150	125/25	0/25	0/100	0/175	125/50	0/50
MS#2	0/75	0/150	125/25	0/25	0/100	0/175	125/50	0/50
MS#3	0/75	0/150	0/150	125/25	0/100	0/175	0/175	125/50
MS#4	0/75	0/150	0/150	125/25	0/100	0/175	0/175	125/50

Basically in each round, the deficit counter is increased by the quantum size, 75. The packets from queues with the deficit counters greater than the head packet size are scheduled. Again, in this particular example, only two packets in each frame are transmitted, and result in 50 unused bytes. The packets being transmitted in each frame are shown in Table 1. Notice that N is always 125 bytes.

3. DEFICIT ROUND ROBIN WITH FRAGMENTATION (DRRF) (FOR BEST EFFORT TRAFFIC)

In the previous section, we described the basic idea of two common scheduling algorithms: GPS and DRR. Although GPS can achieve perfect fairness with frame utilization, the resource allocation results in a large overhead. On the other hand, DRR can mitigate the effect of packet fragmentation. However, again full frame utilization cannot be achieved. In other words, some spaces may be left unused in a Mobile WiMAX frame in case the next full packet does not fit in that space. We will also show the simulation results to illustrate these claims (See Section 7). As a result, we introduce a modification of DRR, the so-called Deficit Round Robin with Fragmentation (DRRF) [10].

DRRF is basically a combination of GPS and DRR. In fact, DRRF is similar to DRR, but allows packet fragmentation for the purpose of achieving full frame utilization, that is, in case there are some left-over spaces within a frame, DRRF allocates those left-over spaces to some mobile stations.

Table 2 shows an example of DRRF. Again, we use the same example as in DRR (See Section 2.2). Four frames are illustrated as an example (in 5 rounds). In each round, the deficit counter is increased by the quantum size, 75. Similar to the DRR, the packets with the deficit counter greater than the packet size are scheduled. However, to achieve full frame utilization, a fragmented packet is allowed as well. In other words, a fragmented packet with the deficit counter greater than the fragment size is also scheduled. In this table, there are 2 rounds in the first two frames. In frame 3, there is no round and only one round for the fourth frame. In each frame, some fragmented packets may be also transmitted (we indicate these by bold numbers.)

Table 2. Updated Deficit Counter for DRRF (Transmitted Packet Size/Deficit Counter) [10].

Frame No.	1st			2nd				3rd	4th		
Round	1	2			3	4				5	
MS#1	0/75	0/150	125/25	0/25	0/100	0/175	**100**/75	**25**/50	0/50	0/125	125/0
MS#2	0/75	0/150	125/25	0/25	0/100	0/175	0/175	125/50	0/50	0/125	**75**/50
MS#3	0/75	0/150	**50**/100	**75**/25	0/100	0/175	0/175	125/50	0/50	0/125	0/125
MS#4	0/75	0/150	0/150	125/25	0/100	0/175	0/175	**25**/150	**100**/50	0/125	0/125

3.1. Throughput Fair Allocation

In traditional wired networks, the simple scheduling algorithms, i.e., GPS, and DRR, have the assumption that the link capacity is constant over time and location. However, it is not always the case in wireless networks, particularly IEEE 802.16e Mobile WiMAX Networks. In Mobile WiMAX, to support a reliable transmission and throughput optimality, various modulation and coding schemes (MCSs) are allowed. The MCS levels are determined by CINR (Carrier to Interference-plus-Noise Ratio), typically sent back over a CQI&ACK channel, and the desired BLock Error Rate (BLER) [1,2].

Thus, to overcome the issue of varying link capacity, we apply MCSs to calculate the proper quantum size and the updated deficit counters for DRRF. In other words, we basically use the number of requested slots not the number of bytes in the queue length (in downlink direction). The derivation is the ceiling of the queue length and its MCS size (in terms of number of bytes).

For example, for downlink PUSC with 10 MHz and 1024 FFT (Fast Fourier Transform), the number of subcarrier×symbol combinations per slot is 56. Of these, 8 combinations are used as pilots leaving 48 combinations for data. With one bit per symbol coding, this results in 48 bits (or 6 bytes) per slot [1,5]. In case the queue length is 125 bytes, the number of requested

slots is or 21 slots.

For uplink PUSC, the requested slots can be derived from the bandwidth request. However, since there is no mechanism to send the information about individual packet sizes to BS, the concept of deficit round robin cannot be applied directly.

For performance comparison, we also apply the MCS levels to calculate the fair share for GPS; and the proper quantum size and updated deficit counters for DRR. In other words, again the resource allocation unit is in terms of number of requested slots.

Notice that in Section 2, if the fair share and the quantum size are measured in slots, in those particular examples, each slot's capacity is one byte. However, different MCSs result in a different number of bytes per slot, and thus result in different fair share and quantum sizes in bytes. For the rest of the paper, we use the number of requested slots instead of requested bytes for all three schedulers in order to achieve throughput fairness for various channel conditions.

3.2. Max-min Fairness

In the previous section, to achieve throughput fair allocation, we assume that all MSs have infinite traffic to send, and can use the resources allocated to them. In other words, at the scheduling stage all MSs always have packets waiting in their queues. However, in practice, the available traffic is finite. Some MSs may not have enough traffic, may not be able to use a fair share, or may have too much traffic.

Our goal is to make a fair allocation among MSs. In the case where some MSs cannot use a fair share, their left-over share should be fairly allocated to other MSs. This leads to a commonly used max-min criterion for fair allocation. An allocation is said to be max-min fair if it maximizes the allocation for the user who received the minimum. In IEEE 802.16e Mobile WiMAX Networks, we apply the max-min concept by maximizing the minimum number of requested slots.

Figure 2 shows steps in computing the max-min fair allocation. The first step is to compute the number of requested slots from the number of bytes requested by an active mobile station and its MCS. Then, we sort the requested slots for all active MSs in ascending order (Step 2). In Step 3, for each active MS the number of fair share slots is derived. Next, the number of granted slots, the minimum of number of fair share, and the number of requested slots are updated. Then, the number of free slots is updated. This loop continues until there are no more free slots, or the number of requested slots for all active MSs is satisfied. The results are used to as the actual quantum for DRR and DRRF, and the actual fair share allocation for GPS. Note that the quantum size and fair share may change over a frame period according to the MCS level.

```
Calculate #requested_slots_i per frame for each active MS_i given its MCS //1st step
Sort active MSs in ascending order of active_MS_requested_slots //2nd step
FOR each active_MS_i DO //3rd step
Calculate #fairshare_slots for active_MS_i
IF #requested_slots_i/frame < #fairshare_slots THEN
#fairshare_slots_i/frame = #requested_slots_i
END IF
Update #granted_slots for active_MS_i
Update #free_slots and exit if #free_slots == 0
END FOR
```

Figure 2. Steps in a Simple Max-Min Fairness Algorithm [10].

(a) Slot Fairness and Byte Fairness	(b) Weighted Fairness	
Slot Fair Allocation Scheme	S_i and S_j	Number of slots allocated to MS_i and MS_j
(temporal fair allocation):	B_i and B_j	Number of bytes allocated to MS_i and MS_j
$Total\_Slots = \sum_{i=1}^{N} S_i$	b_i and b_j	Number of bytes per slot for MS_i and MS_j
$S_i = S_j \| i, j \leq N$	N	Number of active mobile stations
	w	Weight parameter (between 0 and 1)
Byte Fair Allocation Scheme	M	Number of bytes per slot for the highest MCS
(throughput fair allocation):		
$Total\_Slots = \sum_{i=1}^{N} S_i$	$Total\_Slots = \sum_{i=1}^{N} S_i$	
$B_i = B_j \| i, j \leq N$	$wS_i + \frac{(1-w)B_i}{M} = wS_j + \frac{(1-w)B_j}{M} \| i, j \leq N$	
$B_i = b_i \times S_i$	$B_i = b_i \times S_i$	

Figure 3. Fairness Definitions. (a) Slot Fair Allocation and Byte Fair Allocation; (b) Weighted Fair Allocation [11].

4. GENERALIZED WEIGHTED FAIRNESS (GWF)

From the discussion so far, we have described mechanisms on how to achieve throughput fair allocation by applying various modulation and coding schemes (MCSs) to calculate the fair share (as a resource allocation unit). However, in wireless networks, e.g., wireless local area networks, or WLANs, there are two ways to define fairness: temporal fairness and throughput fairness [14].

The former approach is to allocate equal number of service time units to each user. Again, in IEEE 802.16e Mobile WiMAX Networks, the resource allocation is in terms of the number of slots [1,2]. Therefore, here the unit used to achieve the temporal fairness is slot. In other words, the base station will allocate equal number of slots to each user. The users, who are near the base station, and therefore have a very good link, will be able to use good MCSs (multiple bits per symbol), and consequently receive good throughput in bytes per second. This is also called slot fairness or temporal fairness.

On the other hand, the other approach is to allocate equal throughput (or bytes) to each user. The users near the base station will need fewer slots than those who are far away to transmit the same number of bytes. This is called byte fairness or throughput fairness. Figure 3a shows formal descriptions for both slot and byte fairness schemes.

When setting either slot fairness (temporal fairness) or byte fairness (throughput fairness) for each user, it can be argued which definition is better. For example, the users want to run certain applications, such as VoIP (Voice over Internet Protocol), which require a certain bit rate, and so the users would prefer a guaranteed bit rate for their applications or a fair share of the available bit rate in case of best effort service. This is the argument in favor of byte fairness or throughput fairness.

The argument in favor of temporal fairness or slot fairness goes as follows: The service provider has a fixed number of slots, and if the user happens to choose a bad location, e.g., the basement of a building on the edge of the cell, the provider will have to allocate a significant number of slots to provide the same quality of service as to a user who is near the base station. Since the providers have no controls over the user locations, they can argue that they will provide the same resources to all users, and the throughput observed by the user will depend upon their locations.

Some service providers will prefer throughput fairness while the others will favor the slot fairness. The equipment manufacturers should allow both possibilities. One way to do this is to implement the weighted fairness introduced in this paper [11]. The formal definition of the weighted fairness is shown in Figure 3b.

In general, setting the weight, w, to 0 results in byte fairness, and setting it to 1 results in slot fairness. A carrier can select the weight to be either of the two values or any values between 0 and 1. Note that the GWF formulation as defined in Figure 3b uses equal weight for all mobile stations. It is possible to generalize this formulation further by allowing different weights for different mobile users based on their priorities or the price paid.

4.1. DRRF and GWF

To apply the generalized weighted fairness (GWF) concept into IEEE 802.16e Mobile WiMAX Networks, we modified the Deficit Round Robin with Fragmentation (DRRF) algorithm so that the resource allocation follows the GWF criterion. We first calculate the pre-quantum number derived from the GWF equation (See Figure 3b). Then, similar to the quantum derivation in DRRF, the finalized quantum number is updated based on number of free slots, queue length, and its deficit counter (See also Figure 2).

5. DRRF EXTENSIONS (FOR NON REAL-TIME TRAFFIC)

In Section 3 and Section 4, we discussed how the scheduler can achieve (weighted) fair allocation using DRRF. The concept of fair allocation can be applied easily to the best effort (BE) traffic. However, in IEEE 802.16e Mobile WiMAX Networks, there are also other quality of service (QoS) parameters used for non real-time Polling Service (nrtPS), i.e., minimum reserved traffic rate, maximum sustained rate, and traffic priority [1,3]. Therefore, this section describes how to modify DRRF to support those parameters. Note that to simplify the derivation; we assume all these three criteria are based on the throughput requirement. In other words, considering the GWF criterion, the weighted value is set to 0.

5.1. Minimum Reserved Traffic Rate Extension

As described so far, the scheduling scheme ensures that all users will achieve fair throughput with the channel conditions' considerations. However, by defining minimum reserved traffic rate, or r_{min}, some users may need to be favored over others. Note that users without minimum reserved traffic rates are treated as users with a zero guaranteed rate.

We modified DRRF by updating the deficit counter to achieve a minimum guaranteed rate criterion. First, similar to what we described in Section 3, we derive the number of requested slots, requested_slots, from the queue length and its corresponding MCS. Then, we calculate the number of slots required to guarantee the minimum reserved traffic rate, rmin_slots. The minimum of these two numbers, called pre_deficit, is used to update the deficit counter. Then, the number of free slots used to calculate the proper quantum is also updated accordingly.

$$pre\_deficit(i) = Min[requested\_slots(i), rmin\_slots(i)] \quad (1)$$

$$rmin\_slots(i) = \left\lceil \frac{rmin(i) \times t\_frame}{MCS\_size(i)} \right\rceil \quad (2)$$

Equations 1 and 2 show how to derive these numbers. Here, t_frame is Mobile WiMAX frame size (5 ms). MCS_size(i) is the number of bytes per slot for the given MCS level of mobile station i. Note that after these derivations, we also apply the max-min fairness algorithm (See Figure 2) to calculate the proper quantum size to distribute the left-over slots fairly for both two categories: (1) MSs without minimum reserved traffic rate and (2) MSs with minimum reserved traffic rate but still need more slots than the total guarantee.

5.2. Maximum Sustained Traffic Rate Extension

Maximum sustained traffic rate, r_{max}, is typically used to control the peak rate. This rate excludes overheads, such as a MAC header and other subheaders [1,2]. We modified DRRF to achieve this criterion as follows: We derive the maximum number of slots required to meet r_{max}. We call this number rmax_slots. The derivation is shown below (Equation 3):

$$rmax\_slots(i) = \left\lfloor \frac{rmax(i) \times t\_frame}{MCS\_size(i)} \right\rfloor \quad (3)$$

Here, again t_frame is the frame size. MCS_size(i) is the number of bytes per slot for the given MCS level of MS i. Note that instead of the ceiling function in case of rmin_slots, here we use a floor function.

Then, we apply the max-min fairness algorithm first without r_{max} constraint. Consequently, the number of fair share slots is derived. Due to the r_{max} constraint, some MSs may receive the fair share which is over the r_{max} limitation. We limit those fair shares to rmax_slots. Finally, in case there are some left-over slots due to the r_{max} limitation, we then apply another round of the max-min fairness algorithm to distribute the left-over slots fairly among the flows which have no r_{max} constraint.

5.3. Traffic Priority Extension

In IEEE 802.16e Mobile WiMAX Networks, traffic priority is one of the QoS parameters which differentiate the relative importance of flows within the same QoS class in which all other QoS parameters are identical. The standard does not specify how to differentiate between traffic priorities; however, it recommends that the scheduler should give lower delay and/or higher buffering preference for flows with high priority [1,3].

In a priority based system, flows of higher priority are serviced before those of lower priority. Fairness implies that multiple flows of the same priority should get similar service, that is, if the system is overloaded, all flows of the lowest priority should be penalized equally. It is also a good practice to police the higher priority flows so that they do not overload the system, and starve lower priority flows.

6. DRRF AND REAL-TIME TRAFFIC

Based on the discussion so far, we have introduced a Deficit Round Robin with Fragmentation (DRRF) and its extensions including the Generalized Weighted Fairness criterion basically to support both Best Effort (BE) and non real-time Polling Service (nrtPS) classes for IEEE 802.16e Mobile WiMAX. However, Mobile WiMAX also supports a real-time Polling Service (rtPS) class [1,3]. For the formal two classes, the performance metrics are based on fairness and throughput guarantees. For rtPS, an additional QoS parameter is the delay (maximum allowable delay) constraint and optional delay jitter (delay variation).

In this section, we focus on an investigation of a well-known scheduler for real-time traffic in wired networks, i.e., EDF (Earliest Deadline First), applied in IEEE 802.16 WiMAX Networks. Note that a Deficit Round Robin (DRR) scheduling algorithm is commonly used to schedule non real-time traffic basically to maintain the expected throughput and fairness in a downlink direction [3]. We compare the results with DRRF (Deficit Round Robin with Fragmentation), as we described earlier, as well as a combination of the two.

6.1. Earliest Deadline First (EDF)

Given a set of flows, the first algorithm, EDF [12], compares the packets at the head of the flow queues, and schedules the packet the earliest deadline constraint. One additional complication in the case of Mobile WiMAX is that the entire packet may not fit in the current Mobile WiMAX frame, and a fragment may be left-over. This fragment is possibly transmitted first in the next Mobile WiMAX frame. The entire packet is discarded if the fragment does not meet the deadline (or delay constraint).

6.2. Deficit Round Robin with Fragmentation (DRRF)

Traditionally, DRR [9] is used to avoid packet fragmentation by scheduling only a full packet. If a packet will result in exceeding the fair share, the packet is not scheduled, and then the deficit (amount that would have been allocated) is stored. However, to fully utilize a Mobile WiMAX frame, we use a modified version of DRR, proposed in this paper (See Section 3), DRR with fragmentation, or DRRF. In general, if a packet meets the fair share limit, we schedule it in the current Mobile WiMAX frame, and if necessary, allow the part that will not fit in the current Mobile WiMAX frame to be scheduled in the next frame. This ensures that Mobile WiMAX frame capacity is not wasted.

6.3. Earliest Deadline First with DRRF (EDF-DRRF)

For a combination of EDF and DRRF, called EDF-DRRF, we first apply EDF, and then regulate the packet stream with DRRF. In other words, the packets are sorted according to the deadline, and then DRRF is used to decide whether the packet with the earliest deadline is eligible for transmission without exhausting the flow's credits (deficits). Again, we allow fragmented packets to be transmitted for frame utilization purposes.

6.4. Scheduling Algorithms with Enforced Deadline

For real-time traffic, video traffic in particular, received packets with the huge delay or over the deadline are not useful. Since the deadline or average delay is negotiated during the connection setup, we can use the deadline information at the scheduler by dropping the packets that are over the deadline, and then save the bandwidth. Therefore, for all three algorithms described above, packets are dropped if they will not meet the deadline after transmission. We also analyze cases without this option; however, the results showed worse performance with a large fraction of packets being discarded at the destination due to exceeding the deadline. We conclude that given the resource constrained nature of a wireless medium, any reasonable implementation should minimize waste by discarding late packets before transmission.

7. PERFORMANCE EVALUATION

In this section, we present four main simulation results to show the effect of DRRF in both non real-time and real-time traffic: (1) DRRF, (2) DRRF+GWF, (3) DRRF Extensions, and (4) EDF-DRRF. The first simulation configuration is used to show system performance of DRRF in terms of system throughput, percentage of overheads, and fairness index of DRRF compared to GPS and DRR. The second configuration is to show if DRRF can be used to achieve GWF criteria. The third configuration is to show whether DRRF can support other QoS parameters for nrtPS. Finally, the fourth configuration is to show system performance of EDF, DRRF, and EDF-DRRF in terms of system throughput, delay, delay jitter, and fairness.

Table 3. Performance Evaluation Parameters [2,5].

Parameter	Value
PHY	OFDMA
Duplexing mode	TDD
Frame length	5 ms
System bandwidth	10 MHz
FFT size	1024
Cyclic prefix length	1/8
DL permutation zone	PUSC
RTG + TTG	1.6 symbol
DL:UL ratio	2:1 (29: 18 OFDM symbols)
DL preamble	1 symbol-column
MAC PDU size	Variable length
ARQ and packing	Disable
Fragmentation	Enable
DL-UL MAPs	4 symbol-columns

7.1. Simulation Parameters and Topology

The simulation parameters follow the performance evaluation parameters specified in Mobile WiMAX System Evaluation documents and WiMAX profiles [2]. These parameters are briefly summarized in Table 3. With 10 MHz system bandwidth, 5 ms frame, 1/8 cyclic prefix, and a DL:UL ratio of 2:1, the number of downlink symbol-columns per frame is 29 [2] (18 for uplink). Note that 1.6 symbol-columns are used for TTG and RTG. Of these,

1 symbol-column is used for preamble. In the Partial Usage of Subchannels (PUSC) mode, there are 30 subchannels in the downlink, and each slot consists of one channel over two symbol durations. As a result, there are 30 × (24/2) = 360 downlink slots per frame. Each MS has only one flow, and is mapped to a different MCS used in Mobile WiMAX.

Table 4. Data Throughput Analysis (excluding DCD/UCD) [5].

MCS	Bit per symbol	Coding rate	Bytes per slot	Throughput (kbps)
BPSK1/2	1	½	3	1728
QPSK1/2	2	½	6	3456
QPSK3/4	2	¾	9	5184
16QAM1/2	4	½	12	6912
16QAM3/4	4	¾	18	10368
64QAM2/3	6	⅔	24	13824
64QAM3/4	6	¾	27	15552

Of these, Frame Control Header (FCH), DL-MAP, UL-MAP, Downlink Channel Descriptor (DCD), and Uplink Channel Descriptor (UCD) overheads can approximately range from 51 slots to 195 slots (including 31 slots for DCD/UCD in case of five, and fifteen mobile stations, respectively [5,8,10,13]). The repetition of 4 with QPSK1/2 is used for MAPs.

Notice that the total overhead depends on the number of actual burst allocations in both uplink and downlink; and other management messages.

Table 4 lists system throughput for each MCS. We show seven MCSs. The analysis can be easily extended for more flows or MCSs. In our analysis, interference is represented as a change of MCS. To keep it simple, the MCS level is constant over the simulation period.

We consider only downlink resource allocation because we want to analyze the effect of overheads for various scheduling algorithms. For DL, the queue size in number of bytes is translated to the number of requested slots. The analysis for UL is very similar except that the bandwidth requests are used to derive the number of requested slots. However, BS has no information about individual packet sizes. Therefore, both DRR and DRRF cannot be used for uplink scheduling.

We used a modified version of the WiMAX Forum's ns-2 simulator [15] in which a mobile WiMAX module has been added [16].

Consider a topology configuration. Figure 4 shows the simulation topology, a standard point to multipoint configuration. BS is a central control for resource allocation for both uplink and downlink directions. The link between BS and MSs is the only bottleneck link.

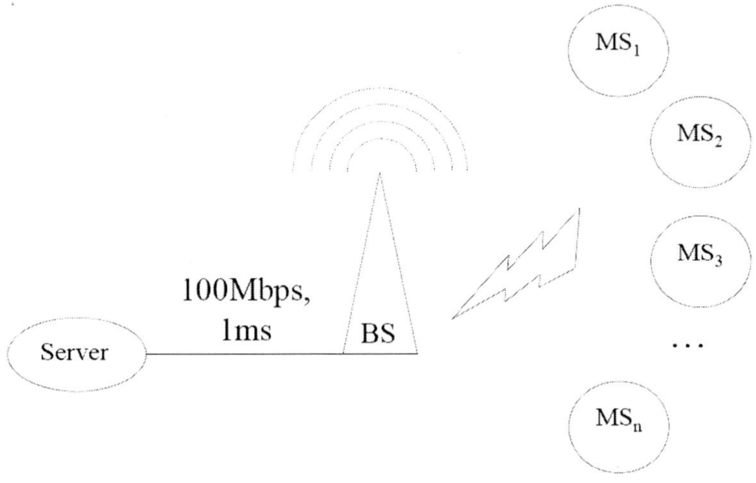

Figure 4. Simulation Topology.

7.2. Simulation Configurations

In this section, we describe the simulation configurations for DRRF in both real-time and non real-time traffic scenarios. The metrics are the throughput (used to achieve three constraints: minimum rate guaranteed, maximum rate, and traffic priority) and percentage of overhead reduction. Note that the percentage of overheads was calculated from a MAC header and a fragmentation subheader. We do not consider an optional packing subheader and other subheaders. In terms of fairness, we used a well-known Jain Fairness index [17], which is computed as follows:

$$f(x_1, x_2, \ldots x_n) = \frac{(\sum_{i=1}^{n} x_i)^2}{n \times \sum_{i=1}^{n} x_i^2} \quad (4)$$

In this equation (Equation 4), x_i is the throughput for the ith MS, and there are n MSs or n flows. For a real-time traffic class, we used the deadline or hard delay constraint as the metric versus throughput fairness.

Again, there are four main configurations: (1) DRRF, (2) DRRF+GWF, (3) DRRF Extensions, and (4) EDF-DRRF (See Table 5). For (1) and (3), the simulations were run from 0 to 10 seconds with 5 seconds of traffic duration. Flows start at 5 seconds, and then end at 10 seconds. There are ranging, registration, and connection setup processes during the first 5 seconds. The packet size was set at 1500 bytes. Each queue is 100 packets long; one queue per flow.

For (2) and (4), the simulations were run from 0 to 22 seconds with 10 seconds of traffic duration. Flows start at 10, 10.02, 10.04, and 10.06 seconds, and then end after 20 seconds. The packet size was set at 500 bytes. At the base station, there is one queue for each mobile station, and each queue is 50 packets long.

Table 5. System Configurations.

Scheme	Number of MSs	Bit Rates	MCS	Note
7.2.1(a) GPS,DRR, and DRRF	5 MSs	CBR; 3 Mbps each	QPSK3/4, 16QAM1/2, 16QAM3/4, 64QAM2/3, 64QAM3/4	
7.2.1(b) GPS,DRR, and DRRF	15 MSs	CBR: 500 Kbps	QPSK3/4	
7.2.2 DRRF and GWF	4 MSs	CBR: 5 Mbps	QPSK3/4	Weight = 0.0, 0.5, and 1.0.
7.2.3(a) DRRF extension	5 MSs	CBR: 3 Mbps	QPSK3/4	One MS is 2 Mbps min rate
7.2.3(b) DRRF extension	5 MSs	CBR: 3 Mbps	QPSK3/4	One MS is 500 kbps max rate
7.2.4(a) EDF, DRRF, EDF-DRRF	3 MSs + 1 MS	3 Videos: 1.35 Mbps average rate 1 CBR: 3 Mbps	QPSK3/4	Underload
7.2.4(b) EDF, DRRF, EDF-DRRF	5 MSs	5 Videos: 1.35 Mbps average rate	QPSK3/4	Overload
7.2.4(c) EDF, DRRF, EDF-DRRF	3 MSs + 1 Ms	2 Videos: 1.35 Mbps average rate 1 Video: 3.3 Mbps average rate 1 CBR: 3 Mbps	QPSK3/4	One ill-behaved flow

7.2.1. GPS, DRR, and DRRF

We illustrate that DRRF outperforms GPS and DRR by showing the fairness and the overhead reduction versus system utilization. To exercise the effect of the scheduler, we used Constant Bit Rate (CBR) traffic flows. We did not use TCP (Transmission Control Protocol) flows because the slow-start feedback mechanism of TCP has a significant impact on the resource usage, and it is difficult to isolate the effect of TCP from that of the scheduling algorithm. In stead, we use CBR traffic so that the results indicate effects on non-TCP applications like mobile video.

First (7.2.1a), to show the effect of throughput fairness, there are 5 MSs in this configuration competing at the same bottleneck link. Each MS has 3 Mbps CBR traffic rate. Each MS has only one flow, and is mapped to one of the MCSs shown in Table 4 (from QPSK3/4 to 64QAM3/4).

Second (7.2.1b), to show the effect of the overhead reduction versus the system utilization, we simulated 15 MSs. Again, each MS has only one flow with 500 kbps CBR traffic. The modulation and coding scheme was fixed to QPSK3/4.

Note that five flows (15 Mbps) and fifteen flows (7.5 Mbps) can saturate the link capacity so that the fairness is exercised; there are dropped packets due to buffer overflow at the base station.

7.2.2. DRRF and GWF

DRRF was chosen to show the effect of GWF. Therefore, the set up is similar to the DRRF simulation configuration. However, we only used 4 MSs with 5 Mbps CBR traffic rate. These 4 MSs can saturate the link capacity so that the fairness is exercised. Although we simulated four CBR flows to demonstrate the effect of GWF, with more mobile stations or number of flows, intuitively the effect of weighted fairness will be more obvious. Moreover, we set the weight values at 0.0, 0.5, and 1.0.

7.2.3. DRRF Extensions

In this setup, we show that DRRF can support three other nrtPS QoS parameters; namely, minimum reserved traffic rate, maximum sustained rate, and traffic priority.

We used Constant Bit Rate (CBR) traffic flows. There are 5 MSs in this configuration competing at the same bottleneck link, and 3 Mbps each. However, we fixed the MCS level to QPSK3/4 and a minimum reserved traffic rate of 2 Mbps for MS_1 (7.2.3a).

With a maximum sustained traffic rate extension (7.2.3b), again there are five MSs with 3 Mbps each used the MCS level of QPSK3/4. In addition, we set the maximum sustained traffic rate of MS_1 to 500 kbps. There is no minimum reserved traffic rate required.

Finally, with a traffic priority extension (7.2.3c), we simulated 5 MSs, 1.5 Mbps each with the MCS level of QPSK3/4. However, we set the traffic priority of MS_1 and MS_2 at 2, and MS_3 to MS_5 at 1. There are no other constraints. The expected fair throughputs of the first two MSs are 1.5 Mbps and the fair share of the left-over capacity for MS_3 to MS_5.

7.2.4. EDF, DRRF, and EDF-DRRF

There are three main simulation configurations in order to show the fairness behavior among all MSs and the delay constraints for all three algorithms.

First (7.2.4a), an under-load scenario with three video flows with 1.35 Mbps average rate each, and one Constant Bit Rate (CBR) flow with 3 Mbps. The purpose of CBR flow is to measure the unused space in the frame. We treated the CBR flow as a lower priority so that the CBR flow acquires transmission opportunity only if there is an unused space within the WiMAX frame. The second case (7.2.4b) is an overload scenario with five video flows with 1.35 Mbps average rate each.

Third (7.2.4c), we used three video flows and one CBR flow; however, one of the video flows is ill-behaved or sending more traffic. Note that because of the effect of link overload, CBR flows do not really get any transmission opportunities in this particular case.

Again, for all video flows, we used the SAM traffic generator [13] to produce a video stream with an average rate of 1.35 Mbps and 3.3 Mbps (for the ill-behaved flow). The SAM traffic generator was based on the video trace from Lord of the Ring movies: Episode I, II, and III. We also randomly chose 10-second scenes for both under-load and overload scenarios. The full trace is available for the research community as shown in the reference [18].

Although we used three to five flows to show the effect of fairness, the results are expected to be similar with a greater number of MSs and higher MCS levels. Note that the video frames were packetized, and so RTP (Real-Time Transport Protocol), UDP (User Datagram Protocol), and IP header overheads were added. All video flows' deadlines were set to 20 ms.

Table 6. System Throughput versus Percentage of Overheads [10].

Algorithm	System Throughput (kbps)	Percentage of Overhead	Jain Fairness Index
GPS	4,658	4.62	1.00
DRR	2,418	0.37	1.00
DRRF	4,651	0.71	1.00

7.3. Simulation Results and Discussions

In this section, we show the simulation results based on the simulation configuration discussed in the previous section to illustrate the effect of DRRF in terms of throughput, fairness, and delay.

7.3.1. GPS, DRR, and DRRF

We show system throughput, percentage of overhead, and fairness index of GPS, DRR, and DRRF. The results show that the throughput is similar to that obtained by our numerical analysis (GPS and DRRF). Note that the actual throughput varies because of the variation of the actual number of burst, Downlink Channel Descriptor (DCD), Uplink Channel Descriptor (UCD), and other management messages being transmitted as well.

For the first scenario, since the number of slots is derived from the queue length, and its MCS level is used, the results show that all three algorithms can achieve perfect throughput fairness (Jain Fairness Index is 1.) The average throughput of GPS and DRRF is around 1.78 Mbps. For DRR, the average throughput is around 1.47 Mbps due to the frame underutilization.

In the second scenario (See also Table 6), the overhead versus system utilization, since DRR only sends a full packet; the percentage of the overhead is the lowest (0.37%). In other words, there are no fragmented packets. Table 6 shows that the overhead of GPS is the highest because GPS has no consideration of the packet size (an allocated portion of the packet is transmitted to each MS in each frame.) The overhead of DRRF is close to that of DRR because it favors full packet transmission with fragmentation allowed.

In terms of system throughput, since the fragmentation is not allowed for DRR, the system throughput of DRR is the lowest, i.e., 2.42 Mbps versus 4.65 Mbps for GPS and DRRF. To sum up, DRRF provides higher system throughput than DRR, and has less overhead than GPS.

7.3.2. DRRF and GWF

Figure 5a, Figure 5b, and Figure 6a show that the throughput follows the GWF criterion with weights of 0, 1, and 0.5. Figure 5a shows that all mobile stations achieve byte fairness with the weight set to 0. Figure 5b shows slot fairness with weight 1. In other words, the mobile station throughput is proportional to its channel condition or modulation and coding scheme (MCS). Figure 6a shows the effect of GWF with weighted fairness of 0.5.

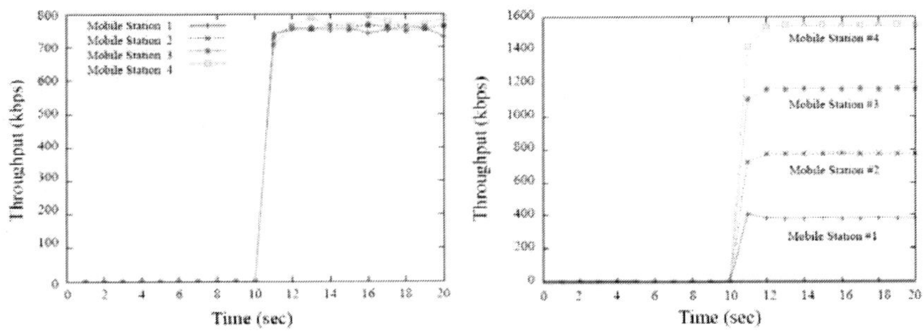

Figure 5. DRRF Throughput. a. (left) 4 flows: with GWF weight 0; b. (right) 4 flows: with GWF weight 1 [11].

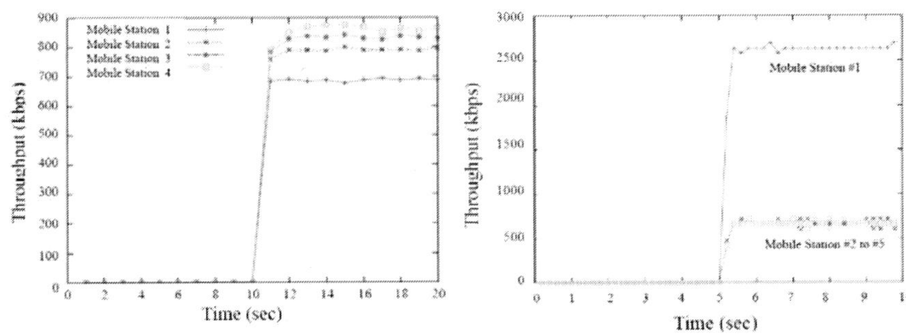

Figure 6. DRRF Throughput. a. (left) 4 flows: with GWF weight 0.5; b. (right) 5 flows: MS_1 with 2Mbps r_{min} [10,11].

Notice that the throughput varies because of Downlink Channel Descriptor (DCD), Uplink Channel Descriptor (UCD), and other management messages being transmitted as well. In addition, since the packing feature is disabled, the anticipated number of a MAC header for each allocation is not precise; the overhead here includes both the MAC header (6 bytes) and the fragmentation subheader (2 bytes). In these and subsequent figures, the variability is low and so the confidence intervals are expected to be small.

7.3.3. DRRF Extensions

We show that DRRF can support three other nrtPS QoS parameters. With a minimum reserved traffic rate extension, the result shows DRRF can support the minimum reserved traffic rate. Figure 6b shows the throughput for all five MSs. In general, MS_1 receives the minimum bandwidth at 2 Mbps, and the left-over bandwidth is distributed fairly among MS_1 and other mobile stations. Observing that to meet the 2 Mbps rate guaranteed, BS also needs to consider two main overheads: a 6-byte MAC header and a 2-byte fragmentation subheader.

For DRRF with a maximum sustained traffic rate extension, Figure 7a shows that the average throughput of MS_1 is 492 kbps, and the left-over bandwidth is distributed fairly among others (fairness indexes are all 1, each flow throughput except MS_1 is around 1.2 Mbps.)

Finally, for DRRF with a traffic priority extension, Figure 7b shows both flows of priority 2 receive equal throughput, i.e., 1.5 Mbps, while the three flows of priority 1 receive a fair share of the left-over capacity, i.e., 0.7 Mbps.

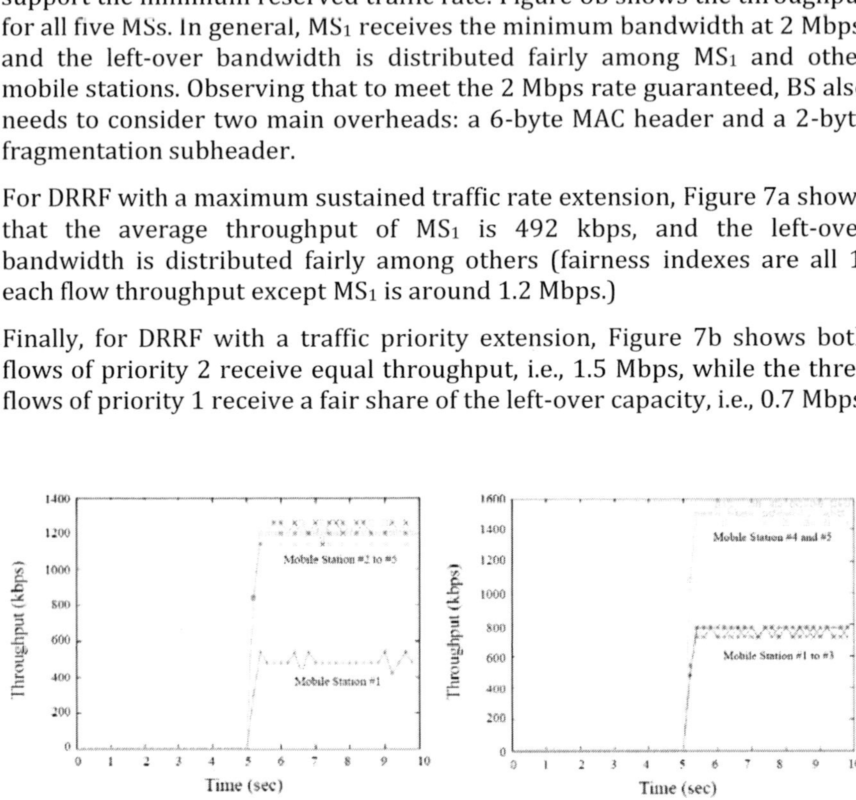

Figure 7. DRRF Throughput. a. (left) with 5 flows: MS_1 with 500 kbps r_{max}; b. (right) 5 flows: MS_1 and MS_1 with traffic priority 2 and MS_3 to MS_5 with traffic priority 1 [10].

7.3.4. EDF, DRRF, and EDF-DRRF

We show system throughput, delay, and delay jitter of EDF, DRRF, and EDF-DRRF with deadline enforcement. For the first scenario (under-load scenario), all three algorithms result in the same throughput, 1.5, 1.24, and 1.49 Mbps with 1.19 Mbps with CBR. There are no dropped packets for video flows. The average delay ranges from 6 to 7 ms.

Table 7 shows the results for the second overload scenario with deadline enforcement. Because of the deadline enforced, average delays for all three algorithms are within the specified deadline of 20 ms plus 5 ms additional transmission delays (duration of the WiMAX frame). EDF is unfair, while DRR and EDF-DRRF are fair. The degree of fairness of DRRF is a bit higher than EDF-DRRF, 0.9998 versus 0.9986, respectively.

Table 8 shows the results for the case with one ill-behaved flow. Deadline is enforced for all flows. Again, EDF cannot maintain the share for well-behaved flows. On the other hand, DRRF and EDF-DRRF can achieve max-min fairness for all flows.

Table 7. System Throughput, Delay, and Delay Jitter with Deadline Enforced (5 Flows in Overload Scenario) (a) EDF, (b) DRRF, and (c) EDF-DRRF [13].

Mobile Station	Sender (Mbps)	Receive (Mbps)	Delay (ms)	Delay Jitter (ms)
(a) EDF				
1	1.49	0.90	21.38	1.80
2	1.18	0.76	20.83	2.07
3	1.53	1.14	21.31	1.79
4	1.24	0.74	21.22	1.83
5	1.47	1.11	21.20	1.75
(b) DRRF				
1	1.49	0.95	18.65	3.84
2	1.18	0.98	16.27	4.30
3	1.53	0.96	19.03	3.46
4	1.24	0.97	16.91	4.31
5	1.47	0.98	18.28	3.86
(c) EDF-DRRF				
1	1.49	0.89	19.63	3.08
2	1.18	0.95	17.07	4.26
3	1.53	0.85	19.90	3.24
4	1.24	0.89	18.04	4.04
5	1.47	0.88	19.63	3.06

The results in Table 7 and Table 8 follow the theoretical expectation in that EDF with a hard deadline constraint dominates the resources resulting in other flows to starve and their packets to be dropped because video traffic is bursty in its nature. Hence, the number of the dropped packets in the EDF scheme is higher than in the other schemes. Bursty packets with deadline constraints results in larger delays for packets in other queues.

Table 8. System Throughput, Delay, and Delay Jitter with Enforced Deadline: 2 Well-Behaved Flows and 1 Ill-Behaved Flow in Overload Scenario (a) EDF, (b) DRRF, and (c) EDF-DRRF [13].

Mobile Station	Send (Mbps)	Receive (Mbps)	Delay (ms)	Delay Jitter (ms)
(a) EDF				
1	1.49	1.19	20.41	1.93
2	1.24	1.03	20.16	1.99
3	3.33	2.66	21.28	1.83
(b) DRRF				
1	1.49	1.49	9.69	3.33
2	1.24	1.24	8.70	3.05
3	3.33	2.34	20.58	2.87
(c) EDF-DRRF				
1	1.49	1.49	9.80	3.72
2	1.24	1.24	9.07	3.41
3	3.33	2.34	20.62	2.77

Consider delay jitter. Since we use bursty traffic which allows a group of continuous packets to be passed in a short time, the results show in a smaller delay jitter value. In comparing DRR and EDF-DRRF, in EDF-DRRF the bursty nature of the traffic restricts the influence of EDF, and allow more dominance of DRRF that results in similar values to the one obtained by using DRRF.

8. CONCLUSIONS

In this paper, we investigated a Deficit Round Robin (DRR) and a simple scheduling algorithm (imitating Generalized Processor Sharing or GPS) in the context of IEEE 802.16e Mobile WiMAX Networks. We introduced a modification of DRR by allowing packet fragmentation (DRRF) to take the best features of both DRR (reducing a fragmentation overhead) and GPS (achieving full WiMAX frame utilization).

To overcome the issue of varying link capacity, we used the slot allocation derived from the queue lengths and MCSs. The results show that all three scheduling algorithms can achieve perfect throughput fairness, but the total utilization of the link and the percentages of overhead are different. DRRF provides the best goodput.

In addition, we formulated a general definition of fairness, the so-called Generalized Weighted Fairness, or GWF. In Mobile WiMAX, this definition provides a compromise between slot fairness or temporal fairness in which the service provider allocates slots fairly, but throughput of the users varies with their locations, and byte fairness or throughput fairness, in which all users get the same throughput, but the service provider has to allocate extra resources for users in poor channel conditions. The GWF criteria allow the service providers to configure the weight parameter to any values between 0 and 1, and achieve any levels of compromise between these two extremes. We also used DRRF as the scheduling discipline to show the effect of GWF.

In general, DRRF can be used as a best effort (BE) scheduler when the fairness is mainly used as the metric. However, to support a non real-time Polling Service (nrtPS) class, a mechanism to derive a proper quantum number was also introduced for DRRF to support users with minimum reserved traffic rate, maximum sustained traffic rate, and traffic priority constraints.

For a real-time Polling Service (rtPS) class, in this paper, video traffic is used as an example. Common well-known scheduling disciplines, Earliest Deadline First (EDF) and DRRF, were investigated. We demonstrated the effect of EDF and DRRF with the deadline enforced. The results show that DRRF outperforms EDF, and maintains fairness under an overload scenario.

Notice that although the simulations show the precise effect of each algorithm, more variations of simulation topologies, configurations, different traffic types including large number of mobile stations, and variable packet sizes can be further investigated. Moreover, with Automatic Repeat reQest (ARQ) and Hybrid ARQ features enabled, the scheduler needs to accommodate scheduling of the retransmission/feedback and the boundary of ARQ blocks [19].

One more requirement for Mobile WiMAX scheduling with Hybrid ARQ disabled is that all downlink allocations being mapped to a rectangular area in the Orthogonal Frequency Division Multiple Access (OFDMA) frame. That restriction can reduce the throughput because some spaces may need to be left unused to make the allocation rectangular [20]. Moreover, the optional packing feature can also be added. These issues need further investigation.

In our analysis, we assumed that the throughput is always non-zero, i.e., the worst case MCS is BPSK. However, it is possible that the channel conditions are so poor that even BPSK will not work. In other words, the base station and/or the mobile station cannot transmit any bits for some time. In this case, the scheduler can simply record the transmitted opportunity of MSs during the unacceptable channel state condition [3]. Then, once the mobile station is in an acceptable channel condition, the MS will receive the transmission opportunity back within some specified thresholds distributed uniformly over some periods to prevent the starvation of other flows.

REFERENCES

1. Institute of Electrical and Electronics Engineers. In IEEE Standard for Local and Metropolitan Area Networks. Part 16: Air Interface for Broadband Wireless Access Systems; IEEE Std-802.16-2009; IEEE: New York, NY, USA, 2009; p. 2082.
2. WiMAX Forum. WiMAX System Evaluation Methodology V2.1. 2008. Available online: http://www.wimaxforum.org/technology/documents (accessed on 25 September 2010).
3. So-In, C.; Jain, R.; Al-Tamimi, A. Scheduling in IEEE 802.16e WiMAX Networks: Key Issues and a Survey. IEEE J. Sel. Area. Commun. **2009**, 27, 156–171.
4. Jain, R.; So-In, C.; Al-Tamimi, A. System Level Modeling of IEEE 802.16e Mobile WiMAX Networks: Key Issues. IEEE Wirel. Commun. Mag. **2008**, 15, 73–79.
5. So-In, C.; Jain, R.; Al-Tamimi, A. Capacity Evaluation for IEEE 802.16e Mobile WiMAX. J. Comput. Syst. Netw. Commun. **2010**, 2010, 279807.
6. Sayenko, A.; Alanen, O.; Hamaainen, T. Scheduling solution for the IEEE 802.16 base station. Int. J. Comp. Telecommun. Netw. **2008**, 52, 96–115.
7. So-In, C.; Jain, R.; Al-Tamimi, A. SWIM: A Scheduler for Unsolicited Grant Service (UGS) in IEEE 802.16e Mobile WiMAX Networks. Lect. Note. Inst. Comp. Sci. **2010**, 37, 40–51.
8. Parekh, A.K.; Gallager, R.G. A Generalized Processor Sharing approach to Flow control in Integrated Services Networks: The Single Node Case. IEEE/ACM Trans. Netw. **1993**, 1, 334–337.

9. Shreedhar, M.; Varghese, G. Efficient fair queueing using deficit round robin. Proc. ACM SIGCOMM Commun. Rev. **1995**, 25, 231–242.

10. So-In, C.; Jain, R.; Al-Tamimi, A. Deficit Round Robin with Fragmentation Scheduler for Mobile WiMAX. In Proceedings of IEEE Sarnoff Symposium, Princeton, NJ, USA, March 2009.

11. So-In, C.; Jain, R.; Al-Tamimi, A. Generalized Weighted Fairness and its Application for Resource Allocation in IEEE 802.16e Mobile WiMAX. In Proceedings of the 2nd International Conference on Computer and Automation Engineering, Singapore, February 2010.

12. Andrews, M. Probabilistic end-to-end delay bounds for earliest deadline first scheduling. In Proceedings of IEEE Conference on Computer Communications, Tel-Aviv, Israel, March 2000.

13. Al-Tamimi, A.; So-In, C.; Jain, R. Modeling and Resource Allocation for Mobile Video over WiMAX Broadband Wireless Networks. IEEE J. Sel. Area. Commun. **2010**, 28, 354–365.

14. Fallah, Y.P.; Alnuweiri, H. Analysis of temporal and throughput fair scheduling in multirate WLANs. Comput. Netw. **2008**, 52, 3169–3183.

15. UCB/LBNL/VINT. Network Simulator-ns-2. Available online: http://www.isi.edu/nsnam/ns/index.html (accessed on 25 September 2010).

16. WiMAX Forum. The Network Simulator NS-2 MAC+PHY Add-On for WiMAX. Available online: http://www.wimaxforum.org (accessed on 25 September 2010).

17. Jain, R.; Chiu, D.M.; Hawe, W. A Quantitative Measure of Fairness and Discrimination for Resource Allocation in Shared Systems. Available online: http://www.cse.wustl.edu/~jain/papers/fairness.htm (accessed on 25 September 2010).

18. Al-Tamimi, A.; So-In, C.; Jain, R. Video Modeling and Generation Tools. Available online: http://www.cse.wustl.edu/~jain/sam/index.html (accessed on 25 September 2010).

19. Sayenko, A.; Alanen, O.; Hamalainen, T. ARQ aware scheduling for the IEEE 802.16 base station. In Proceedings of IEEE International Conference on Communications, Beijing, China, May 2008.

20. So-In, C.; Jain, R.; Al-Tamimi, A. OCSA: An Algorithm for Burst Mapping in IEEE 802.16e Mobile WiMAX Networks. In Proceedings of the 15th Asia-Pacific Conference on Communications, Shanghai, China, October 2009.

12

CHAPTER

A SIGNALING ARCHITECTURE FOR MULTIMEDIA MBS OVER WIMAX

Mauro Femminella*, Roberto Francescangeli and Gianluca Real

ABSTRACT

In this article, we design and demonstrate a signaling architecture for multicast and broadcast services over a laboratory-emulated worldwide interoperability for microwave access (WiMAX) network. WiMAX is a broadband wireless access technology which includes different quality of service levels. Currently, a significant research effort focuses on data transmission optimization and mobility support for multicast and broadcast services. Our proposal is fully IP-based, and the relevant signaling architecture is modular in all its entities and extensible to other, non-WiMAX, IP access networks. We specifically address the non-trivial support of multicast services, which has been demonstrated through a prototype implementation of our proposal for an Internet protocol television (IPTV) service. This prototype has been implemented by using open source technologies and its signaling is orchestrated by a JAIN SLEE server. The results obtained show that the implemented system is scalable, can achieve both high signaling throughput and low service latency, and has a signaling overhead lower than similar IMS-based solutions.

Keywords: MBS JAIN SLEE WiMAX signaling SIP IPTV

1. INTRODUCTION

The worldwide interoperability for microwave access (WiMAX) is a wireless broadband technology. Its architecture includes a set of flexible capabilities which enable the convergence of mobile and fixed broadband networks. By using suitable communication technologies, such as multiple input multiple output smart antennas and orthogonal frequency division multiple access, WiMAX can provide high throughput, efficient data multiplexing, and low transmission latency in both fixed and mobile communications. These features, together with the capability of covering large areas with relatively few base stations (BSs), are fundamental to deploy effective wireless broadband network services able to support many applications, such as data exchange, audio/video streaming, and VoIP services, with different levels of quality of service (QoS) [1]. Leveraging on these network services, service providers can also implement the so-called Multicast Broadcast Service (MBS) over WiMAX, which consists of providing mechanisms for delivering the same contents to multiple users who share radio resources. In particular, a single-frequency operating mode is envisaged, through which a single radio channel is used to distribute information to all users subscribed to the same service [2]. The MBS is typically used for delivering multimedia contents, such as mobile Internet protocol television (IPTV) and audio/video file casting, and for implementing massive software updates [1, 2]. Implementing MBS over WiMAX is not trivial for service providers, since customers may also need a return channel for interacting with a management system in order to dynamically subscribe/unsubscribe to any available service. Moreover, multicast services are more complex to be designed and maintained than broadcast services, due to additional signaling, content delivery authorization and management, and required support of multicast streams in WiMAX subscriber stations (SSs) and mobile stations (MSs). Finally, an MBS architecture must be scalable with the number of users, requires a central management system, should both include an easy to use management software application and be implemented through standard technologies so as to simplify MBS service development and maintenance.

In this paper, we illustrate an original design, and the relevant prototype implementation, of a simple and effective end-to-end service architecture that extends the MBS Network Reference Model (NRM) [3] and provides MBS over WiMAX.

Our aim was to create a flexible and efficient signaling system for MBS architecture, with a low architectural complexity, so as to reduce the initial investment for service providers. In fact, an MBS system has a quite complex set of requirements, such as the need of using a return channel and a central service orchestration point that needs a suitable design. This design has to guarantee scalability, manageability, efficiency and maintainability. Most the recent literature on MBS focuses mainly on data transport over WiMAX networks. Nevertheless, in order to achieve the MBS requirements mentioned above, a suitable signaling support is also essential.

Our proposal addresses these features by resorting to available standard and open interfaces that speed up implementation process and reduce time to market. In fact, the use of development frameworks providing high-level APIs may be extremely helpful to service creation, since suitable APIs can mask the complexity of underlying network and transport layers. In addition, the use of standard open interfaces, rather than proprietary ones, has the additional advantages of both supporting multiple access networks, devices, and protocols, and attracting a larger number of developers. For these reasons, some key implementation issues of our MBS architecture have been faced by resorting to the Service Logic Execution Environment (SLEE), which is a hosting environment for advanced telecom services. The SLEE logic is designed to fulfill the specifications of telecom services, which are typically asynchronous and require high throughput and low signaling latency. Thus, a SLEE server can be useful to effectively orchestrate the overall signaling architecture of MBS over WiMAX.

Our MBS solution is IP-based and adopts standard IP signaling protocols, such as SIP (Session Initiation Protocol) [4] and IGMP (Internet Group Multicast Protocol) [5]. We have implemented the core signaling elements by using an open source JAIN SLEE application server (AS), named Mobicents [6]. Being based on standard protocols and open interfaces, our architecture is not limited to WiMAX, but it may both include other existing networks, and be accessed by most existing terminal devices. As a proof of concept, we have implemented an IPTV multicast service on top of our MBS prototype, which is often indicated as the main application for MBS architectures [2]. We show that this service is scalable with the number of users, flexible in the deliverable contents, modular in all its entities, and centrally manageable. Latency and throughput are evaluated experimentally, and an in-depth analysis of the signaling overhead and its impact on a WiMAX network is presented.

The paper is organized as follows. The next background section shows a brief overview of the SLEE container, including the open source SLEE package that we have used in our implementation; it also illustrates the

MBS over WiMAX issues and some recently published results, including a related work about an alternative signaling architectures for IPTV. Section 3 describes the design of our service architecture and the end-to-end signaling flows between network entities. Section 4 illustrates the test-bed implementation. Section 5 shows the achieved performance, including a specific analysis of the signaling overhead. Some final remarks are reported in Section 6.

2. BACKGROUND AND RELATED WORK

2.1 JAIN SLEE specifications

A SLEE is a service container made of several abstraction layers. Its multi-tier architecture simplifies the development of complex services by providing the non-functional features needed for their execution, such as object lifecycle management, persistence management, thread management, object pooling, and so on.

The Java APIs for Integrated Networks (JAIN) activity has delivered the JAIN SLEE (JSLEE) specifications, which delineate a Java-based, event-oriented, and protocol-agnostic container suitable for hosting carrier-grade telecom services [7]. It includes different Java Enterprise Edition (J2EE) components, adapted to the needs of telecom applications.

In a JSLEE container, a service logic is organized and implemented in system components called Service Building Blocks (SBBs). In operation, a JSLEE server creates a pool of SBB objects and manages them according to a well-defined lifecycle. SBBs operate asynchronously by receiving, processing, and triggering events. They are attached to one or more data streams called Activities, by which they can exchange events and share state variables (Activity Context) regarding the Activity itself with other JSLEE system entities. SBBs may be hierarchically structured according to parent-child relations, in order to enhance service logic modularity. Events are internally managed and routed by a functional entity called Event Router, which delivers each event to the appropriate recipient SBB. With regard to the JSLEE performance, an Event Router plays a critical role, since it processes all system events.

External network events, such as a SIP message arrival, are translated into JSLEE events by means of specific entities called Resource Adaptors (RAs). Thus, the set of available RAs constitutes an abstract interface layer that allows a JSLEE to access external resources. If service interoperability with a particular protocol stack is needed, it can be achieved easily by including the relevant RA interface in service code (e.g. methods to access SIP header fields) and deploy the appropriate RA Java archive (JAR) file within the JSLEE container.

Our MBS signaling server has been developed by using the Mobicents Communication Platform [6] that is an open source project currently owned by Red Hat. Mobicents includes a JSLEE, a Media server, a Presence server, and a SIP Servlet server. At the time of writing, the Mobicents JSLEE (MSLEE) is under active development and it is the first and only open source implementation of the latest JSLEE v.1.1 specifications [7]. The MSLEE includes several J2EE components, such as Container Managed Persistence (CMP) fields, which enable data persistence for SBB objects, Java Database Connector (JDBC) drivers, Java Management Extensions, SNMP interfaces, and Java Naming and Directory Interface (JNDI), which provides lookup functions for services and variables. In order to leverage the aforementioned J2EE technologies, the MSLEE is deployed and executed within the JBoss AS [8], which plays the role of container of containers, offering advanced capabilities such as JSLEE service deployment, thread pooling, logging, etc.

2.2 MBS over WiMAX

A broadcast service may be defined as a unidirectional point-to-multipoint (PMP) service in which data is transmitted from a single source to all terminals (Subscriber Units, SUs, including either MSs or SSs) included in its broadcast service area. Thus, broadcast services are often referred to as *push*-type services [9]. Multicast services are unidirectional PMP services in which data is transmitted from a single source to a multicast group of SUs located in the multicast service area. Only SUs that have subscribed to a specific multicast service and joined the relevant multicast group can receive data relevant to that service. Whilst a broadcast service can be accessed without any explicit request from customers [10], multicast users need a return channel for interacting with a management system, in order to dynamically subscribe/unsubscribe available services. Thus, designing an end-to-end multicast service architecture is not trivial since it must fulfill all service requirements and be compliant with the underlying network architecture.

The IEEE 802.16-2009 [11] standard specifies the MBS deployment over WiMAX networks. MBS provides an efficient transmission of multimedia streams from one or more BSs to multiple MSs through a shared radio resource [2]. Multiple BSs compose an MBS zone, where BSs synchronously transmit over the air multicast data belonging to the same service flow, by using a common connection identifier (CID) and through the same security association. Since BSs located within the same MBS zone use a common Multicast CID (MCID) for the same MBS multimedia stream transmission, MSs do not require registration to any specific BS and the handover functions are transparent to a MS moving within the same MBS zone.

The MBS service can be provided over the WiMAX air interface [12], since:

- high data rate and large coverage using a single frequency network;
- flexible radio resource allocation that enables dynamic radio resource management in MBS areas;
- low MS power consumption;
- low channel switching time, in order to support multichannel services such as mobile IPTV.

MBS over WiMAX is suitable for applications that need to simultaneously provide large subscriber groups with the same data. Usage of broadcast/multicast bearers allows multiple receiving SUs to share downlink radio resources and receive the same data in parallel. Hence, this technology allows network operators reducing costs for delivering multimedia services, such as mobile IPTV, audio/video file-casting, mass software, and configuration updates. Further, multicast/broadcast transmission allows such services to scale well with the number of users, without modifying the network capacity even in case of extremely high service demand. Multicast services are more complex to be designed and maintained than broadcast services due to additional signaling exchange and content delivery authorization and management. For this reason, the implemented service, shown in what follows as a proof-of-concept of our architecture, focuses on multicast transmission of IPTV to WiMAX users.

2.3 Related studies on MBS over WiMAX

Some of the recent literature on the MBS-enabled WiMAX focuses on the extension of the baseline MBS architecture in order to improve the performance of video broadcasting over the air [2, 13]. Other articles are related to video QoS provisioning [14, 15] and handover management [16]. Nevertheless, the end-to-end signaling support for IPTV provisioning over WiMAX is not investigated. In [17], the multicast service management is optimized in terms of channel switching time, uplink bandwidth, and power consumption. Nevertheless, this proposal requires SU modifications, which do not operate in a standard way, especially for what concerns IGMP message handling. We believe the need of modifying SUs to be a strong limitation.

Differently, we prefer the solution of using a dedicated service delivery entity [2, 13–15, 17] and enhance it by adding signaling and authorization capabilities. In addition, the solutions proposed in [2, 14, 15] can easily be integrated in our architecture, since we do not require any specific procedures for handling multicast data flows.

In regard to signaling for IPTV, the articles [18, 19] present an approach based on the IP Multimedia Subsystem (IMS) [20]. In [18], the authors present a detailed SIP exchange for live TV and Video On Demand, both for content request and content switch, and analyze the relevant message delays. In [19], an IMS-based architecture for providing converged IPTV services, compliant with the ETSI TISPAN NGN release 2 specifications, is illustrated. Extensions to IMS session setup for supporting multicast-based many-to-many services are discussed in [21]. Proposals in [22, 23] enhance the IMS-based service architecture illustrated in [19], by adding scenarios aiming to a future interactive TV experience, illustrated by a SIP signaling exchange.

The SIP-based IMS architecture is the most largely used in the IPTV signaling proposals [18, 19, 21–24]. Nevertheless, IMS is a very complex reference architecture, and its complexity often leads service providers to resort to simplified architectures, often proprietary, to provide users with advanced services. Due to its complexity, IMS typically exhibits worse performance than a pure SIP-based service architecture [25]. Our aim is to create a simple and effective architecture that both require a very small initial investment and provides advanced MBS services without changing existing network entities and configurations.

Finally, with reference to IPTV over general network architectures, it is worth to mention a very interesting survey in [24], whereas quality of experience of an IPTV service is analyzed in [26].

3. SYSTEM ARCHITECTURE

To facilitate end-to-end interoperability, the WiMAX forum has specified the WiMAX NRM [3], that is, a logical representation of the network architecture. The NRM is based on all-IP core and a packet-switched air interface. The main NRM advantage is that the network entities are agnostic of the IEEE 802.16 radio specifications. In [2], a new entity, called MBS Controller (MBSC), has been included within NRM. However, as mentioned above, the MBSC has been defined only for improving service data delivery, without specifying signaling entities and signaling flow management.

Figure 1 shows our proposed NRM. The top-level scheme includes the following logical entities: client MS/SS, network access provider (NAP), connectivity service network (CSN), and MBS. MS/SS includes the generalized collection of functions used to provide connectivity between mobile/subscriber equipment and BS. The NAP includes access service network (ASN) entities. Each ASN includes an ASN gateway (ASN-GW) and BSs. ASN is a node aggregation defined as a logical boundary in a mobile WiMAX radio access network. Typically, an ASN consists of multiple BSs that perform radio-related functions, and a gateway node (ASN-GW), that both interfaces with a CSN (where the key functional entities for WiMAX operators reside) and provides IP connectivity services to the WiMAX MS/SS. A CSN includes the MBSC, which in turn includes three separate entities, the MBSC enhanced transport layer [15] (MBSC-ETL) and two further entities that we propose to include in this architecture, the MBSC signaling server (MBSC-SS) and the MBSC database (MBSC-DB). The MBSC-ETL is used for scheduling and transmitting MBS data and for chopping contents of both announcements and data. It is also used to manage MBS zones, MCID, IP addresses, and security at MAC layer [15]. Some key reference points, R1 and R6, also are shown in Figure 1. The MBSC-SS and the MBSC-DB manage service announcements, user subscriptions, security at service level, and session state management for charging and policing. In more detail, the MBSC-SS handles all signaling messages related to providing multicast services to subscribers, whereas the MBSC-DB is the database where the status of each subscriber is stored (e.g., subscribed packages, joined multicast flows, current user activities using multicast services).

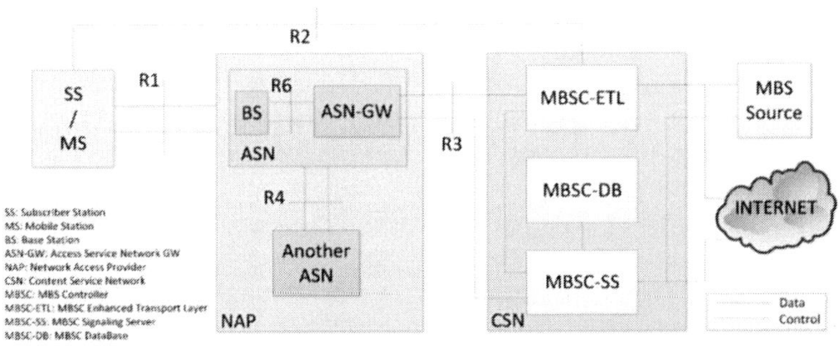

Figure 1. PROPOSED END-TO-END ARCHITECTURE FOR MBS OVER WIMAX.

The MBS is used to offer data services, including real-time and non-real-time multimedia streaming, together with relevant service announcements. Since the NRM is IP-based, streams can be transmitted from either a dedicated multicast server (e.g., a content provider directly connected to the CSN) or an Internet streaming server.

3.1 Signaling

As mentioned above, providing MBS over WiMAX is not trivial for many reasons, such as the need of using a return channel for interacting with the network entities and dynamically subscribe/unsubscribe services. Thus, in what follows we focus on the design and the evaluation of a suitable signaling architecture for multicast services. A simplified version of our architecture can also be used to provide broadcast or *push*-type services.

Our solution is IP-based and it includes standard IP signaling protocols only, such as SIP and IGMP. This way, the scope of the signaling architecture is not limited to WiMAX access networks, but it includes most of existing data networks and is accessible by most existing terminal devices.

SIP is the main signaling protocol of IP multimedia services [24] due to both its flexibility in supporting different signaling scenarios and its reliability, also when messages are transported by UDP segments. In addition, SIP is the *de facto* standard protocol for VoIP communications and it is central in the 3GPP IMS [20]. However, it is worth noting that in this context the concept of MBS session is very different from the call session typically used in the Internet telephony field. In fact, in our case signaling is used to set up a unidirectional point-to-multipoint media flow, whereas in the Internet telephony it is typically used to set up a bidirectional, point-to-point media flow. Also the pricing scheme is different. In Internet telephony each call is typically charged on a per-second basis [27], and the use of SIP calls handled by SIP back-end servers on the signaling path is necessary to correctly apply charging policies. Instead, in the scenario proposed here, it is much more suitable a per-session charging. Thus, we have decided to use simple, session stateless, SIP MESSAGE requests to wrap data exchanged between MBS entities [4, 28], instead of the classic SIP INVITE used in IMS-based systems [18, 19, 21, 29]. This approach simplifies signaling management (e.g., see [21]) and maintains user state consistent by keeping its information updated in the database running in the MBSC-DB.

Finally, IGMP is used to manage multicast groups. Clients use IGMP to report their multicast group memberships to any immediately neighboring multicast router. There are essentially three IGMP message types: Report, Query, and Leave Group. A Report message is used in two situations. When a client wants to receive a specific multicast stream, it sends out an unsolicited Join Report message to the local router in order to join the group. Another situation is when the client passively generates a Membership Report for its interested groups in response to a Query message. A Leave message is sent by a client when it leaves a multicast group. It allows group membership termination to be quickly reported to the router in order to release resources.

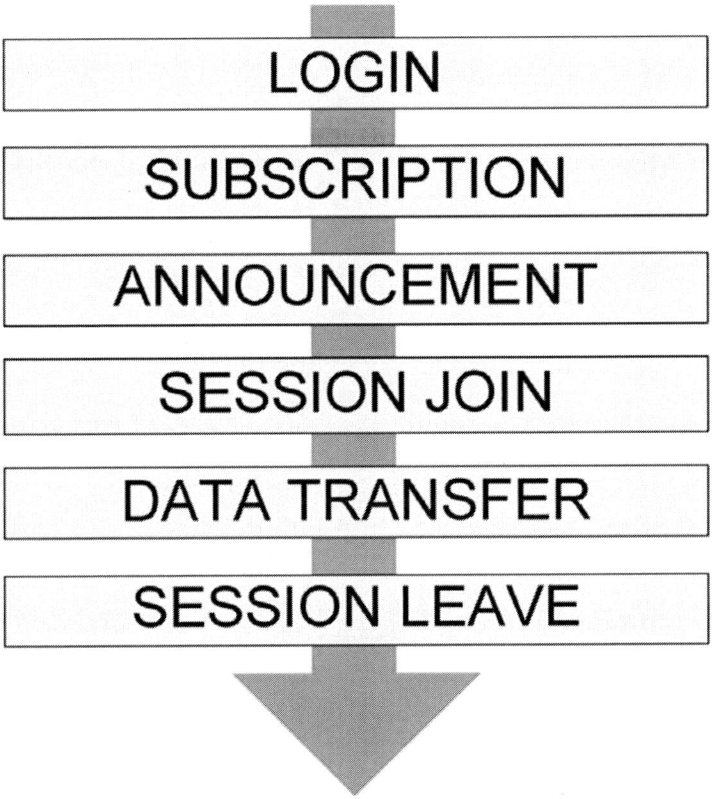

Figure 2. MBS SERVICE FLOW.

Our MBS architecture has been designed according to the service flow shown in Figure 2 (see also [10]). First of all, a user must authenticate its software client with the MBSC-SS by using its credentials, which implies that he has to log in to the system. User's credentials (user name and password or digital certificates) are usually obtained during the initial service registration, which is typically performed through a web page advertising the multicast/broadcast service (i.e., an IPTV web portal). Clearly, the registration step to activate a service has to be executed only once.

When the *first* user access to the MBS system is gained, he is asked to subscribe to categories of interest. For example, in an IPTV service user can subscribe to a package containing his favorite TV channels, whereas in a software update service user can select the software type to be updated (e.g., anti-virus or system updates). Users can add or remove subscriptions at any time, after being authenticated by the MBS system through the *login* phase described above.

In the *announcement* phase, users are informed about the subscribed MBS sessions, such as list of TV channels and the relevant contents. An announcement can be distributed in either multicast or unicast fashion. We have implemented both approaches. In case of multicast distribution, we have provided a specific multicast channel, for each package or channel offered by the service provider, to send announcements from the MBS source to clients. In case of unicast distribution, a user periodically receives announcements relevant to its subscribed packages or channels from an announcement server, co-located with the MBS source. In both cases, either the multicast IP address of the announcement multicast group, or the IP address of the announcement server, is communicated to the client during his subscription to the relevant channel or package. After receiving an announcement, a user can join the multicast group (*session join*) relevant to the selected content and, after being authorized by the MBSC-SS, the *multicast data transfer* phase from the MBS source starts. Finally, if either a data stream ends or a user leaves the current session, the *session leave* signaling exchange closes the network bearers towards the specific subscriber and, if that user is the last one within an MBS zone, network resources are released.

To implement the aforementioned service phases, our proposal includes two main signaling flows: one for the login-subscription-announcement chain and one for the join/leave a multicast data session. In what follows, when we refer to user credentials, we assume that the considered user has already registered with the service provider prior to use the proposed MBS architecture.

In order to preserve generality, we keep client and WiMAX SU entities separated. For example, the SU could be the WiMAX interface of a vehicular gateway, and the client could run within a notebook/smartphone connected to the gateway through a WiFi connection). Clearly, in case of mobile WiMAX terminals, they are co-located. Note that being the MBSC-ETL entity part of the data plane it is not involved in the following signaling exchanges of the control plane.

3.1.1 Login, subscription, and announcement

The login, subscription, and announcement signaling exchanges are implemented by the message exchange shown in Figure 3. The initial exchange is used to authenticate the user with the MBSC-SS. A normal *SIP Register* request is sent by the client device to the signaling server. In this case, the server acts as a SIP registrar and, upon receiving the SIP request, it issues an authentication challenge to the user by sending a 401 *Unauthorized SIP* response back containing a nonce. At this point, the client can send another SIP Register request including the received nonce and the credentials requested. Since SIP is easily extendable, the second Register request could include additional headers and also a body part to help conveying authentication data to the MBSC-SS. For example, additional information needed could be SU MAC address, service id, public key certificate, and so on. When the server receives the second request, it checks the user identity against the MBSC-DB that contains a list of eligible users (those that have previously subscribed to the service). Finally, it sends a 200 OK response back to the client, so closing the login phase. Once a user has successfully logged in, the subscription phase follows, as depicted in Figure 3. The subscription signaling messages are used to allow customers to subscribe to multiple packages, each containing one or more MBS contents. For example, an IPTV service may allow selecting different channel packages, or a software update service may offer different software types. The user sends a SIP Message request to the MBSC-SS in order to subscribe/unsubscribe to a specific package. The signaling server MBSC-SS checks on the MBSC-DB whether the user is allowed to receive the specified service and responds by sending another SIP Message back to the client. Each SIP Message request is followed by an empty 200 OK response on the opposite direction [28], shown in Figure 3. Clearly, the subscription phase is not executed at each log in, but only when the user wants to change the list of subscribed packages.

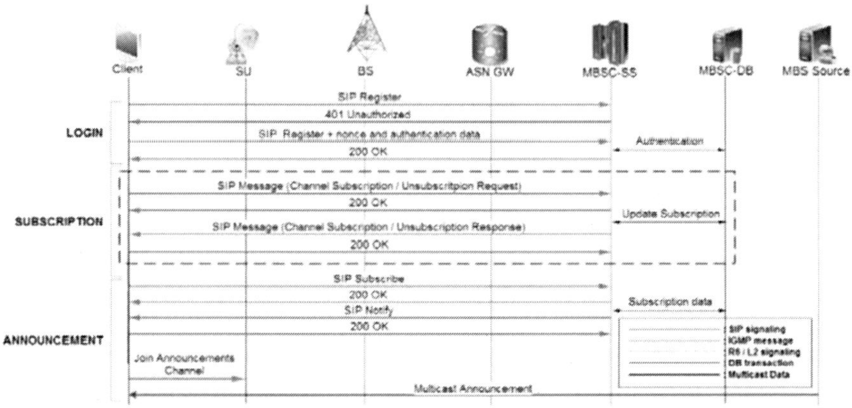

Figure 3 SIGNALING FLOW FOR USER LOGIN, PACKAGE SUBSCRIPTION, AND ANNOUNCEMENTS.

Once logged in and subscribed to a service package, the client device needs to discover the multicast address of the announcement stream in order to receive the relevant service announcements. This is achieved by sending a *SIP Subscribe* request to the MBSC-SS. The signaling server sends a 200 OK response back and checks on the DB the appropriate announcement channel(s) for that user. In fact, the announcement channel could vary depending on the content of the service being advertised and on the user type or permissions. Once the multicast IP addresses and ports used to transport announcements are selected, the signaling server can send them to the client within a *SIP Notify* message. The client then acknowledges the notification by sending a 200 OK response back and then proceeds in joining the announcement multicast groups. When the user requests to join an announcement multicast stream, the client device sends an unsolicited IGMP report, shown as an orange arrow in Figure 3, in order to join the multicast group and receive announcement data. More details on how to join a multicast group in our MBS over WiMAX architecture will be given in the next section and illustrated in Figure 4. Announcement data can also be made available in a unicast fashion, for instance by downloading a file from a web server using HTTP; in this case the *SIP Notify* request sent by the MBSC-SS to the client includes a set of URLs used by the client to access the announcement file. At this point the client is able to receive the announcement data of the service provider from the MBS source. Thus, the user has all the information needed to join a specific MBS session.

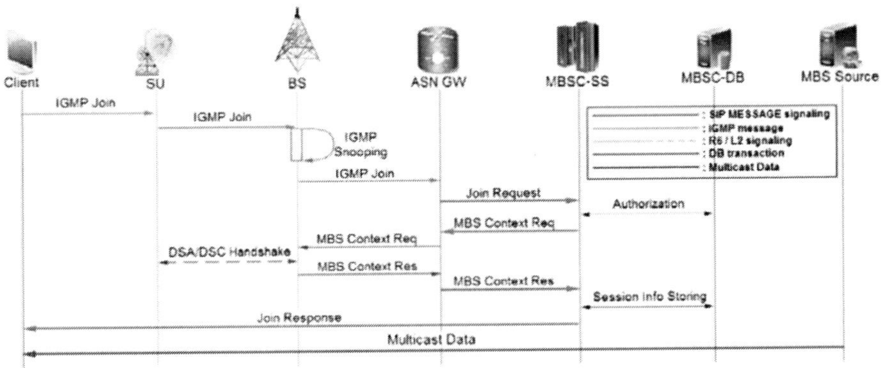

Figure 4 SIGNALING FLOW FOR SESSION JOIN.

In our solution, we have encapsulated signaling messages into the body of *SIP Message* requests. The reason of this approach is that SIP Message requests can be issued at any time and they do not require any SIP session to be established and managed by a signaling server, thus decreasing the overall MBSC-SS load. In addition, a SIP Message can contain different payload types, including contents encoded also in binary format and not only in the text-based format typical of SIP, even though the original design purpose of SIP messages was to carry instant messages in VoIP communications [28].

3.1.2 Session join/leave and session refresh

The signaling flow for a client joining or leaving an MBS session over WiMAX is shown in Figure 4. Before going into details, we point out that all the SIP protocol messages used in Figure 4 are *SIP Message* requests. For the sake of neatness, all SIP 200 OK responses are not shown in Figure 4.

In order to join a multicast group, the client becomes aware of the multicast address and port through the *announcement* phase and issues an unsolicited IGMP report. This report is forwarded by the SU to the BS, and from the BS to the closest multicast capable router which, in our scenario, is the ASN-GW. In order to optimize traffic delivery, the BS *can* "snoop" the IGMP report [30] and send it to the ASN-GW only. Once the latter receives the IGMP report, a new SIP *Message* request, called *Join/Leave Request*, is created and sent to the MBSC-SS. The message body of the *Join/Leave* request contains both user information and the IGMP report that has triggered it. The signaling server parses such an information and uses it to check whether the user is allowed to join the desired session. For example, the MBSC-SS can check on the MBSC-DB if the user has subscribed to that session. If the authorization check is successful, the MBSC-SS issues an MBS Context Request towards the ASN-GW in order to either trigger the creation of a new multicast context or reuse an already existing one for the new client. When the ASN-GW receives the MBS *Context* request, a message exchange between ASN-GW and BS is initiated through the R6 reference point. If this signaling exchange is successful, it triggers a BS-initiated three-way Dynamic Service Addition handshake, carried out at MAC layer, between the BS and the SU of the client. Once the multicast context is created, the BS notifies the ASN-GW by sending it another R6 message, called MBS Context Response. The ASN-GW, in turn, creates a SIP Message to notify the MBSC-SS that it can save the session information on the MBSC-DB.

The final step of the procedure consists of another SIP message, called *Join/Leave Response*, sent by the MBSC-SS to the client (see Figure 4), and used to notify it whether his request was successfully processed by the signaling entities. It is worth noting that this message is not mandatory, and could be avoided in order to save bandwidth, unless it is used to transport some information needed by the application to receive the content. For example, it could transport a cryptographic key needed by the client software to decode the streamed data, or custom data enabling more options for service providers, such as live advertisements or notifications. When the packets of the joined session are routed to the BS, they are delivered to the SU through the established multicast connection over the air.

When the client leaves the multicast group, the same signaling flow as shown in Figure 4 is triggered by sending an IGMP *Leave* message to the neighboring multicast router (ASN-GW). This signaling might cause the deletion of the over-the-air connection by means of BS-initiated *Dynamic Service Deletion* messages, transmitted on the Primary Management Connection of the MS/SS, if the leaving SU is the last multicast user for that multicast connection on the BS, and the consequent release of the unused wireless resources.

To refresh an already established session, we make use of the IGMP soft-state refresh, as shown in Figure 5. By default, every 125 s, the first multicast-enabled router on path (in our case the ASN-GW) issues a multicast IGMP query message. All users interested to the advertised multicast group, must respond to this message. Every IGMP report sent by the users is in turn encapsulated by the ASN-GW into a SIP Message containing a *Join* request and sent to the MBSC-SS. When the message arrives at the MBSC-SS, the signaling server becomes aware that it represents a session refresh by checking the MBSC-DB, and it updates the session information accordingly. Finally, the refresh may be acknowledged to the user by a *Join* response, if this message is requested by the application.

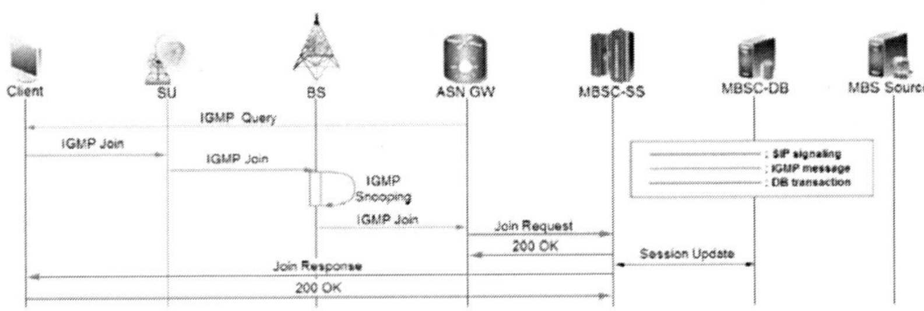

Figure 5 SIGNALING FLOW FOR SESSION REFRESH.

Another option consists of defining a timer inside the MBS client that sends unsolicited IGMP messages towards the ASN-GW. In this second case, the timer for the IGMP query sent by the ASN-GW could be relaxed or disabled at all, since clients autonomously refresh their status. It is easy to show that the second solution does not imply additional overhead with respect to the first one, which is analyzed in detail in the article. In this second option, when the ASN-GW receives the IGMP message, and encapsulates it inside the payload of a SIP message towards the MBSC-SS, it includes additional information such as the BS serving the MBS client, which, in turn, has an its own identifier. When this information is received by the MBSC-SS, it checks the data stored in the MBSC-DB to determine if the previous status of the MBS client is changed. This step allows the MBSC-SS to realize if the client has changed the serving BS, and thus to trigger, if necessary, a signaling procedure to update the system by using exactly the same messages already defined (i.e., MBS Context Req and MBS Context Res). Thus, this way of performing session refresh could allow an easy way to maintain active the multicast session in case of client mobility.

Differently from [17], where the BS infers its subordinate interest of SU in a multicast group from the MAC-layer management messages, in order to avoid periodic exchange of IGMP reports, we prefer to not include additional, mandatory functions in the BS. Thus, the IGMP exchange occurs on the wireless channel in a standard way, since the overhead of exchanging IGMP message is small (see Section 5.2).

If either a user leaves the session abnormally, thus without issuing an IGMP leave message, or the client software crashes, such a situation is handled at the MBSC-SS level. Within the signaling server, we have designed a further simple service, which periodically checks the MBSC-DB to make sure that there are no users with active sessions that have not been refreshed for more than a fixed amount of time. If it happens, the MBSC-SS issues an MBS Context Request that acts as an unsolicited, MBSC-SS-initiated leave procedure (see Figure 4). This periodical check is very useful to release resources on the wireless interface when the user that got stuck was the last one in his multicast session. A good choice for the value of this timer is an integer multiple of the IGMP refresh period (the default value is 125 s).

4. PROTOTYPE IMPLEMENTATION

In this section, we illustrate the implementation of the testbed used for validating the proposed service architecture, in particular the end-to-end signaling architecture for MBS over WiMAX. Thus, we focus on the entities involved in the signaling process only and exclude those belonging to the data path, such as the MBSC-ETL, that has already been described in the literature [2, 14, 15].

The testbed has been designed to provide a multicast IPTV service as an instance of a typical MBS over WiMAX service. In what follows, "TV channel" and "multicast session" share the same meaning.

Five main entities are involved: MBSC-SS, which is the core of the proposed signaling service architecture, MBSC-DB, ASN-GW, client, and MBS Source.

4.1 MBSC-SS

The MBSC-SS is the key functional entity that orchestrates all the components used to enable, create, and manage the MBS IPTV streaming service. In order to fulfill the requirements illustrated in Section 2.3, the Mobicents JSLEE (MSLEE) has been selected to implement the central signaling server. It handles all the incoming SIP signaling messages from the MBS clients and the ASN-GW. In addition, it performs user authentication, service subscription, session join/leave, periodic check of active users, and their authorizations.

The MSLEE service, running in the same MBSC-SS where the signaling logic resides, has been implemented in a single SBB, called MessageSbb. As mentioned above, all the SIP sessions used in our architecture are stateless, differently from the typical INVITE-initiated VoIP call dialogs [31]. Thus, the SBB has been configured to process all incoming SIP events fired by the SIP RA as a root SBB. Specifically, each incoming SIP request is processed by a different MessageSbb entity, which is identified by the Call-ID value of the originating SIP request. As a design choice, requests belonging to the same transaction (e.g., MBS Context REQ--RES) share the same Call-ID value and are processed by the same SBB entity, thus increasing the MSLEE efficiency.

The format of data encapsulated into the body part of the SIP Message requests is based on the Cisco® R6 interface [32], which consists of a set of control and bearer plane communication protocols between BS and ASN-GW [33]. In short, in our implementation each data packet is encoded as a binary Type-Length-Value (TLV) data structure and located within the body field of a SIP Message request. After receiving such a request, the recipient replies with a final SIP response with an empty body (i.e., 200 OK) [28]. The ASN-GW encapsulates the R6 message into the body of SIP Message requests. In this way, the modifications needed in the ASN-GW are marginal, since it only has to encapsulate the relevant R6 and IGMP protocol messages into a SIP message body. Thus, they affect the implementation effort and the complexity of the signaling architecture on the MBSC-SS only. If needed, given the SLEE flexibility and extendibility, the MBSC-SS can easily be modified, so as to support other R6 protocols, different from the Cisco® version.

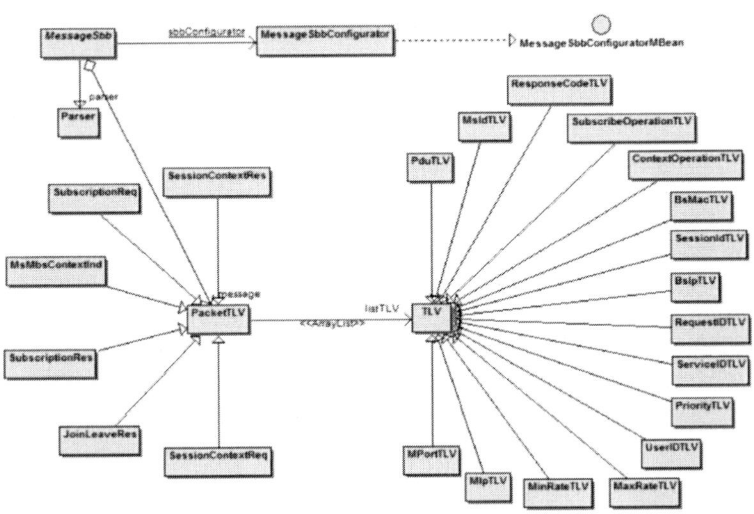

Figure 6 UML CLASS DIAGRAM OF THE JSLEE SERVICE.

When a MESSAGE is received, its content is parsed by using the *parser* Java class that decodes and creates Java objects from each TLV (e.g., BsMacTLV, BsIpTLV, SessionIdTLV, etc.). Each TLV field and each packet exchanged have been modeled by using a different Java class, as shown in the UML class diagram shown in Figure 6. The parsed data, wrapped into a Java object, are then processed inside the SBB class, and it is used to both query/update the database and send subsequent requests.

The signaling management service runs on top of Mobicents JSLEE v.1.2.4.GA, which is executed in the JBoss AS v.4.2.2. The designed service makes an extensive use of the external database (MBSC-DB), hence the MySQL Connector/J v.5.1.6 JDBC, which is the official JDBC driver for MySQL, has been deployed onto JBoss and configured to access MBSC-DB from the MBSC-SS service class.

The *MessageSbbclass* has also been equipped with a management interface by means of a Management Bean (MBean) class called MessageSbbConfigurator (Figure 6). Through the management interface, an administrator can change configuration parameters of the MBS service at runtime without stopping or even pausing the JSLEE server. The MBean exploits the Java Management eXtension (JMX) interface provided by JBoss, thus the parameters are easily accessible through a simple JMX web interface.

4.2 MBSC-DB

All the service-related information, such as user subscription and registration data, channels, packages, and configuration variables, are stored and maintained in the MBSC-DB.

The MBSC-DB is implemented by a Linux box running the open source MySQL database server v.5.0.75 as our Data Base Management System (DBMS). This database stores user data, package composition, channel information, ongoing sessions, subscriptions, and scheduled streams by using different tables. Each table has been designed by optimizing the memory space occupied by each field. The database can be managed using, e.g., MySQL Administrator or the phpMyAdmin free software tool [34].

4.3 ASN-GW

ASN-GW, which is placed at the edge of the ASN, is the entity that connects the ASN to the CSN. The ASN-GW assists mobility and security in the control plane and handles IP forwarding. In our architecture, the ASN-GW also connects the WiMAX entities with the MBS service entities. It translates the protocol messages from the R6 protocol to the SIP protocol.

The testbed design includes a custom-built C++ software module, the WiMAX Access Network Emulator (WANE) that emulates the functions of the WiMAX access network entities (including the SUs, the BSs, and ASN-GW) which are significant for our use case. In more detail, our WANE is in charge of (i) capturing IGMP reports coming from the client by using a raw socket, (ii) encapsulating them into R6 messages, and (iii) encapsulating the latter messages together with some additional information within SIP message requests, which are then sent to the MBSC-SS. Conversely, the WANE generates IGMP queries and sends them to clients. Upon receiving MBS Context Request messages from the MBSC-SS, it replies with MBS Context Response messages emulating the overall ASN. Finally, it is in charge of handling and generating 200 OK SIP messages as specified in [28]. Figure 7 shows the role of the WANE in the WiMAX access network.

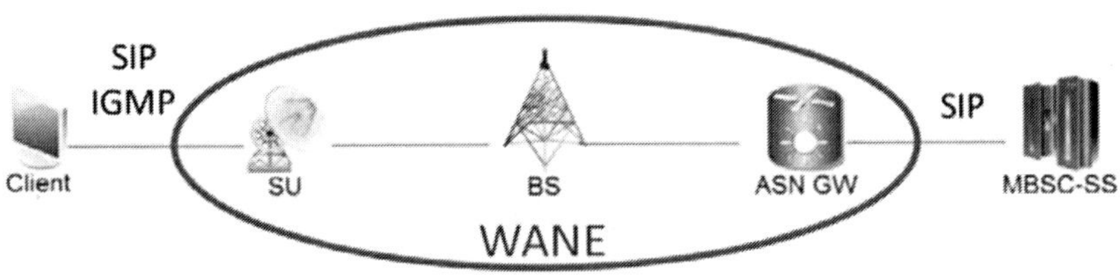

Figure 7 WANE: EMULATED NETWORK ENTITIES AND SUPPORTED PROTOCOLS.

The MSLEE server providing the MBSC-SS is then in charge of parsing and interpreting encapsulated data. Thus, the service running in the MSLEE is always aware of the protocol being used within the WiMAX network [32].

Given our focus on the IP part of the MBS signaling architecture, we treat the WiMAX network as a black box. Thus, mobility effects, modulation parameters, and wireless loss rates and retransmissions are not emulated by our WANE module and are not included in our performance evaluation results.

Finally, it is worth citing that we also executed functional tests of the overall system in the laboratories of a WiMAX manufacturer, running our software on a real ASN-GW, able to dialogite with real WiMAX BSs and our MBSC.

4.4 MBS software client

The MBS software client handles the signaling messages at the client side and provides a graphical user interface (GUI) of the implemented IPTV service. The system architecture that we have designed is modular and allows developers to implement an MBS client as both a standalone application and a Web 2.0 client.

The standalone version consists of a standard Java application using the open source JAIN SIP stack v.1.2 [35] to enable SIP communication with the MBSC-SS. The same SIP stack is used also internally by the SIP RA in the MSLEE signaling server. The client is also equipped with a multimedia player compliant with the used multimedia flow formats and able to send the IGMP unsolicited report message to join a multicast group. The join message triggers all the signaling exchange described in Section 3.

The Web 2.0 version of the MBS client consists of a multi-tier Web-based application developed by using Microsoft Silverlight™ 3.0 together with PHP, Javascript, ActiveX, NPAPI, and CSS, all embedded in a Web page accessible through the Web portal of the MBS over WiMAX service (an additional entity in the CSN beyond those depicted in Figure 1). In this case, the announcements are transferred in a unicast fashion, and they are retrieved by periodically downloading a Really Simple Syndication (RSS) file through a simple HTTP unicast connection. An ActiveX library, for Internet Explorer®, and an NPAPI plug-in, for Mozilla Firefox®, have been included to handle SIP and IGMP messages and fire events towards the Silverlight application.

In both standalone and web cases, the GUI displays user information about subscribed and available channels and packages, announcements, and multicast streaming contents.

4.5 MBS source

The MBS Source is one of the key entities for deploying an MBS over WiMAX architecture. Nonetheless, thanks to our modular architecture, the service administrator is allowed to select the technology that better fits his interests, without modifying the other service entities. In our IPTV experiment, this is achieved by isolating the actual streaming server technology from the service logic. The only requirement for the MBS Source is to support multicast streaming endpoints. We have used a simple VideoLAN client (VLC), running in server mode in a Linux machine, and configured some multicast endpoints to stream a sample video. Then, we have recorded the URLs of the created endpoints in the appropriate database table of the MBSC-DB. We have also created channel packages that the user can subscribe to, such as sport, movies, etc. Users can subscribe to packages and/or to a single channel. The MBS Source has also to create and send multicast announcement data on a dedicated multicast group when a standalone client is used. This is done by using the Session Announcement Protocol (SAP, [36]) functionality of the mini-SAP server bundled with the VLC software. On the other hand, if a Web 2.0 MBS client is used, then a Web server is also necessary. The Web server is used to provide users with the Web portal through which they can access the MBS Service. The Web server has been deployed by installing the open source LAMP (Linux Apache, MySQL, PHP) package on a Linux machine executing Ubuntu Server 9.1 x86. The Web portal also provides an administrative Web interface by which service providers can add, remove, and manage channels, users, and subscriptions. The Web server also dynamically creates and sends the unicast announcement data for the channels subscribed by a user. In this case, data are used to update the web page of the service with real-time information. In fact, each user, once subscribed and authorized by the MSLEE server, periodically polls the web server for announcements encoded as an RSS news file. The client GUI is then in charge of parsing XML files and visualizing data in a user-friendly way. Even if not implemented in our prototype yet, the web client could have the option of using multicast announcements distributed through the SAP protocol, as it happens in the standalone client approach.

5. PERFORMANCE EVALUATION

In this section, we show the performance metrics used to evaluate the MBS signaling. A number of experiments have been realized by using the system testbed described in Section 4. For this purpose, we have deployed the Mobicents JAIN SIP RA v.1.2.4.GA in the MSLEE server that acts as MBSC-SS. We have optimized the SIP RA load by increasing the number of threads, in order to maximize the number of simultaneous processed events. During test execution, we have stopped recording logs in order to improve MSLEE and JBoss AS performance. For what concerns the operating system, we have used the 32-bit Ubuntu server 10.04, running in a virtual machine (VM) equipped by three CPU cores and 3 GB of RAM, and managed by the ESXi 4.1 VMWare hypervisor [37]. The server running the hypervisor is equipped with a dual Intel Xeon E5410 with clock frequency equal to 2.33 GHz and 16 GB of RAM. The MSLEE has been configured with 2.5 GB Java heap, using the Parallel Garbage Collector scheme to periodically clean the memory allocated to the Java server. The choice of deploying the MBSC-SS in a VM allows using computing resources efficiently, implementing energy efficient policies, simplifying replication of ASs along with their failover management, and so on. An interesting reading about virtualization advantages can be found in [38]. The DBMS has been deployed in a different VM having the same characteristics of the MBSC-SS VM and executed in the same server.

In our experiments, clients have been emulated by using the well-known SIPp traffic generator [39]. This choice is due to the unavailability of high performance IGMP message generators. Thus, by using SIPp, the client generates SIP messages with the IGMP report information already encoded in the payload, thus emulating the output of the ASN-GW towards the MBSC-SS. In turn, the WANE has been modified with respect to its base version, in order to simply forward the SIPp-generated messages to the MBSC-SS, without modifying them. By using this configuration, we have collected significant statistics, such as the number of lost SIP messages and the signaling latency, by using the built-in logging capabilities of the SIPp. In particular, we have measured the throughput and latency of the core component of our architecture, the MBSC-SS. We have checked that the WANE was never the system performance bottleneck in our testbed implementation. In addition, we have measured the average signaling protocol overhead of each message exchanged between the MBS entities, and compared it with a typical IMS-based system. In all service lifecycle phases, per-user and per-BS frequency overhead have been evaluated. Finally, we have evaluated the maximum number of users that our testbed can support while keeping a good QoS level in terms of signaling latency and service availability. In particular, we consider a maximum tolerable channel switching time of 100 ms and service availability equal to, at least, 99%. As mentioned in Section 4.3, it is worth to remember that our focus is the performance of the IP part of the MBS signaling architecture, thus the WiMAX network may be simply regarded as a "black box" emulated by our WANE module.

5.1 Throughput and latency

We define a *service transaction* as the signaling exchange which starts with an IGMP message sent by the client, and ends when a SIP reply is sent by the MBSC-SS back to the client (see Figure 4). Therefore, this signaling exchange corresponds to a Join or a Leave operation. We use the service transaction as the basic message exchange to evaluate performance in terms of signaling throughput and latency. Since it is the most frequent and complex signaling exchange of our signaling architecture, it is a meaningful benchmark for evaluating the processing capability of the MBSC-SS. As for the service transaction latency, this is a significant parameter, since it represents the contribution of the core elements of the signaling architecture to the channel switching time. It is defined as the delay between the IGMP message generation (directly encapsulated within a SIP message) at the client side and the reception of the Join Response message.

The SIPp client, which emulates the MBS client, has been configured to generate SIP Messages according to a deterministic arrival process, at a fixed rate λ. Each test has been executed by using the same λ value for 60 min. Both TCP and UDP protocols have been used for transmitting SIP messages. In case of TCP usage, SIPp is configured to work in a mono socket mode.

Figure 8 shows the achieved throughput versus offered load, both expressed in terms of service transactions/s. The first observation is that TCP outperforms UDP significantly. In particular, TCP can reach 210 service transactions/s with a loss rate lower than 1%, whereas UDP can reach only 70 service transactions/s with the same desired maximum loss performance (75 service transactions/s if the loss target is increased by 5%). This large difference is due to the fact that, by using TCP, retransmissions do not occur at the application layer (i.e., SIP) as it happens by using UDP. In fact, if UDP is used, retransmissions are managed by the SIP RA of the MSLEE server, and this can cause avalanche restart effects immediately after a full garbage collection (i.e., after the Java heap cleanup). Note that this operation is quite different from the classical SIP call management in IP telephony where the three way SIP handshake (INVITE-200 OK-ACK) is used, since in that case the SIP layer manages the 200 OK retransmissions directly, even if TCP is used (see also [31]). When using TCP, retransmissions are managed more efficiently by the transport layer at the kernel level and not by the SIP layer in user space. This means that when TCP is used, if the MBSC-SS server get overloaded, it experiences packet losses in the internal queues of the JSLEE application server (e.g., during the garbage collection process). However, since TCP is used, SIP disables retransmission timers, thus the sender knows, from the TCP ACK, that the packet has correctly been received and will not retransmit it. Nevertheless the receiver, which is overloaded, has lost the packet and there is no means of recovering it, and the service transaction fails. Thus, when this process begins, a large number of service transactions will fail and the harsh behavior shown in Figure 8 happens. Note that we have assumed a perfect wireless network channel, meaning that there are no packet losses at the MAC layer, so as to evaluate the MBSC-SS worst-case signaling scenario.

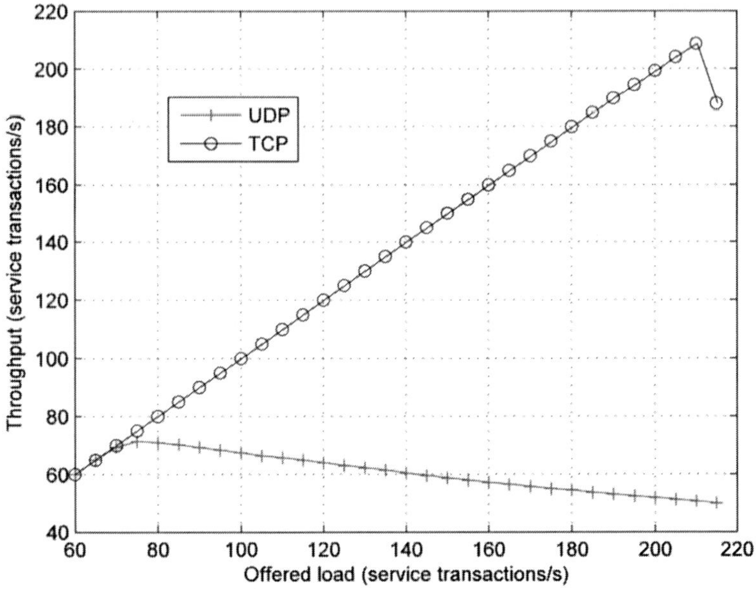

Figure 8 THROUGHPUT VERSUS OFFERED LOAD.

In UDP, a packet lost in the internal queues of the JSLEE application server will always be retransmitted with increased timer value. Thus, overall throughput is much lower, but the decrease is smoother.

Figure 9 illustrates the average latency, expressed in seconds, associated with a service transaction, as a function of the offered load (service transactions/s). The ordinate axis has a logarithmic scale. Again, TCP exhibits superior performance, since it is possible to maintain the latency, which affects the IPTV channel switching time, below 100 ms even if the offered load grows up to 210 service transactions/s. This latency value is compliant with the specifications provided in [26]. On the other hand, by using UDP, a similar latency performance can be obtained only if the offered load is lower than 60 service transactions/s. This is due to the need of managing SIP message retransmissions by the MSLEE RA. The slope for UDP is much steeper than that of TCP since, when the MBSC-SS becomes overloaded; it is not able to process all SIP messages before timer expiration. This implies an increased number of SIP packets to be managed (original packets and retransmissions), which first experience higher delays (in the order of hundreds of ms, since first retransmission occur after at least 500 ms), and then are discarded (service transaction failure), thus having a nearly constant delay independently on the offered load. This almost flat delay of about 20 s is due to a maximum value of SIP timer equal to 32 s. When TCP is used, retransmissions are managed at a lower layer and the process is much more efficient. However, this is a special case relevant to SIP messages only, since in case of SIP sessions with INVITE messages they are managed differently.

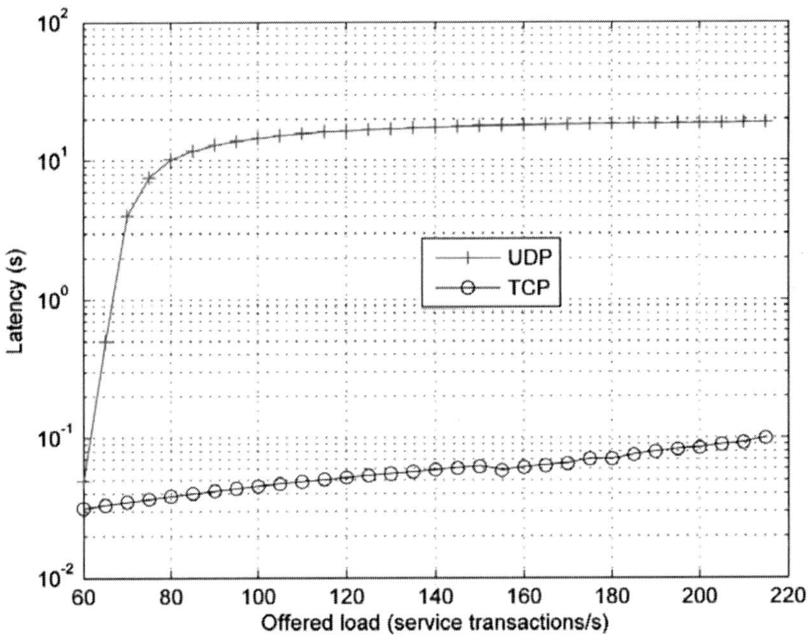

Figure 9 LATENCY VERSUS OFFERED LOAD.

Consider also that the additional latency due to the over-the-air data transmission is not included in our tests, thus it is really important to keep the measured latency value as small as possible in order to have an acceptable service in operation.

By using the maximum throughput values, expressed in service transactions/s, obtained by using either TCP or UDP, we have calculated the maximum number of clients that our MBSC-SS can simultaneously support. We assume that all users are already logged in. This means that they need to issue one service transaction in order to be authorized to join the multicast data flow, an additional one to notify their intention to leave the multicast stream, and a number of service transactions are used to refresh the current multicast session, by answering the Query sent by the ASN-GW with an IGMP Join message (the relevant signaling exchange is shown in Figure 5). Clearly, in this case the only parameter that can influence the number of supported users is the IGMP query period value, since it determines the number of service transactions per user. Note that during a session refresh (see Figure 5), the number of SIP messages exchanged by the MBSC-SS is lower than those exchanged during a service transaction (see Figure 4). Thus, our evaluation is a lower bound of the real capacity of the proposed implementation of the signaling server.

To calculate the maximum number of users N_a supported by our system, and considering movie TV channels as use case, we assume an average flow duration FD = 5400 s (i.e., one and a half hour, which is a typical movie duration). The maximum number of session refresh transactions (*NSTF*) for a multicast session is given by Equation (1). Table 1 reports all the parameters used in the following performance evaluation equations.

Table 1. Definition of symbols used in overhead formulas

Symbol	Definition	Value
MAX_{STS}	Maximum number of service transactions per second supported by an MBSC-SS instance	210 (TCP) 70 (UDP)
FD	Average multicast flow duration, expressed in seconds	5400
$IGMP_{T1}$	IGMP timer value, in seconds, for queries refreshing membership to multicast groups transmitting data flows	125
$IGMP_{T2}$	IGMP timer value, expressed in seconds, for queries refreshing membership to multicast groups transmitting announcement data	1250
R_t	SIP Registration expiration value, expressed in seconds	3600
NSTF	Number of service transactions per user needed to maintain the membership to a multicast data flow of duration FD	-
N_p	Number of available subscription packages (each one with likely more than one channel)	20
N_a	Maximum number of users receiving a multicast data flow supported by a MBSC-SS instance	-
N_{BS}	Number of users served by a BS (or BS sector) on a single WiMAX channel	200
R_{OH}	Overhead per user evaluated at the IP layer associated to a SIP registration	-
A_{OH}	Overhead per user evaluated at the IP layer for maintaining the membership to N_p multicast announcement groups (one for each subscribed package of TV channels)	-
S_{OH}	Overhead per user evaluated at the IP layer for maintaining the membership to a multicast data flow	-
C_{TCP}	TCP connection setup/teardown overhead, expressed in bytes	268 (TCP) 0 (UDP)

By definition, the *NSTF* value depends only on the timer of the IGMP queries and it is equal to

$$\mathrm{NSTF} = 1 + \left\lceil \frac{\mathrm{FD}}{\mathrm{IGMP}_{T1}} \right\rceil . \tag{1}$$

Given the *NSTF* value and the maximum throughput of the signaling server (MAX_{STS}) for both TCP and UDP (see Figure 8), the maximum number of users supported by our system is given by

$$N_a = MAX_{STS} \cdot \frac{FD}{NSTF}. \tag{2}$$

Figure 10 shows the N_a values for different IGMP query interval values ($IGMP_{T1}$). Note that both the abscissa and the ordinate values are expressed in a logarithmic scale. We considered $IGMP_{T1}$ values ranging from 10 s (about 10 times lower than the default value 125 s) to 15 min (about 7 times larger than the default value). It appears that by changing the value of the IGMP query period, the maximum number of users supported by the system varies in a nearly linear fashion, spanning more than two orders of magnitude. In the worst case, by using UDP with $IGMP_{T1}$ = 10 s, only about 800 users are supported. Nevertheless, by using the IGMP default value, $IGMP_{T1}$ = 125 s, and the TCP transport protocol, the number of supported users increases significantly, up to about 26,000 (about 8,000 when UDP is used). With a lower query rate, it is possible to reach a number of users very high, up to 300,000 for TCP and 90,000 for UDP.

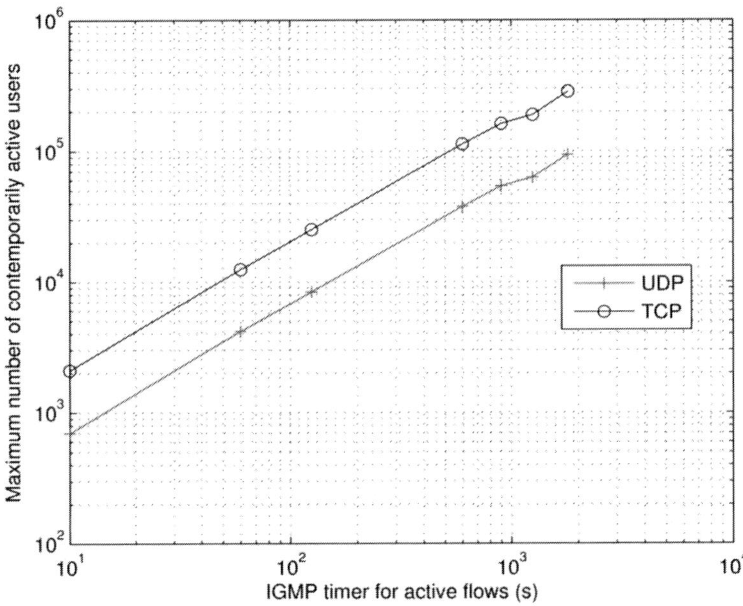

Figure 10 MAX NUMBER OF USERS VERSUS IGMP SESSION REFRESH TIMER.

In this evaluation, we have done the realistic assumption that each user joins only one multicast session. In addition, we have assumed that the ASN-GW sends Group-Specific multicast Queries, which are queries relevant to the membership to a *single, specific* multicast group [5]. In this way, we avoid handling messages from users not involved in such a multicast group, i.e., in the specific case, from inactive users.

In the above evaluation, we have neglected the effect of the queries needed to maintain the membership to multicast groups used to transport announcement data. Clearly, this effect is present only when announcements are multi-casted. We suggest two possible approaches to take this effect into account and limit the performance penalty to a negligible level. The first one is to use an $IGMP_{T2}$ timer value larger than $IGMP_{T1}$ (e.g., 10 times larger), so as to have a lower rate, and thus a lower impact of the responses to queries on the signaling server. The other approach consists of both using $IGMP_{T2} >> IGMP_{T1}$ and sending a General Query upon expiration of the $IGMP_{T2}$ timer at ASN-GW. In this way, irrespective of the number of the multicast groups the client has joined, it will answer by means of a single Report message, listing all multicast groups [5]. In both approaches, the additional resource consumption is negligible.

5.2 Signaling overhead

Each signaling protocol produces some overhead due to control messages. Transmission of such messages requires an additional bandwidth consumption and a processing load in all involved entities. In this section, we determine the average signaling protocol overhead of our proposed MBS over WiMAX architecture. Table 2 summarizes the overall message overhead, for all messages described in Section 3.1, when no compression is used. Since we have proposed an IP-based signaling, we will evaluate the overhead at the IP layer only, thus the additional MAC overhead calculation (see [40] for details) at the R1 interface is not considered. It can be observed that the largest contribution to the signaling overhead is given by the message exchange needed by the join/leave process. In fact, when a user joins a new multicast service session, the procedure requires an exchange of five different messages: an IGMP *join* and a Join *response* between the user and the MBSC-SS on the wireless interface, a Join *request*, an MBS *context request*, and an MBS *context response* between the ASN-GW and the MBSC-SS.

Table 2. Message overhead by protocol layer (expressed in bytes)

Message type	Label	IP layer	Transport layer	Signaling overhead	Overall overhead
IGMP query	I_q	24	-	16	40
IGMP report	I_r	20	-	16	36
Register	$R1$	20	8 (UDP), 20 (TCP)	320	348 (UDP), 360 (TCP)
401–unauthorized	$R2$	20	8 (UDP), 20 (TCP)	450	478 (UDP), 490 (TCP)
Register w/auth data	$R3$	20	8 (UDP), 20 (TCP)	500	528 (UDP), 540 (TCP)
200 OK	OK	20	8 (UDP), 20 (TCP)	250	278 (UDP), 290 (TCP)
Subscription req	S_{req}	20	8 (UDP), 20 (TCP)	350	378 (UDP), 390 (TCP)
Subscription res	S_{res}	20	8 (UDP), 20 (TCP)	350	378 (UDP), 390 (TCP)
Subscribe	$S1$	20	8 (UDP), 20 (TCP)	250	378 (UDP), 290 (TCP)
Notify[a]	$S2$	20	8 (UDP), 20 (TCP)	350[a]	378[a] (UDP), 390[a] (TCP)
Join/leave req	J_{req}	20	20	370	410
Join/leave res	J_{res}	20	8 (UDP), 20 (TCP)	360	388 (UDP), 400 (TCP)
MBS context req	MC_{req}	20	20	370	410
MBS context res	MC_{res}	20	20	360	400

[a]Additional 20 bytes for each subscribed package are necessary.

Therefore, the total average amount of signaling overhead for a session setup is $I_r + J_{\text{req}} + MC_{\text{req}} + MC_{\text{res}} + J_{\text{res}} = 36 + 410 + 410 + 400 + 388 = 1644$ bytes, plus any possible retransmissions when UDP is used to deliver SIP messages. If only TCP is used, there are no SIP retransmissions at the application layer, but the average overhead increases by 12 bytes per packet plus the TCP handshake overhead when the connection is not yet established. Note that only the first *join* and the final *leave* messages produce this amount of overhead, whereas the periodic *join* messages, triggered by the IGMP queries, used to refresh data or announcement multicast sessions, only use $I_q + I_r + J_{\text{req}} + J_{\text{res}} = 40 + 36 + 410 + 388 = 874$ bytes (886 bytes in TCP) including the IGMP query sent by the ASN-GW.

Similarly, the registration process involves four different messages, even if their size depends on the type of security support implemented in the selected user authentication procedure. In fact, if public key encryption is used, then the size of the Register messages can increase significantly. Even though the registration could easily become the largest contribution to the signaling overhead, in an operational service scenario we believe that the session *join* and *refresh* exchanges are those to be kept under control in terms of size, latency, and rate. To this aim, we will refer to the quantities reported in Tables 1 and 2.

Since the wireless link is typically the bandwidth bottleneck, in what follows we analyze the overhead only in the radio interface.

We start evaluating the average overhead rate for each user as follows. The user registration overhead, expressed in bits, is composed of five components, as follows:

$$R_{OH} = (R1 + R2 + R3 + OK + C_{TCP}) \cdot 8. \tag{3}$$

The initial four quantities in Equation (3) are the size of the signaling messages used in the registration phase (see Table 2) and C_{TCP} is the TCP connection setup and teardown contribution to the overhead (see Table 1).

Similarly, the overhead of both refreshing active data and announcing multicast sessions is given by

$$A_{OH} = S_{OH} = (I_q + I_r + J_{req} + J_{res} + C_{TCP}) \cdot 8. \tag{4}$$

The overhead rate of the registration phase for each user is given by

$$Rr_{user} = \frac{R_{OH}}{R_t}, \tag{5}$$

where R_t represents the expiration value of the SIP registration, which is typically equal to 3600 s, and R_{OH} is the registration overhead calculated by Equation (3).

The overhead rates for each user needed to maintain the membership to both announcement and existing data sessions in steady state, are given by Ar_{user} and Sr_{user}, respectively

$$Ar_{user} = \frac{A_{OH}}{IGMP_{T2}}, \tag{6}$$

$$Sr_{user} = \frac{S_{OH}}{IGMP_{T1}}. \tag{7}$$

In fact, as shown in Figures 4 and 5, the announcement multicast session and the data multicast session are both joined and refreshed by similar procedures, which start with an IGMP message. Thus, each time they are performed, they produce the same amount of overhead on the wireless interface, even if membership refresh requires a lower number of signaling messages for the signaling server. Given the greater importance of data sessions queries than announcement ones,[a] two different IGMP query periods can be used in the ASN-GW, $IGMP_{T1}$, and $IGMP_{T2}$, which represent the refresh period of both the data and the announcement multicast sessions, respectively. Their values used in subsequent calculations are reported in Table 1. We have used a value for $IGMP_{T2}$ 10 times larger than the value of $IGMP_{T1}$. These values can be adjusted for different multicast groups in the ASN-GW configuration, and the overhead rate can therefore be optimized.

Figure 11 shows the average per-user signaling overhead rate in the wireless interface. All service phases are evaluated and compared from a transport protocol point of view. When a user watches a TV channel, the largest contribution is given by the membership refresh process for the active multicast data flow. In this case, a rate of 48 bit/s is reached by using UDP, whereas the TCP usage provides 67 bit/s, which is 40% higher. A value of $IGMP_{T1}$ = 125 s has been used as IGMP query interval for TV channel multicast groups. The announcement and registration overhead is much smaller than the active flow management one, and is lower than 10 bit/s per user in both cases. In fact, for the former we have defined $IGMP_{T2}$ = 1250 s and for the latter we use a registration refresh time R_t = 3600 s. The overall amount of per-user overhead is thus about 80 and 60 bit/s, by using TCP and UDP, respectively. If we assume that a user receives a multicast video streaming flow in low resolution (e.g., 500 kbit/s), the ratio between the overall signaling overhead and the data traffic is about 1.6×10^{-4} for TCP and 1.2×10^{-4} for UDP, respectively. The overall comment is that the per-user overhead required by our solution is definitely affordable in a WiMAX network. Also increasing the rate of queries, the overhead is very low.

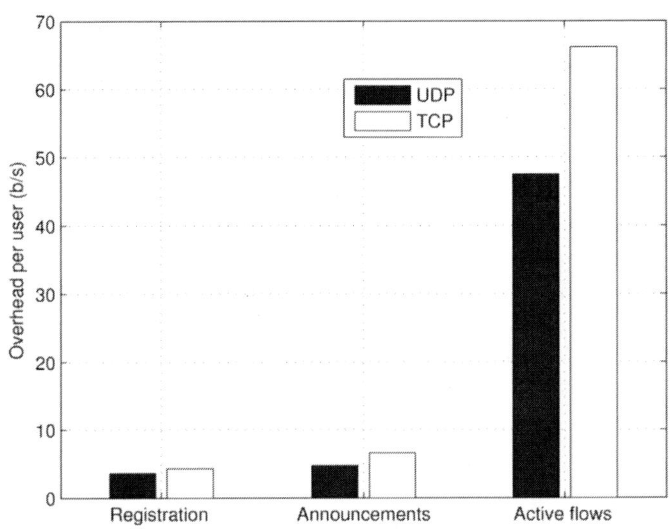

Figure 11 PER-USER SIGNALING OVERHEAD IN BIT/S USING TCP OR UDP FOR EACH SERVICE PHASE WITH $\mathbf{R}_t$ = 3600 S, $IGMP_{T1}$ = 125 S, AND $IGMP_{T2}$ = 1250 S.

By using the single user overhead rate, we can estimate the amount of aggregate overhead rate, consumed by a group of 200 users being served by a single MBS frequency under a single BS. We assume that each user is already registered and is watching a TV channel. In addition, all users are subscribed to $N_p = 20$ different packages and receive the relevant announcement data.

The aggregate overhead bit rate, relevant to all users served by a single BS on an MBS frequency required for keeping the registration state active within the MBSC-SS, is given by

$$Rr_{\text{freq}} = \frac{R_{\text{OH}} \cdot N_{\text{BS}}}{R_t}, \tag{8}$$

where N_{BS} is the number of users served by a single BS on an MBS frequency.

The aggregate overhead required for maintaining the membership to both announcement and active sessions in steady state, considering all users served by a single BS with an MBS frequency, is given by, respectively

$$Ar_{\text{user}} = \frac{(A_{\text{OH}} - I_q \cdot 8) \cdot N_{\text{BS}} + I_q \cdot 8}{\text{IGMP}_{T2}} \cdot N_p, \tag{9}$$

$$Sr_{\text{freq}} = \frac{(S_{\text{OH}} - I_q \cdot 8) \cdot N_{\text{BS}} + I_q \cdot 8}{\text{IGMP}_{T1}}. \tag{10}$$

In Equations (9) and (10), we consider a single IGMP query triggering a number N_{BS} of IGMP reports, as in the standard IGMP settings. This happens N_p times in the case of announcements (we assume, as worst case, that all users subscribed all N_p packages) and only once for the TV channel that the user is currently watching. Similar to Equations (6) and (7), also in Equations (9) and (10), we assume that the IGMP query periods of announcement multicast channels are longer than for those of normal data channels ($\text{IGMP}_{T2} = 10 \times \text{IGMP}_{T1}$).

In Figure 12, we can observe that when a large group of users is considered, the largest contribution to signaling is the membership update of the announcement channel sessions. In fact, by using UDP this contribution is equal to 18 kbit/s, whereas using TCP it grows up to 25 kbit/s, i.e., 40% more than using UDP. In both cases, it is quite low. Keeping the TV channel session open for all users under the same BS and frequency requires less than 10 kbit/s for UDP usage, and 13 kbit/s for TCP. For this reason, it is very important to differentiate the IGMP query interval timer. For updating a registration, the aggregate overhead rate is below 2 kbit/s for both TCP and UDP.

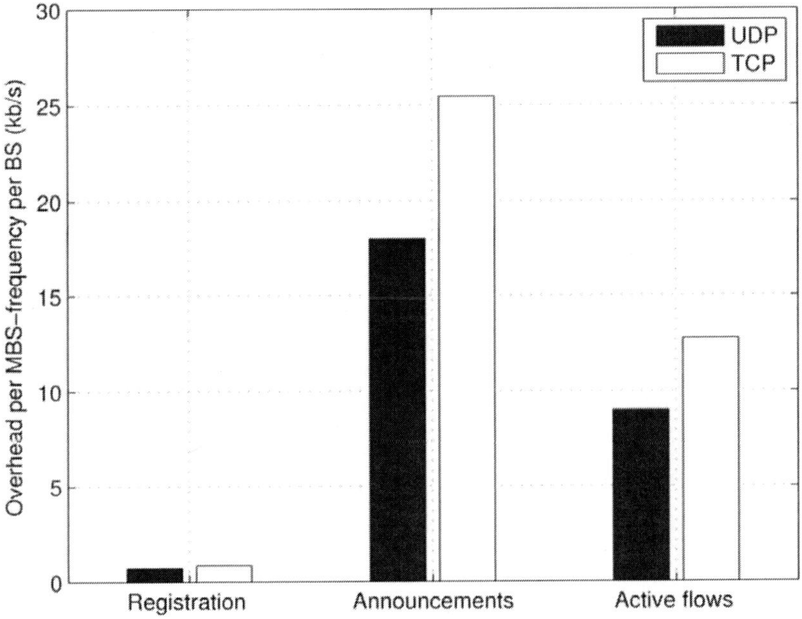

Figure 12 AGGREGATE SIGNALING OVERHEAD IN KBIT/S WITH $\mathbf{N}_{BS} = 200$, $\mathbf{N}_{p} = 20$, $IGMP_{T1} = 125$ S, $IGMP_{T2} = 1250$ S, AND $\mathbf{R}_{t} = 3600$ S.

The overall aggregate overhead rate for 200 users is then around 40 kbit/s for TCP and 28 kbit/s for UDP. If we consider that each announcement channel, using the SAP protocol, is limited to a maximum of 4 kbit/s [36], and that the streaming TV channel can easily use more than 1 Mbit/s, then the overall overhead percentage of our signaling architecture is very low. In more detail, assuming the *worst* case for the signaling overhead, in which 200 users subscribed to N_p packages, all watching the same channel with a data rate equal to 500 kbit/s (e.g., low-quality video), the ratio between the maximum overhead (if TCP is used) and aggregate data rate is given by the ratio 40/500 = 0.08, without any compression. Clearly, this value is definitely affordable.

As already mentioned in Section 2, in the IMS-based architectures the use of the SIP INVITE message is mandatory to create a session state and to provide an IPTV service [18, 19, 22, 23]. Even other proposals [21] try to extend the IMS session setup for multipoint-to-multipoint service type, using again INVITE messages. Conversely, the signaling overhead of our architecture is significantly lower than that of an IMS-based one, due to the usage of simple Message requests. In [29], the overhead of the IMS architecture is analyzed together with the signaling latency, and in both cases our architecture proves to save bandwidth and provides lower latency values for initiating a service session.

5.3 A deployment scenario

In designing our solution, in order to minimize the signaling architecture complexity, we have used only messages that do not require a SIP session to be established. In fact, by both using simple SIP Message requests and by saving and recovering the transaction state on the MBSC-DB, our architecture is easily deployable in a carrier-grade scenario, as shown in Figure 13.

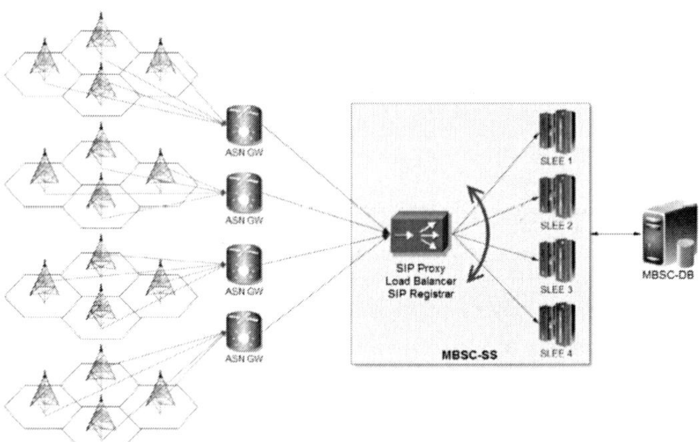

Figure 13 SERVICE DEPLOYMENT IN A CARRIER-GRADE SCENARIO.

The MBSC-SS is an AS and thus it is usually deployed on a cluster of servers in the CSN. It serves a number of ASN-GWs, each providing Internet connectivity to a set of BS. By using a cluster of servers in the MBSC-SS, critical situations, such as failures and overload, can easily be managed without compromising the user's perceived grade of service. Moreover, it is also possible to increase throughput and decrease latency by using different small servers in parallel rather than using a single powerful node processing all messages. In fact, recent studies (see [41] and references therein) show that Java-based ASs, especially SIP servers, perform better if they are executed in a VM with a limited number of CPU cores (e.g., two or three CPU cores), and scalability is efficiently obtained by deploying multiple VMs on the same physical server, each one hosting a replica of the AS. In this deployment, the entry point of all SIP signaling messages in the MBSC-SS is a SIP proxy load balancer, in charge of dispatching SIP messages among AS replicas. This task can easily be accomplished by using a SIP proxy such as SIP Express Router, which can provide high throughput also on low end hardware [42]. In addition, it can also act as a SIP registrar during the registration phase of the service [42].

In a cluster scenario, like the one shown in Figure 13, the session-oriented nature of SIP has important implications for load balancing. In a typical VoIP scenario, all transactions relevant to the same call must be routed to the same server; otherwise, the server will not recognize the call. Session-aware request assignment is the process through which a system allocates requests to servers such that sessions are properly recognized by that server, and subsequent requests, corresponding to the same session, are allocated to the same server [43]. In our case, in which no SIP INVITEs are used, there is no need of complex algorithms. On the contrary, a simple static approach to load balancing (such as a call-id hash algorithm [43]) is enough to ensure that messages with the same call-id, hence belonging to the same MBS service transaction (as defined in Section 5.1), are routed to the same server within the cluster. In addition, when the MBS service transaction (e.g., the Join procedure) is completed, in our architecture any future update of the user state can be processed by another replica of the MBSC-SS, since the session status of each user is stored in the DB. This allows managing computing resources very efficiently by employing simple load-balancing and failover techniques, since ASs do not have to maintain session states and can be powered off without any problem.

For what concerns the use of UDP or TCP for exchanging SIP messages, the best solution is to use UDP between the MBS clients and the SIP load balancer, so as to minimize the overhead on the wireless interface, and to use TCP for all other exchanges, that is between the ASN-GW and the SIP load balancer and between the SIP load balancer and the MSLEE instances (see Figure 13). In this way, it is possible to maximize the achievable throughput and to control the system latency.

Finally, we point out that the solution proposed, although based on Java, has not been tailored exclusively to the JAIN SLEE service framework. In fact, another very interesting approach to the deployment of our MBSC-SS entails the usage of the Java-based NSF NetServ in network container [44]. By using NetServ, a dynamic approach, in which MBSC-SS replicas can be created and installed closer to the ASN-GW lowering signaling latencies and dynamically adapting the number of servers to the load, is made possible with no significant development effort.

6. CONCLUSION

In this article, we have illustrated the novel design and implementation of a modular, open source, and IP-based MBS over WiMAX service signaling architecture. The modularity of the proposed architecture allows service providers to select the best technology for each service entity without any need of modifying the other service entities or the signaling interfaces. The presented service architecture is service independent, and allows creating services such as file-casting, media streaming, mass software updates, and data push services without any need of modifying the signaling messages or network entities. Any change of the service type requires only the change of the user GUI of the MBS client and the MBS source content provider technology, both outside the core network elements.

Being full IP-based, our service architecture is independent from the access network technology, thus it is possible to extend it to other fixed or mobile access networks beyond WiMAX.

The signaling messages exchanged between system entities are based on encoded binary data encapsulated within the payload of SIP messages. In the proposed solution, we have chosen to shift implementation complexity from the ASN-GW towards the JSLEE server orchestrating all signaling. Thanks to the open nature of the JSLEE specifications and the Java capability to ease the development process of telecommunication services, our approach allows supporting a number of different vendors/protocols without major changes.

To test the performance of the service architecture, we have focused on the non-trivial case of a multicast IPTV service offered through an MBS over WiMAX architecture. The signaling server has been implemented by using a single *service building block* deployed over the open source Mobicents JSLEE server. Also the other signaling entities have been implemented by using open source technologies, thus reducing costs and leveraging the community of developers. Our results show that a simple virtual machine hosting the MBSC signaling server, which can be regarded as a common mid-level machine in terms of memory size and CPU processing power can provide up to 210 service transaction/s (and thus supporting tens of thousands of users) and a signaling latency as low as 100 ms when the TCP transport protocol is used by the MBSC-SS.

Although the experiments shown in this article and in some referenced articles are quite different, it could be interesting to compare session set-up times reported by Munir and Gordon-Ross [29] (well above 1 s in the best case) and our switching time (below 100 ms by using TCP as transport protocol). This should be enough to show an order of magnitude on the advantage of deploying our system with respect to IMS-based ones, which are much more complex. As for the comparison with other MBS systems, let us consider the one defined for the UMTS illustrated in [45]. Comparing that proposal with our solution, two differences emerge. The first is that the solution proposed in [45] is tailored exclusively to the UMTS system, whereas our proposal is quite general and could be adapted to other systems with minimal software changes in the MBSC-SS service logic, maintaining the same protocols. The second is that, analyzing the complexity of the signaling exchange, our solution is definitely simpler.

A future work goal is the definition of a standard high level interface to be used between the ASN-GW and the MBSC-SS based on the emerging Open WiMAX standard [46]. Using such a standard interface, the JSLEE service logic becomes independent from the specific protocol implementation used in the WiMAX network, and thus portable over different vendor devices.

ENDNOTE

[a]As explained in last part of Section 3.1.2, missing replies to IGMP query refresh messages could cause service interruption, thus IGMP queries relevant to data sessions are much more important than those relevant to announcements.

DECLARATIONS

Competing interests

The authors declare that they have no competing interests.

REFERENCES

1. Mobile WiMAX--Part I: A Technical Overview and Performance Evaluation[http://www.wimaxforum.org/technology/downloads/Mobile_WiMAX_Part1_Overview_and_Performance.pdf]
2. Jiang T, Xiang W: Multicast broadcast services support in OFDMA-based WiMAX system. IEEE Commun Mag 2007, 45(8):78-86.
3. WiMAX Forum, WiMAX end-to-end network systems architecture, Stage 2: architecture tenets, reference model and reference points2006. [http://www.wimaxforum.org/]
4. Rosenberg J, Schulzrinne H, Camarillo G, Johnston A, Peterson J, Sparks R, Handley M, Schooler E: SIP: session initiation protocol. IETF RFC 3261 2002.
5. Cain B, Deering S, Kouvelas I, Fenner B, Thyagarajan A: Internet group management protocol, Version 3. IETF RFC 3376 2002.
6. Mobicents Web Site[http://www.mobicents.org]
7. JSR 240, JAIN SLEE v1.1. Web Site[http://jcp.org/en/jsr/detail?id=240]
8. Boss JWeb Site, available at: www.jboss.com
9. Etemad K, Wang L: Multicast and broadcast multimedia services in mobile WiMAX networks. IEEE Commun Mag 2009, 47(10):84-91.
10. 3GPP TS 22.146. Multimedia Broadcast/Multicast Service (MBMS); Stage 1 (Release 9)
11. IEEE 802.16-2009, IEEE Standard for Local and metropolitan area networks, Part 16: Air Interface for Broadband Wireless Access Systems 2009.
12. WiMAX Forum, Mobile WiMAX--Part II: A Comparative Analysis2006. [http://www.wimaxforum.org/]
13. Bhatnagar S: Multicast data delivery in WiMAX by reusing L2 handoff procedure. Annu Rev Commun 2008, 61: 277-283.
14. Wang J, Venkatachalam M, Fang Y: System architecture and cross-layer optimization of video broadcast over WiMAX. IEEE J Sel Areas Commun 2007, 25(4):712-723.

15. Lee JM, Park H-J, Choi SG, Choi JK: Adaptive hybrid transmission mechanism for on-demand mobile IPTV over WiMAX. IEEE Trans Broadcast 2009, 55(2):468-477.

16. Lee JH, Kwon TT, Choi Y, Pack S: Location management area (LMA)-based MBS handover in mobile WiMAX systems. In IEEE International Conference on Communications System Software and Middleware (COMSWARE) 2008. Bangalore, India; 2008:341-348.

17. Liao N, Shi Y, Chen J, Li J: Optimized multicast service management in a mobile WiMAX TV system. In 6th IEEE CCNC. Las Vegas, NV (USA); 2009:1-5. February 2009

18. Riede C, Al-Hezmi A, Magedanz T: Session and media signaling for IPTV via IMS. In Mobilware '08. Innsbruck, Austria; 2008.

19. Mikoczy E, Sivchenko D, Xu B, Rakocevic V: IMS based IPTV services--architecture and implementation. In ACM MobiMedia'07. Nafpaktos, Greece, August; 2007.

20. ETSI ES 282 007. Telecommunications and Internet converged services and protocols for advanced networking (TISPAN); IP multimedia subsystem (IMS); functional architecture, V2.0.0, 2008-05 2008.

21. Vidal I, Garcia-Reinoso J, Soto I, Valera F: Evaluating extensions to IMS session setup for multicast-based many-to-many services. Comput Netw 2011, 55(3):600-621. 10.1016/j.comnet.2010.09.011

22. Riede C, Friederich O, Seeliger R, Arbanowski S: Interactive IMS-based IPTV--plunge into the reality game show. In IEEE IMSAA 2008. Bangalore, India; 2008:1-6.

23. Hung N, Thanh N: Intelligent features for IMS-based IPTV. In IEEE ATNAC 2010. Auckland, New Zealand; 2010:130-134.

24. Xiao Y, Du X, Zhang J, Hu F, Guizani S: Internet protocol television (IPTV): the killer application for the next-generation Internet. IEEE Commun Mag 2007, 126-134.

25. Kellokoski J, Tukia E, Wallenius E, Hamalainen T, Naarmala J: Registration performance comparison between IP multimedia subsystem and session initiation protocol networks. In International Congress on Ultra Modern Telecommunications and Control Systems and Workshops (ICUMT), 2010. Moscow, Russia; 2010:20-25. 18-20

26. Lloret J, Garcia M, Atenas M, Canovas A: A QoE management system to improve the IPTV network. Int J Commun Syst 2011, 24(1):118-138. 10.1002/dac.1145

27. Di Sorte D, Femminella M, Reali G: A QoS index for IP services to effectively support usage-based charging. IEEE Commun Lett 2004, 8(11):686-688. 10.1109/LCOMM.2004.837661

28. Campbell B, Rosenberg J, Schulzrinne H, Huitema C, Gurle D: Session initiation protocol (SIP) extension for instant messaging. IETF RFC 3428 2002.

29. Munir A, Gordon-Ross A: SIP-based IMS signaling analysis for WiMAX-3G interworking architectures. IEEE Trans Mob Comput 2010, 9(5):733-750.

30. Christensen M, Kimball K, Solensky F: Considerations for Internet group management protocol (IGMP) and multicast listener discovery (MLD) snooping switches. IETF RFC 4541 2006.

31. Femminella M, Francescangeli R, Giacinti F, Maccherani E, Parisi A, Reali G: Scalability and performance evaluation of a JAIN SLEE-based platform for VoIP services. ITC-21 Paris, France; 2009.

32. Cisco Broadband Wireless Gateway[http://www.cisco.com/en/US/prod/collateral/wireless/wirelssw/ps8738/product_data_sheet0900aecd806cdb72.html]

33. Li B, Qin Y, Low CP, Gwee CL: A survey on Mobile WiMAX. IEEE Communications Magazine 2007, 45(12):70-75.

34. phpMyAdmin web page[http://www.phpmyadmin.net/home_page/index.php]

35. JAIN SIP stack Web Site[http://jsip.java.net]

36. Handley M, Perkins C, Whelan E: Session announcement protocol. IETF RFC 2974 2000.

37. VMware ESXi Web Site[http://www.vmware.com/products/esxi/]

38. Singh A: An introduction to virtualization.2004. [http://www.kernelthread.com/publications/virtualization/]

39. SIPp Traffic Generator[http://sipp.sourceforge.net/]

40. Pareit D, Deruyck M, Tangle E, Joseph W, Moerman I, Martens L, Demeester P: Detailed modeling of MAC throughput and ranges for mobile WiMAX. IEEE Commun Lett 2011, 158(8):839-841.

41. Femminella M, Maccherani E, Reali G: Performance management of java-based SIP application servers. In IEEE/IFIP IM'11. Dublin, Ireland; 2011:493-500. 23-27

42. Nahum EM, Tracey J, Wright CP: Evaluating SIP proxy server performance. In NOSSDAV '07. Urbana, IL; 2007.

43. Jiang H, Iyengar A, Nahum E, Segmuller W, Tantawi A, Wright CP: Load balancing for SIP server clusters. In IEEE INFOCOM 2009. Rio de Janeiro, Brazil; 2009:2286-2294. 19-25

44. Femminella M, Francescangeli R, Reali G, Lee J, Schulzrinne H: An enabling platform for autonomic management of the future internet. IEEE Netw 2011, 25(6):24-32.

45. Xylomenos G, Vogkas V, Thanos G: The multimedia broadcast/multicast service. Wirel Commun Mob Comput 2008, 8(2):255-265. 10.1002/wcm.463

46. Open WiMAX whitepaper, The disruptive approach of open WiMAX[http://www.wimax-industry.com/sp/wcm/avr/dl/wp/Disruptive_Approach_wp.pdf]

INDEX

V

W

Z